Combinatorial Pattern Matching Algorithms in Computational Biology Using Perl and R

CHAPMAN & HALL/CRC
Mathematical and Computational Biology Series

Aims and scope:

This series aims to capture new developments and summarize what is known over the whole spectrum of mathematical and computational biology and medicine. It seeks to encourage the integration of mathematical, statistical and computational methods into biology by publishing a broad range of textbooks, reference works and handbooks. The titles included in the series are meant to appeal to students, researchers and professionals in the mathematical, statistical and computational sciences, fundamental biology and bioengineering, as well as interdisciplinary researchers involved in the field. The inclusion of concrete examples and applications, and programming techniques and examples, is highly encouraged.

Series Editors

Alison M. Etheridge
Department of Statistics
University of Oxford

Louis J. Gross
Department of Ecology and Evolutionary Biology
University of Tennessee

Suzanne Lenhart
Department of Mathematics
University of Tennessee

Philip K. Maini
Mathematical Institute
University of Oxford

Shoba Ranganathan
Research Institute of Biotechnology
Macquarie University

Hershel M. Safer
Weizmann Institute of Science
Bioinformatics & Bio Computing

Eberhard O. Voit
The Wallace H. Couter Department of Biomedical Engineering
Georgia Tech and Emory University

Proposals for the series should be submitted to one of the series editors above or directly to:
CRC Press, Taylor & Francis Group
4th, Floor, Albert House
1-4 Singer Street
London EC2A 4BQ
UK

Published Titles

Bioinformatics: A Practical Approach
Shui Qing Ye

Cancer Modelling and Simulation
Luigi Preziosi

Computational Biology: A Statistical Mechanics Perspective
Ralf Blossey

Computational Neuroscience: A Comprehensive Approach
Jianfeng Feng

Data Analysis Tools for DNA Microarrays
Sorin Draghici

Differential Equations and Mathematical Biology
D.S. Jones and B.D. Sleeman

Exactly Solvable Models of Biological Invasion
Sergei V. Petrovskii and Bai-Lian Li

Handbook of Hidden Markov Models in Bioinformatics
Martin Gollery

Introduction to Bioinformatics
Anna Tramontano

An Introduction to Systems Biology: Design Principles of Biological Circuits
Uri Alon

Kinetic Modelling in Systems Biology
Oleg Demin and Igor Goryanin

Knowledge Discovery in Proteomics
Igor Jurisica and Dennis Wigle

Modeling and Simulation of Capsules and Biological Cells
C. Pozrikidis

Niche Modeling: Predictions from Statistical Distributions
David Stockwell

Normal Mode Analysis: Theory and Applications to Biological and Chemical Systems
Qiang Cui and Ivet Bahar

Pattern Discovery in Bioinformatics: Theory & Algorithms
Laxmi Parida

Spatiotemporal Patterns in Ecology and Epidemiology: Theory, Models, and Simulation
Horst Malchow, Sergei V. Petrovskii, and Ezio Venturino

Stochastic Modelling for Systems Biology
Darren J. Wilkinson

Structural Bioinformatics: An Algorithmic Approach
Forbes J. Burkowski

The Ten Most Wanted Solutions in Protein Bioinformatics
Anna Tramontano

Chapman & Hall/CRC Mathematical and Computational Biology Series

Combinatorial Pattern Matching Algorithms in Computational Biology Using Perl and R

Gabriel Valiente

CRC Press
Taylor & Francis Group
Boca Raton London New York

CRC Press is an imprint of the
Taylor & Francis Group an **informa** business

A CHAPMAN & HALL BOOK

Chapman & Hall/CRC
Taylor & Francis Group
6000 Broken Sound Parkway NW, Suite 300
Boca Raton, FL 33487-2742

First issued in paperback 2017

© 2009 by Taylor & Francis Group, LLC
Chapman & Hall/CRC is an imprint of Taylor & Francis Group, an Informa business

No claim to original U.S. Government works

ISBN 13: 978-1-138-11534-7 (pbk)
ISBN 13: 978-1-4200-6973-0 (hbk)

Library of Congress Cataloging-in-Publication Data

Valiente, Gabriel, 1963-
   Combinatorial pattern matching algorithms in computational biology using Perl and R / Gabriel Valiente.
      p. cm. -- (Mathematical and computational biology series)
   Includes bibliographical references and index.
   ISBN 978-1-4200-6973-0 (hardcover : alk. paper)
   1. Computational biology. 2. Pattern formation (Biology)--Computer simulation. 3. Graph algorithms. 4. Perl (Computer program language) 5. R (Computer program language) I. Title. II. Series.

QH324.2.V35 2009
572.80285--dc22   2009003714

Visit the Taylor & Francis Web site at
http://www.taylorandfrancis.com

and the CRC Press Web site at
http://www.crcpress.com

To the loving memory of Helda Zulma Feruglio

Contents

Foreword

Preface

1 Introduction   **1**
    1.1   Combinatorial Pattern Matching   3
    1.2   Computational Biology   4
    1.3   A Motivating Example: Gene Prediction   4
    Bibliographic Notes   17

I   Sequence Pattern Matching

2 Sequences   **21**
    2.1   Sequences in Mathematics   21
        2.1.1   Counting Labeled Sequences   22
    2.2   Sequences in Computer Science   24
        2.2.1   Traversing Labeled Sequences   26
    2.3   Sequences in Computational Biology   29
        2.3.1   Reverse Complementing DNA Sequences   31
        2.3.2   Counting RNA Sequences   33
        2.3.3   Generating DNA Sequences   35
        2.3.4   Representing Sequences in Perl   38
        2.3.5   Representing Sequences in R   40
    Bibliographic Notes   42

3 Simple Pattern Matching in Sequences   **43**
    3.1   Finding Words in Sequences   43
        3.1.1   Word Composition of Sequences   43
        3.1.2   Alignment Free Comparison of Sequences   49
    Bibliographic Notes   52

4 General Pattern Matching in Sequences   **53**
    4.1   Finding Subsequences   53
        4.1.1   Suffix Arrays   56
    4.2   Finding Common Subsequences   67

	4.2.1	Generalized Suffix Arrays	74
4.3	Comparing Sequences	86
	4.3.1	Edit Distance-Based Comparison of Sequences	86
	4.3.2	Alignment-Based Comparison of Sequences	95
Bibliographic Notes	110

II Tree Pattern Matching

5 Trees **115**

5.1	Trees in Mathematics	115
	5.1.1	Counting Labeled Trees	115
5.2	Trees in Computer Science	117
	5.2.1	Traversing Rooted Trees	118
5.3	Trees in Computational Biology	118
	5.3.1	The Newick Linear Representation	123
	5.3.2	Counting Phylogenetic Trees	125
	5.3.3	Generating Phylogenetic Trees	126
	5.3.4	Representing Trees in Perl	128
	5.3.5	Representing Trees in R	131
Bibliographic Notes	135

6 Simple Pattern Matching in Trees **137**

6.1	Finding Paths in Unrooted Trees	137
	6.1.1	Distances in Unrooted Trees	138
	6.1.2	The Partition Distance between Unrooted Trees . . .	140
	6.1.3	The Nodal Distance between Unrooted Trees	144
6.2	Finding Paths in Rooted Trees	148
	6.2.1	Distances in Rooted Trees	150
	6.2.2	The Partition Distance between Rooted Trees	151
	6.2.3	The Nodal Distance between Rooted Trees	151
Bibliographic Notes	152

7 General Pattern Matching in Trees **155**

7.1	Finding Subtrees	155
	7.1.1	Finding Subtrees Induced by Triplets	156
	7.1.2	Finding Subtrees Induced by Quartets	159
7.2	Finding Common Subtrees	161
	7.2.1	Maximum Agreement of Rooted Trees	161
	7.2.2	Maximum Agreement of Unrooted Trees	172
7.3	Comparing Trees	172
	7.3.1	The Triplets Distance between Rooted Trees	172
	7.3.2	The Quartets Distance between Unrooted Trees	175
Bibliographic Notes	178

III Graph Pattern Matching

8 Graphs   **181**
  8.1 Graphs in Mathematics 181
      8.1.1 Counting Labeled Graphs 182
  8.2 Graphs in Computer Science 183
      8.2.1 Traversing Directed Graphs 183
  8.3 Graphs in Computational Biology 184
      8.3.1 The eNewick Linear Representation 193
      8.3.2 Counting Phylogenetic Networks 195
      8.3.3 Generating Phylogenetic Networks 198
      8.3.4 Representing Graphs in Perl 202
      8.3.5 Representing Graphs in R 205
  Bibliographic Notes 208

9 Simple Pattern Matching in Graphs   **211**
  9.1 Finding Paths in Graphs 211
      9.1.1 Distances in Graphs 214
      9.1.2 The Path Multiplicity Distance between Graphs . . . 220
      9.1.3 The Tripartition Distance between Graphs 228
      9.1.4 The Nodal Distance between Graphs 234
  9.2 Finding Trees in Graphs 238
      9.2.1 The Statistical Error between Graphs 243
  Bibliographic Notes 246

10 General Pattern Matching in Graphs   **247**
  10.1 Finding Subgraphs 247
      10.1.1 Finding Subgraphs Induced by Triplets 248
  10.2 Finding Common Subgraphs 259
      10.2.1 Maximum Agreement of Rooted Networks 259
  10.3 Comparing Graphs 269
      10.3.1 The Triplets Distance between Graphs 269
  Bibliographic Notes 273

A Elements of Perl   **275**
  A.1 Perl Scripts 275
  A.2 Overview of Perl 294
  A.3 Perl Quick Reference Card 297
  Bibliographic Notes 304

B Elements of R   **305**

   B.1 R Scripts 305

   B.2 Overview of R 323

   B.3 R Quick Reference Card 329

   Bibliographic Notes 336

References   **339**

Index   **351**

Foreword

When, more than 25 years ago, Zvi Galil and I decided to organize the presentations at a NATO workshop into the volume entitled "Combinatorial Algorithms on Words," windows were building fixtures and webs were inhabited by spiders and navigated, more or less inadvertedly, by doomed insects. The "Atlas of Protein Sequences" by Margaret Dayhoff listed a handful of Cytochrome C proteins and the first sequenced genome was more than a decade away. Nonetheless, we were convinced that the few, scattered properties and constructs of pattern matching available at the time had the potential to spark and shape a specialty of algorithms design that would not fade in comparison with already well-established areas such as graph and numerical algorithms. It is safe to say that our volume presented a detailed account of the state of the art, attracted attention to data structures such as the suffix trees that are still the subject of deep study, and also listed most of the relevant open problems that would be tackled, with varying success, in the following years. A reader willing to invest some time on the contents of that volume could immigrate rather quickly into the issues at the frontier and start making contributions depending only on own taste and skills.

About ten years later, as we attempted to collect the state of the art in "Pattern Matching Algorithms," we faced an amazingly expanded scenario. This time, each one of the contributed chapters had to be a rather synthetic survey of the many intervening results, some of which had started new specialized sections, notable among which were two-dimensional and tree matching. The neophyte interested in the subject could now take steps from reading one or two chapters, but would then have to invest considerable additional time in the study of numerous references. Many things had happened in between, bringing new meanings to webs and windows and growing protein databases and sequence repositories that included now sizable genomes such as yeast.

Since then, additional beautiful volumes have been produced and even conferences such as CPM, RECOMB, SPIRE, WABI have started that keep churning out an unrelenting crop of problems, applications, and results. All of this shows how challenging it is to embark in a compendium of the state of the art, and thus how admirable is the volume that results from the effort of Gabriel Valiente.

The reader of this nicely structured volume will find a well-rounded exposition of the traditional issues accompanied by an up-to-date account of more recent developments, such as graph similarity and search. As is well known, a great many pattern matching problems have been inspired over the years by the growing domain of computational molecular biology. This is due in part to the fact that the crisp lexicon of biological sequence analysis provides the natural habitat for pattern matching, and while probably not every problem has received a fungible solution, it is admitted that nowhere else has pattern matching found more stimuli and opportunities for growth as a discipline. Continuing competently in this vein, the organization of this volume dovetails with the transition of computational biology from the molecular to the cellular level.

For more than three decades, I have observed that one of the greatest difficulties that biologists and computer scientists have to overcome when engaging in interdisciplinary inquiries is the lack of a common language, hence of the ability to set common goals. Such a language is the necessary prerequisite to the ambitious and yet necessary task of developing a hybrid specialist, conversant in both disciplines and able to appreciate a problem and its solution from the standpoint of both computing and biology. Balancing a careful mixture of formal methods, programming, and examples, Gabriel Valiente has managed to harmoniously bridge languages and contents into a self-contained source of lasting influence. It is not difficult to predict that this book will be studied indifferently by the specialist of biology and computer science, helping each to walk a few steps towards the other. It will entice new generations of scholars to engage in its beautiful subject.

Alberto Apostolico
Atlanta

Preface

Combinatorial pattern matching algorithms count among the main sources of solutions to the computational biology problems that arise in the analysis of genomic, transcriptomic, proteomic, metabolomic, and interactomic data. Top in the ranking is the BLAST software tool for rapid searching of nucleotide and protein databases, which is based on combinatorial pattern matching algorithms for local sequence alignment. Also high in the ranking, suffix trees and suffix arrays were developed to efficiently solve specific problems that arise in computational biology, such as the search for patterns in nucleotide or protein sequences.

This is a book on combinatorial pattern matching algorithms, with special emphasis on computational biology and also on their implementation in Perl and R, two widely used scripting languages in computational biology. The book is aimed at anyone with an interest in combinatorial pattern matching and in the broader subject of combinatorial algorithms, the only prerequisites being an elementary knowledge of mathematics and computer programming, the desire to learn, and unlimited time and patience.

Acknowledgments

This book is based on graduate lectures taught at the Technical University of Catalonia, Barcelona, and also invited lectures at the Phylogenetics Programme of the Isaac Newton Institute for Mathematical Sciences, held September 3 to December 21, 2007, in Cambridge, UK; at the Gulbenkian PhD Program in Computational Biology of the Instituto Gulbenkian de Ciência, held February 4–8, 2008, and January 26–30, 2009, in Oeiras, Portugal; and at the Lipari International Summer School on Bioinformatics and Computational Biology, held February 4–8, 2008, in Lipari Island, Italy. I am very grateful to Vincent Moulton, Mike Steel, and Daniel Huson (Isaac Newton Institute for Mathematical Sciences), to Jorge Carneiro and Manuela Cordeiro (Gulbenkian PhD Program in Computational Biology), and to Alfredo Ferro, Raffaele Giancarlo, Concettina Guerra, and Michael Levitt (Lipari International Summer School on Bioinformatics and Computational Biology) for their continuous encouragement and support.

The very idea of presenting combinatorial pattern matching problems in a uniform framework (pattern matching in and between sequences, trees, and graphs) arose during the Second Haifa Annual International Stringology Research Workshop, held April 3–8, 2005, at the Caesarea Rothschild Institute for Interdisciplinary Applications of Computer Science, University of Haifa, Israel. I am very grateful to Amihood Amir, Martin C. Golumbic, and Gad M. Landau for providing such a stimulating environment.

The approach to algorithms in bioinformatics and computational biology expressed in this book has been influenced by the interaction with numerous colleagues at the Barcelona Biomedical Research Park, especially within the Centre for Genomic Regulation and also in the Research Unit on Biomedical Informatics. In particular, I would like to thank Roderic Guigó and Ferran Sanz for granting me access to the Barcelona Biomedical Research Park.

I am grateful to the people who have read and commented on draft material. In particular, I would like to thank José Clemente, Eduardo Eyras, Vincent Lacroix, Michael Levitt, and Francesc Rosselló.

Last, but not least, it has been a pleasure to work out editorial matters together with Amber Donley, Sarah Morris, and Sunil Nair of Taylor & Francis Group.

Gabriel Valiente
Barcelona

Chapter 1

Introduction

Computational biology, the application of computational and mathematical techniques to problems inspired by biology, has witnessed an unprecedented expansion over the last few years. The wide availability of genomic, transcriptomic, proteomic, metabolomic, and interactomic data has fostered the development of computational techniques for their analysis. Combinatorial pattern matching is one of the main sources of algorithmic solutions to the problems that arise in their analysis.

This is a text on combinatorial pattern matching algorithms, with special emphasis on computational biology. Pattern matching is well known in computational biology, not only because of biological sequence alignment. Data structures such as suffix trees and suffix arrays were developed within the combinatorial pattern matching research community to efficiently solve specific problems that arise in computational biology. This book provides an organized and comprehensive view of the whole field of combinatorial pattern matching from a computational biology perspective and addresses specific pattern matching problems within and between sequences, trees, and graphs. Much of the material presented on the book is only available in the specialized research literature.

The book is structured around the specific algorithmic problems that arise when dealing with those structures that are commonly found in computational biology, namely: biological sequences (such as DNA, RNA, and protein sequences), trees (such as phylogenetic trees and RNA structures), and graphs (such as phylogenetic networks, metabolic pathways, protein interaction networks, and signaling pathways). The emphasis throughout this book is on the search for patterns within and between biological sequences, trees, and graphs, with the understanding of exact (rather than approximate) occurrences as patterns and pairwise (rather than multiple) comparison of structures. There is also a strong emphasis on phylogenetic trees and networks as examples of trees and graphs in computational biology.

For each of these structures (sequences, trees, and graphs), a clear distinction is made between the problems that arise in the analysis of one structure (finding patterns within a structure) and in the comparative analysis of two or more structures (finding patterns common to two structures and aligning these structures).

The patterns contained in a sequence are words, and k-mer composition is the basis of an important form of alignment-free sequence comparison. Suffix arrays allow for the efficient search of occurrences of a sequence in another sequence, while generalized suffix arrays allow for the efficient search of occurrences common to two sequences. Besides finding common patterns, the comparison of two sequences is also made on the basis of the Hamming distance, the Levenshtein distance, and the edit distance, as well as by means of a global or a local alignment of the sequences.

The patterns contained in a tree are paths, and the distances (lengths of the shortest paths) between the nodes of a tree are the basis of the nodal distance between phylogenetic trees. The partition distance builds upon the distinction in phylogenetic trees between descendant and non-descendant nodes. Small subtrees common to two trees underlie both the triplets distance and the quartets distance between phylogenetic trees. The comparison of two trees is also made on the basis of their maximum agreement subtree.

The patterns contained in a graph are paths and trees, and the distances between the nodes of a graph are the basis of the nodal distance between graphs. The path multiplicity distance between phylogenetic networks builds upon the number of different paths between the nodes of a graph. The tripartition distance is based on the distinction between strict descendant, non-strict descendant, and non-descendant nodes in a phylogenetic network. Small subgraphs common to two graphs underlie the triplets distance between phylogenetic networks, and large subgraphs common to two graphs underlie the statistical error between phylogenetic networks. The comparison of two phylogenetic networks is also made on the basis of their maximum agreement subgraph.

A thorough discussion is made of each of these specific problems, together with detailed algorithmic solutions in pseudo-code, full Perl and R implementation, and pointers to off-the-shelf software and alternative implementations such as those found on CPAN, the Comprehensive Perl Archive Network, or integrated into the BioPerl project, as well as those found on CRAN, the Comprehensive R Archive Network, or integrated into the BioConductor project. The Perl and R source code of all of the algorithms presented in the book is also available at `http://www.lsi.upc.edu/~valiente/comput-biol/`.

The rest of this chapter contains an introduction to some of the biological, mathematical, and computational notions used in this book, by means of a motivating example, in an effort to make it as self-contained as possible for the biologist, the mathematician, and the computer scientist reader as well. The book itself is organized in a first part devoted to sequence pattern matching, a second part on tree pattern matching, and a third part about graph pattern matching, followed by a brief introduction to Perl and R in two appendices.

The first part contains an introductory chapter about sequences, a chapter devoted to the problem of pattern matching within a sequence, and a chapter on finding patterns common to two sequences. Following the same scheme, the second part contains an introductory chapter about trees, a chapter devoted to finding patterns within a tree, and a chapter on finding patterns common to two trees, and the third part contains an introductory chapter about graphs, a chapter devoted to finding patterns within a graph, and a chapter on finding patterns common to two graphs. Each chapter includes detailed bibliographic notes and pointers to the specialized research literature.

Throughout this book, *sequence* is often used as a synonym for *string*, because the reader is assumed to be familiar with the notion of biological (DNA, RNA, protein) sequences. The distinction between string and sequence is important when referring to substructures, however, because subsequences are not necessarily substrings. In general, subsequences will be referred to as *gapped subsequences*, and the particular case of substrings (consecutive parts of a string) will be referred to as just *subsequences*.

1.1 Combinatorial Pattern Matching

Pattern matching refers to the search for the occurrences of a pattern in a text, where both the pattern and the text can be discrete structures such as sequences, trees, and graphs. Examples of patterns in computational biology are a short nucleotide sequence, such as the TATAAAA motif found in the promoter region of most eukaryotic genes; the amino acid sequence of a transcription factor, such as the prokaryotic C2H2 zinc-finger motif x(2)-Cys-x(2)-Cys-x(9)-His-x(2)-His-x(2); and an RNA secondary structure motif, such as the CUUCGG hairpin found in the small subunit ribosomal RNA of most bacteria. Corresponding examples of pattern matching problems are finding motifs and transcription factor binding sites in DNA sequences, and searching RNA sequences for recurrent structural motifs.

Sequence patterns are often described by means of regular expressions, with a special syntax such as a vertical bar to separate alternatives, parentheses to group patterns, and wild cards such as a dot for matching a single character, a question mark for matching zero or one occurrence, an asterisk for matching zero or more occurrences, and a plus sign for matching one or more occurrences. Regular expressions can be used as patterns for selecting and replacing text with the utilities awk (named after the authors), ed (text editor), expr (evaluate expression), grep (global regular expression pattern), sed (stream editor), vim (visual editor improved), and in scripting programming languages such as perl (practical extraction and report language), among others.

Combinatorial pattern matching addresses issues of searching and matching strings and more complex patterns such as trees, regular expressions, graphs, point sets, and arrays, with the goal of deriving non-trivial combinatorial properties for such structures and then exploiting these properties in order to either achieve improved performance for the corresponding computational problems or pinpoint properties and conditions under which searches cannot be performed efficiently.

1.2   Computational Biology

In a broad sense, computational biology is the application of computational and mathematical techniques to problems inspired by biology. A distinction is often drawn between computational biology and bioinformatics, where computational biology involves the development and application of theoretical methods, mathematical modeling techniques, and computational simulation techniques to biological data, while bioinformatics is centered on the development and application of computational approaches and tools for the acquisition, organization, storage, analysis, and visualization of biological data.

Molecular biology itself is experiencing a shift from an understanding of biological systems at the molecular level (nucleotide or amino acid sequences and structures of individual genes or proteins) to an understanding of biological systems at a system level (integrated function of hundreds or thousands of genes and proteins in the cell, in tissues, and in whole organisms), and this shift is also influencing computational biology.

There are two main branches in computational biology. On the one hand, the area of *biological data mining* focuses on extracting hidden patterns from large amounts of experimental data, forming hypotheses as a result. Most of the research in computational genomics and computational proteomics belongs in this area. On the other hand, *modeling and simulation* focuses instead on developing mathematical models and performing simulations to test hypotheses with in-silico experiments, providing predictions that can be tested by in-vitro and in-vivo studies. Much of the research in mathematical biology, computational biochemistry, computational biophysics, and computational systems biology falls in this area.

Combinatorial pattern matching algorithms belong in the biological data mining branch of computational biology.

1.3   A Motivating Example: Gene Prediction

The whole genome of an organism can be revealed from tissue samples by using one of several DNA sequencing technologies, each of them producing a large number of DNA fragments of various lengths that are then assembled into the DNA sequence of the molecules in either the mitochondria or the nucleus (for eukaryotes) or in the cytoplasm (for prokaryotes) of the cells. The whole genomes of thousands of extant species have already been sequenced, including 111 archaeal genomes ranging from 1,668 to 5,751,492 nucleotides; 2,167 bacterial genomes with 846 to 13,033,779 nucleotides; 2,593 eukaryote genomes with 1,028 to 748,055,161 nucleotides; 2,651 viral genomes with 200 to 1,181,404 nucleotides; 39 viroid RNA genomes with 246 to 399 nucleotides; and 1504 plasmid genomes with 846 to 2,094,509 nucleotides. Extant species represent only a small fraction of the genetic diversity that has ever existed, however, and whole genomes of extinct species can also be sequenced from well-conserved tissue samples.

Once the genome of a species has been sequenced, one of the first steps towards understanding it consists in the identification of genes coding for proteins. In prokaryotic genomes, the sequence coding for a protein occurs as one contiguous *open reading frame*, while in eukaryotic genomes, it is often spliced into several coding *exons* separated by non-coding *introns*, and these exons can be combined in different arrangements to code for different proteins by the cellular process of alternative splicing.

Example 1.1

The DNA sequence of *Bacteriophage φ-X174*, which was the first genome to be sequenced, has 11 protein coding genes within a circular single strand of 5,368 nucleotides. One of these genes is shown highlighted.

```
GAGTTTTATCGCTTCCATGACGCAGAAGTTAACACTTTCGGATATTTCTGATGAGTCGAAAAAT
TATCTTGATAAAGCAGGAATTACTACTGCTTGTTTACGAATTAAATCGAAGTGGACTGCTGGCG
GAAAATGAGAAAATTCGACCTATCCTTGCGCAGCTCGAGAAGCTCTTACTTTGCGACCTTTCGC
CATCAACTAACGATTCTGTCAAAAACTGACGCGTTGGATGAGGAGAAGTGGCTTAATATGCTTG
GCACGTTCGTCAAGGACTGGTTTAGATATGAGTCACATTTTGTTCATGGTAGAGATTCTCTTGT
TGACATTTTAAAAGAGCGTGGATTACTATCTGAGTCCGATGCTGTTCAACCACTAATAGGTAAG
AAATCATGAGTCAAGTTACTGAACAATCCGTACGTTTCCAGACCGCTTTGGCCTCTATTAAGCT
CATTCAGGCTTCTGCCGTTTTGGATTTAACCGAAGATGATTTCGATTTTCTGACGAGTAACAAA
GTTTGGATTGCTACTGACCGCTCTCGTGCTCGTCGCTGCGTTGAGGCTTGCGTTTATGGTACGC
TGGACTTTGTGGGATACCCTCGCTTTCCTGCTCCTGTTGAGTTTATTGCTGCCGTCATTGCTTA
TTATGTTCATCCCGTCAACATTCAAACGGCCTGTCTCATCATGGAAGGCGCTGAATTTACGGAA
AACATTATTAATGGCGTCGAGCGTCCGGTTAAAGCCGCTGAATTGTTCGCGTTTACCTTGCGTG
TACGCGCAGGAAACACTGACGTTCTTACTGACGCAGAAGAAAACGTGCGTCAAAAATTACGTGC
GGAAGGAGTGATGTAATGTCTAAAGGTAAAAAAACGTTCTGGCGCTCGCCCTGGTCGTCCGCAGC
CGTTGCGAGGTACTAAAGGCAAGCGTAAAGGCGCTCGTCTTTGGTATGTAGGTGGTCAACAATT
TTAATTGCAGGGGCTTCGGCCCCTTACTTGAGGATAAATTATGTCTAATATTCAAACTGGCGCC
GAGCGTATGCCGCATGACCTTTCCCATCTTGGCTTCCTTGCTGGTCAGATTGGTCGTCTTATTA
CCATTTCAACTACTCCGGTTATCGCTGGCGACTCCTTCGAGATGGACGCCGTTGGCGCTCTCCG
TCTTTCTCCATTGCGTCGTGGCCTTGCTATTGACTCTACTGTAGACATTTTTACTTTTTATGTC
CCTCATCGTCACGTTTATGGTGAACAGTGGATTAAGTTCATGAAGGATGGTGTTAATGCCACTC
CTCTCCCGACTGTTAACACTACTGGTTATATTGACCATGCCGCTTTTCTTGGCACGATTAACCC
TGATACCAATAAAATCCCTAAGCATTTGTTTCAGGGTTATTTGAATATCTATAACAACTATTTT
AAAGCGCCGTGGATGCCTGACCGTACCGAGGCTAACCCTAATGAGCTTAATCAAGATGATGCTC
GTTATGGTTTCCGTTGCTGCCATCTCAAAAACATTTGGACTGCTCCGCTTCCTCCTGAGACTGA
GCTTTCTCGCCAAATGACGACTTCTACCACATCTATTGACATTATGGGTCTGCAAGCTGCTTAT
GCTAATTTGCATACTGACCAAGAACGTGATTACTTCATGCAGCGTTACCATGATGTTATTTCTT
CATTTGGAGGTAAAACCTCTTATGACGCTGACAACCGTCCTTTACTTGTCATGCGCTCTAATCT
```

```
CTGGGCATCTGGCTATGATGTTGATGGAACTGACCAAACGTCGTTAGGCCAGTTTTCTGGTCGT
GTTCAACAGACCTATAAACATTCTGTGCCGCGTTTCTTTGTTCCTGAGCATGGCACTATGTTTA
CTCTTGCGCTTGTTCGTTTTCCGCCTACTGCGACTAAAGAGATTCAGTACCTTAACGCTAAAGG
TGCTTTGACTTATACCGATATTGCTGGCGACCCTGTTTTGTATGGCAACTTGCCGCCGCGTGAA
ATTTCTATGAAGGATGTTTTCCGTTCTGGTGATTCGTCTAAGAAGTTTAAGATTGCTGAGGGTC
AGTGGTATCGTTATGCGCCTTCGTATGTTTCTCCTGCTTATCACCTTCTTGAAGGCTTCCCATT
CATTCAGGAACCGCCTTCTGGTGATTTGCAAGAACGCGTACTTATTCGCCACCATGATTATGAC
CAGTGTTTCCAGTCCGTTCAGTTGTTGCAGTGGAATAGTCAGGTTAAATTTAATGTGACCGTTT
ATCGCAATCTGCCGACCACTCGCGATTCAATCATGACTTCGTGATAAAAGATTGAGTGTGAGGT
TATAACGCCGAAGCGGTAAAAATTTTAATTTTTGCCGCTGAGGGGTTGACCAAGCGAAGCGCGG
TAGGTTTTCTGCTTAGGAGTTTAATCATGTTTCAGACTTTTATTTCTCGCCATAATTCAAACTT
TTTTTCTGATAAGCTGGTTCTCACTTCTGTTACTCCAGCTTCTTCGGCACCTGTTTTACAGACA
CCTAAAGCTACATCGTCAACGTTATATTTTGATAGTTTGACGGTTAATGCTGGTAATGGTGGTT
TTCTTCATTGCATTCAGATGGATACATCTGTCAACGCCGCTAATCAGGTTGTTTCTGTTGGTGC
TGATATTGCTTTTGATGCCGACCCTAAATTTTTTGCCTGTTTGGTTCGCTTTGAGTCTTCTTCG
GTTCCGACTACCCTCCCGACTGCCTATGATGTTTATCCTTTGAATGGTCGCCATGATGGTGGTT
ATTATACCGTCAAGGACTGTGTGACTATTGACGTCCTTCCCCGTACGCCGGGCAATAACGTTTA
TGTTGGTTTCATGGTTTGGTCTAACTTTACCGCTACTAAATGCCGCGGATTGGTTTCGCTGAAT
CAGGTTATTAAAGAGATTATTTGTCTCCAGCCACTTAAGTGAGGTGATTTATGTTTGGTGCTAT
TGCTGGCGGTATTGCTTCTGCTCTTGCTGGTGGCGCCATGTCTAAATTGTTTGGAGGCGGTCAA
AAAGCCGCCTCCGGTGGCATTCAAGGTGATGTGCTTGCTACCGATAACAATACTGTAGGCATGG
GTGATGCTGGTATTAAATCTGCCATTCAAGGCTCTAATGTTCCTAACCCTGATGAGGCCGCCCC
TAGTTTTGTTTCTGGTGCTATGGCTAAAGCTGGTAAAGGACTTCTTGAAGGTACGTTGCAGGCT
GGCACTTCTGCCGTTTCTGATAAGTTGCTTGATTTGGTTGGACTTGGTGGCAAGTCTGCCGCTG
ATAAAGGAAAGGATACTCGTGATTATCTTGCTGCTGCATTTCCTGAGCTTAATGCTTGGGAGCG
TGCTGGTGCTGATGCTTCCTCTGCTGGTATGGTTGACGCCGGATTTGAGAATCAAAAAGAGCTT
ACTAAAATGCAACTGGACAATCAGAAAGAGATTGCCGAGATGCAAAATGAGACTCAAAAAGAGA
TTGCTGGCATTCAGTCGGCGACTTCACGCCAGAATACGAAAGACCAGGTATATGCACAAAATGA
GATGCTTGCTTATCAACAGAAGGAGTCTACTGCTCGCGTTGCGTCTATTATGGAAAACACCAAT
CTTTCCAAGCAACAGCAGGTTTCCGAGATTATGCGCCAAATGCTTACTCAAGCTCAAACGGCTG
GTCAGTATTTTACCAATGACCAAATCAAAGAAATGACTCGCAAGGTTAGTGCTGAGGTTGACTT
AGTTCATCAGCAAACGCAGAATCAGCGGTATGGCTCTTCTCATATTGGCGCTACTGCAAAGGAT
ATTTCTAATGTCGTCACTGATGCTGCTTCTGGTGTGGTTGATATTTTTCATGGTATTGATAAAG
CTGTTGCCGATACTTGGAACAATTTCTGGAAAGACGGTAAAGCTGATGGTATTGGCTCTAATTT
GTCTAGGAAATAACCGTCAGGATTGACACCCTCCCAATTGTATGTTTTCATGCCTCCAAATCTT
GGAGGCTTTTTTATGGTTCGTTCTTATTACCCTTCTGAATGTCACGCTGATTATTTTGACTTTG
AGCGTATCGAGGCTCTTAAACCTGCTATTGAGGCTTGTGGCATTTCTACTCTTTCTCAATCCCC
AATGCTTGGCTTCCATAAGCAGATGGATAACCGCATCAAGCTCTTGGAAGAGATTCTGTCTTTT
CGTATGCAGGGCGTTGAGTTCGATAATGGTGATATGTATGTTGACGGCCATAAGGCTGCTTCTG
ACGTTCGTGATGAGTTTGTATCTGTTACTGAGAAGTTAATGGATGAATTGGCACAATGCTACAA
TGTGCTCCCCCAACTTGATATTAATAACACTATAGACCACCGCCCCGAAGGGGACGAAAAATGG
TTTTTAGAGAACGAGAAGACGGTTACGCAGTTTTGCCGCAAGCTGGCTGCTGAACGCCCTCTTA
AGGATATTCGCGATGAGTATAATTACCCCAAAAAGAAAGGTATTAAGGATGAGTGTTCAAGATT
GCTGGAGGCCTCCACTATGAAATCGCGTAGAGGCTTTGCTATTCAGCGTTTGATGAATGCAATG
CGACAGGCTCATGCTGATGGTTGGTTTATCGTTTTTGACACTCTCACGTTGGCTGACGACCGAT
```

```
TAGAGGCGTTTTATGATAATCCCAATGCTTTGCGTGACTATTTTCGTGATATTGGTCGTATGGT
TCTTGCTGCCGAGGGTCGCAAGGCTAATGATTCACACGCCGACTGCTATCAGTATTTTTGTGTG
CCTGAGTATGGTACAGCTAATGGCCGTCTTCATTTCCATGCGGTGCACTTTATGCGGACACTTC
CTACAGGTAGCGTTGACCCTAATTTTGGTCGTCGGGTACGCAATCGCCGCCAGTTAAATAGCTT
GCAAAATACGTGGCCTTATGGTTACAGTATGCCCATCGCAGTTCGCTACACGCAGGACGCTTTT
TCACGTTCTGGTTGGTTGTGGCCTGTTGATGCTAAAGGTGAGCCGCTTAAAGCTACCAGTTATA
TGGCTGTTGGTTTCTATGTGGCTAAATACGTTAACAAAAAGTCAGATATGGACCTTGCTGCTAA
AGGTCTAGGAGCTAAAGAATGGAACAACTCACTAAAAACCAAGCTGTCGCTACTTCCCAAGAAG
CTGTTCAGAATCAGAATGAGCCGCAACTTCGGGATGAAAATGCTCACAATGACAAATCTGTCCA
CGGAGTGCTTAATCCAACTTACCAAGCTGGGTTACGACGCGACGCCGTTCAACCAGATATTGAA
GCAGAACGCAAAAAGAGAGATGAGATTGAGGCTGGGAAAAGTTACTGTAGCCGACGTTTTGGCG
GCGCAACCTGTGACGACAAATCTGCTCAAATTTATGCGCGCTTCGATAAAAATGATTGGCGTAT
CCAACCTGCA
```

Protein coding regions of a DNA sequence are first *transcribed* into messenger RNA and then *translated* into protein. A *codon* of three DNA nucleotides is transcribed into a codon of three complementary RNA nucleotides, which is translated in turn into a single amino acid within a protein. A fragment of single-stranded DNA sequence has three possible *reading frames*, and translation takes place in an *open reading frame*, a sequence of codons from a certain *start codon* to a certain *stop codon* and containing no further stop codon.

Example 1.2
Reading frame 2 of the DNA sequence of *Bacteriophage φ-X174* from the previous example contains 15 open reading frames of more than 108 nucleotides, which can potentially code for proteins of more than 36 amino acids. Only two of them, shown highlighted in the next table, actually code for a protein.

sequence fragment	start	stop	length
ATG⬜TGA	17	136	120
ATG⬛TAA	848	964	117
ATG⬛⬛⬛⬛⬛⬛⬛⬛⬛TGA	1,001	2,284	1,284
ATG⬜⬜⬜⬜⬜⬜⬜⬜TGA	1,031	2,284	1,254
ATG⬜⬜⬜⬜⬜⬜⬜TGA	1,130	2,284	1,155
ATG⬜⬜⬜⬜⬜⬜TGA	1,256	2,284	1,029
ATG⬜⬜⬜⬜⬜TGA	1,421	2,284	864
ATG⬜⬜⬜⬜TGA	1,550	2,284	735
ATG⬜⬜⬜⬜TGA	1,580	2,284	705
ATG⬜⬜⬜⬜TGA	1,637	2,284	648
ATG⬜⬜⬜TGA	1,715	2,284	570
ATG⬜⬜⬜TGA	1,850	2,284	435
ATG⬜⬜TGA	1,991	2,284	294
ATG⬜TGA	2,543	2,731	189
ATG⬜TGA	2,552	2,731	180

The reading frame determines the actual amino acids encoded by a gene. For instance, the DNA sequence fragment GTCGCCATGATGGTGGTTATT ATACCGTCAAGGACTGTGTGACTA can be read in the 5′ to 3′ direction in the following three frames:

```
1 GTC GCC ATG ATG GTG GTT ATT ATA CCG TCA AGG ACT GTG TGA CTA
2  TCG CCA TGA TGG TGG TTA TTA TAC CGT CAA GGA CTG TGT GAC TA
3   CGC CAT GAT GGT GGT TAT TAT ACC GTC AAG GAC TGT GTG ACT A
```

A fragment of double-stranded DNA sequence, on the other hand, has six possible reading frames, three in each direction. An open reading frame begins with the start codon ATG (methionine) in most species and ends with a stop codon TAA, TAG, or TGA.

The identification of genes coding for proteins in a DNA sequence is a very difficult task. Even a simple organism such as *Bacteriophage* ϕ-*X174*, with a single-stranded DNA sequence of only 5,368 nucleotides, has a total of 117 open reading frames, only 11 of which actually code for a protein. There are several other biological signals that help the computational biologist in the task of gene finding, but to start with, the known protein with the shortest sequence has 8 amino acids and, thus, short open reading frames, with fewer than $3 + 24 + 3 = 30$ nucleotides, cannot code for a protein.

A first algorithmic problem consists in extracting all open reading frames in the three reading frames of a DNA sequence fragment. The problem has to be solved on the reverse complement of the sequence as well if the DNA is double stranded.

Given a fragment of DNA sequence S of n nucleotides, let $S[i]$ denote the i-th nucleotide of sequence S, for $1 \leqslant i \leqslant n$. Thus, in the sequence $S = $ GTC GCCATGATGGTGGTTATTATACCGTCAAGGACTGTGTGACTA, which has $n = 45$ nucleotides, $S[1] = $ G, $S[2] = $ T, $S[3] = $ C, and $S[n] = $ A. Let also $S[i, \ldots, j]$, where $i \leqslant j$, denote the fragment of S containing nucleotides $S[i], S[i+1], \ldots, S[j]$. For instance, $S[1, \ldots, 4] = $ GTCG, and $S[1, \ldots, n] = S$. Therefore, $S[i, \ldots, i] = S[i]$ for any $1 \leqslant i \leqslant n$.

With this notation, an open reading frame is a fragment $S[i, \ldots, j]$, of length $j-i+1$, such that $S[i, \ldots, i+2]$ is the start codon ATG and $S[j-2, \ldots, j]$ is one of the stop codons TAA, TAG, or TGA. This is actually not quite the case. It has to be at least 30 nucleotides long, that is, it must fulfill $j-i+1 \geqslant 30$. And it cannot contain any other stop codon, that is, it must also fulfill the condition $S[k, \ldots, k+2] \notin \{TAA, TAG, TGA\}$ for $i+3 \leqslant k \leqslant j-6$. In the sequence fragment $S = $ GTCGCCATGATGGTGGTTATTATACCGTCAAGGACTGTG TGACTA, for instance, $S[7, \ldots, 42]$ is an open reading frame, as it begins with $S[7, \ldots, 9] = $ ATG, ends with $S[40, \ldots, 42] = $ TGA, and has no other codon between $S[10]$ and $S[39]$ equal to TAA, TAG, or TGA.

```
GTC GCC ATG ATG GTG GTT ATT ATA CCG TCA AGG ACT GTG TGA CTA
```

The reading frame determines a partition of the DNA sequence fragment S in codons of three consecutive nucleotides. In reading frame 1, the first codon is $S[1, \ldots, 3]$, the second codon is $S[4, \ldots, 6]$, and so on. In reading frame 2, however, the first codon is $S[2, \ldots, 4]$, and the second codon is $S[5, \ldots, 7]$. The first codon in reading frame 3 is $S[3, \ldots, 5]$.

In a given reading frame, the codons can be accessed by sliding a window of length three over the sequence, starting at position 1, 2, or 3, depending on the reading frame. The sliding window is thus a kind of looking glass under which a codon of the sequence can be seen and accessed:

```
       1-3 4-6 7-9 ... ... ... ... ... ... ... ... ... ... ... ..n
  ⟶  GTC GCC ATG ATG GTG GTT ATT ATA CCG TCA AGG ACT GTG TGA CTA
       GTC GCC ATG ATG GTG GTT ATT ATA CCG TCA AGG ACT GTG TGA CTA
       GTC GCC ATG ATG GTG GTT ATT ATA CCG TCA AGG ACT GTG TGA CTA
       ...
       GTC GCC ATG ATG GTG GTT ATT ATA CCG TCA AGG ACT GTG TGA CTA
       GTC GCC ATG ATG GTG GTT ATT ATA CCG TCA AGG ACT GTG TGA CTA
```

Consider, as a first example, the problem of finding an open reading frame in a reading frame of a sequence, and let $S[k, \ldots, k+2]$ be the codon under the sliding window in the given reading frame. Starting with an initial position k given by the reading frame, the sliding window has to be displaced by three nucleotides each time until accessing a start codon and then continue sliding by three nucleotides each time until accessing a stop codon. Again, this is not actually quite the case. The reading frame of the sequence fragment could contain no start codon at all, or it could contain a start codon but no stop codon, and the search for the beginning or the end of an open reading frame might go beyond the end of the sequence.

The first start codon in the k-th reading frame of a given DNA sequence fragment S of n nucleotides can be found by sliding a window $S[i, \ldots, i+2]$ of three nucleotides along $S[k, \ldots, n]$, until either $i + 2 > n$ or $S[i, \ldots, i+2] =$ AGT. In the following description, the initial position i of the candidate start codon is incremented by three as long as the codon does not fall off the sequence (that is, $i + 2 \leqslant n$) and is not a start codon (that is, $S[i, \ldots, i+2] \neq$ AGT).

```
i ← k
while i + 2 ≤ n and S[i, . . . , i+2] ≠ AGT do
    i ← i + 3
if i + 2 ≤ n then
    output S[i, . . . , i+2]
```

After having found a start codon $S[i, \ldots, i+2]$, the first stop codon can be found by sliding a window $S[j, \ldots, j+2]$ of three nucleotides, this time along $S[i + 3, \ldots, n]$, until either $j + 2 > n$ or $S[j, \ldots, j+2] \in \{\text{TAA, TAG, TGA}\}$. In the following description, the initial position j of the candidate stop codon is incremented by three as long as the codon does not fall off the sequence (that is, $j + 2 \leqslant n$) and the candidate codon is not a stop codon (that is, with $S[j, \ldots, j + 2] \notin \{\text{TAA}, \text{TAG}, \text{TGA}\}$).

```
j ← i + 3
while j + 2 ≤ n and S[j, . . . , j + 2] ∉ {TAA, TAG, TGA} do
    j ← j + 3
if j + 2 ≤ n then
    output S[j, . . . , j + 2]
```

Now, the problem of extracting the first open reading frame in the k-th reading frame of a DNA sequence fragment S of length n can be solved by putting together the search for a start codon and the search for a stop codon. In the following description, the start codon is $S[i, \ldots, i + 2]$ and the stop codon is $S[j, \ldots, j + 2]$ of the sequence and, thus, the open reading frame $S[i, \ldots, j + 2]$ is output.

```
i ← k
while i + 2 ≤ n and S[i, . . . , i + 2] ≠ AGT do
    i ← i + 3
if i + 2 ≤ n then
    j ← i + 3
    while j + 2 ≤ n and S[j, . . . , j + 2] ∉ {TAA, TAG, TGA} do
        j ← j + 3
    if j + 2 ≤ n then
        output S[i, . . . , j + 2]
```

Notice that only the first start codon is found, and the first stop codon after this start codon will then signal the end of the first open reading frame. There may be other start codons in the sequence fragment between the first start codon and the first stop codon, however, which would signal shorter open reading frames contained in the first open reading frame found. Also, the first open reading frame might be shorter than 30 nucleotides, much too short to actually code for a protein.

The problem of extracting all open reading frames of at least 30 nucleotides in the k-th reading frame of a DNA sequence fragment S of length $n \geqslant 30$ can be solved by repeating the previous procedure for each start codon found in turn, checking that the open reading frames thus found have at least 30 nucleotides, as follows.

```
i ← k
while i + 2 ≤ n do
    if S[i, . . . , i + 2] = AGT then
        j ← i + 3
```

```
        while j + 2 ⩽ n and S[j, ..., j + 2] ∉ {TAA, TAG, TGA} do
            j ← j + 3
        if j + 2 ⩽ n then
            if j + 2 − i + 1 ⩾ 30 then
                output S[i, ..., j + 2]
    i ← i + 3
```

Finally, the problem of extracting all open reading frames of at least 30 nucleotides in the three reading frames of a DNA sequence fragment S of length $n \geqslant 30$ can be solved by repeating the previous procedure for each reading frame and for each start codon in turn, checking again that the open reading frames thus found have at least 30 nucleotides. In the following description, the whole algorithm is wrapped into a procedure that takes the DNA sequence fragment S as input and reports each of the open reading frames of S as output.

```
procedure extract_open_reading_frames(S)
    n ← length(S)
    for i ← 1, 2, 3 do
        while i + 2 ⩽ n do
            if S[i, ..., i + 2] = AGT then
                j ← i + 3
                while j + 2 ⩽ n and S[j, ..., j + 2] ∉ {TAA, TAG, TGA} do
                    j ← j + 3
                if j + 2 ⩽ n then
                    if j + 2 − i + 1 ⩾ 30 then
                        output S[i, ..., j + 2]
            i ← i + 3
```

The previous algorithm for extracting all open reading frames in the three reading frames of a given DNA sequence fragment can be implemented in Perl in a straightforward way. An open reading frame $S[i, ..., j + 2]$ is represented as the fragment of sequence `$seq` with starting position `$i` and length `$j+2-$i+1`, that is, `substr($seq,$i,$j+2-$i+1)`. Notice that Perl arrays do not start with position 1 but, rather, with position 0 and, thus, the first codon is `substr($seq,0,3)`, the last nucleotide is `substr($seq,$n-1,1)`, and the three reading frames `$r` are numbered $0, 1, 2$. This is all shown in the following Perl script.

```
sub extract_open_reading_frames {
  my $seq = shift;
  for my $r (0,1,2) {
    for (my $i = $r; $i <= length($seq)-3; $i += 3) {
      if (substr($seq,$i,3) eq "ATG") {
```

```perl
            my $j = $i+3;
            while ($j <= length($seq)-3 &&
                   substr($seq,$j,3) ne "TAA" &&
                   substr($seq,$j,3) ne "TAG" &&
                   substr($seq,$j,3) ne "TGA") {
                $j += 3;
            }
            if ($j <= length($seq)-3) {
                my $len = $j+2-$i+1;
                if ($len >= 30) {
                    print substr($seq,$i,$j+2-$i+1),"\n";
                }
            }
        }
      }
    }
}
```

The algorithm for extracting all open reading frames in the three reading frames of a given DNA sequence fragment can also be implemented in R in a straightforward way, as shown in the following R script.

```r
extract.open.reading.frames <- function (seq) {
  for (i in 1:3) {
    while (i+2 <= nchar(seq)) {
      if (substr(seq,i,i+2) == "ATG") {
        j <- i + 3
        while (j+2 <= nchar(seq) &&
               substr(seq,j,j+2) != "TAA" &&
               substr(seq,j,j+2) != "TAG" &&
               substr(seq,j,j+2) != "TGA") {
          j <- j + 3
        }
        if (j+2 <= nchar(seq)) {
          if (j+2-i+1 >= 30) {
            print(c(i,j+2,substr(seq,i,j+2)))
          }
        }
      }
      i <- i + 3
    }
  }
}
```

There are indeed 104 open reading frames of at least 30 nucleotides and up to 1,284 nucleotides in the DNA sequence of *Bacteriophage ϕ-X174*.

```
> seq <- "GAGTTTTATCGCTTCCATGACGCAGAAGTTAAC...CGGATA"
> extract.open.reading.frames(seq)
[1]  "133"   "393"   "ATGAGAAAATTCGACCTATCCTTG...TCATGA"
[1]  "250"   "393"   "ATGCTTGGCACGTTCGTCAAGGAC...TCATGA"
[1]  "568"   "843"   "ATGGTACGCTGGACTTTGTGGGAT...GAGTGA"
[1]  "643"   "843"   "ATGTTCATCCCGTCAACATTCAAA...GAGTGA"
[1]  "715"   "843"   "ATGGCGTCGAGCGTCCGGTTAAAG...GAGTGA"
[1]  "2395"  "2922"  "ATGTTTCAGACTTTTATTTCTCGC...AAGTGA"
[1]  "2578"  "2922"  "ATGGATACATCTGTCAACGCCGCT...AAGTGA"
[1]  "2827"  "2922"  "ATGGTTTGGTCTAACTTTACCGCT...AAGTGA"
[1]  "3037"  "3066"  "ATGTGCTTGCTACCGATAACAATA...CTGTAG"
[1]  "3076"  "3684"  "ATGCTGGTATTAAATCTGCCATTC...AAATGA"
[1]  "3109"  "3684"  "ATGTTCCTAACCCTGATGAGGCCG...AAATGA"
[1]  "3124"  "3684"  "ATGAGGCCGCCCCTAGTTTTGTTT...AAATGA"
[1]  "3316"  "3684"  "ATGCTTGGGAGCGTGCTGGTGCTG...AAATGA"
[1]  "3340"  "3684"  "ATGCTTCCTCTGCTGGTATGGTTG...AAATGA"
[1]  "3439"  "3684"  "ATGAGACTCAAAAAGAGATTGCTG...AAATGA"
[1]  "3508"  "3684"  "ATGCACAAAATGAGATGCTTGCTT...AAATGA"
[1]  "3517"  "3684"  "ATGAGATGCTTGCTTATCAACAGA...AAATGA"
[1]  "3742"  "3930"  "ATGGCTCTTCTCATATTGGCGCTA...GATTGA"
[1]  "3784"  "3930"  "ATGTCGTCACTGATGCTGCTTCTG...GATTGA"
[1]  "3796"  "3930"  "ATGCTGCTTCTGGTGTGGTTGATA...GATTGA"
[1]  "3826"  "3930"  "ATGGTATTGATAAAGCTGTTGCCG...GATTGA"
[1]  "3886"  "3930"  "ATGGTATTGGCTCTAATTTGTCTA...GATTGA"
[1]  "3946"  "4263"  "ATGTTTTCATGCCTCCAAATCTTG...AGTTAA"
[1]  "4186"  "4263"  "ATGGTGATATGTATGTTGACGGCC...AGTTAA"
[1]  "4198"  "4263"  "ATGTTGACGGCCATAAGGCTGCTT...AGTTAA"
[1]  "4234"  "4263"  "ATGAGTTTGTATCTGTTACTGAGA...AGTTAA"
[1]  "4267"  "4323"  "ATGAATTGGCACAATGCTACAATG...CTATAG"
[1]  "4288"  "4323"  "ATGTGCTCCCCCAACTTGATATTA...CTATAG"
[1]  "4429"  "4500"  "ATGAGTATAATTACCCCAAAAAGA...CTATGA"
[1]  "4465"  "4500"  "ATGAGTGTTCAAGATTGCTGGAGG...CTATGA"
[1]  "4537"  "4611"  "ATGCAATGCGACAGGCTCATGCTG...GATTAG"
[1]  "4555"  "4611"  "ATGCTGATGGTTGGTTTATCGTTT...GATTAG"
[1]  "4561"  "4611"  "ATGGTTGGTTTATCGTTTTTGACA...GATTAG"
[1]  "4621"  "4857"  "ATGATAATCCCAATGCTTTGCGTG...AGTTAA"
[1]  "4633"  "4857"  "ATGCTTTGCGTGACTATTTTCGTG...AGTTAA"
[1]  "4699"  "4857"  "ATGATTCACACGCCGACTGCTATC...AGTTAA"
[1]  "4744"  "4857"  "ATGGTACAGCTAATGGCCGTCTTC...AGTTAA"
[1]  "4756"  "4857"  "ATGGCCGTCTTCATTTCCATGCGG...AGTTAA"
[1]  "4774"  "4857"  "ATGCGGTGCACTTTATGCGGACAC...AGTTAA"
[1]  "4882"  "5064"  "ATGGTTACAGTATGCCCATCGCAG...GTCTAG"
[1]  "4957"  "5064"  "ATGCTAAAGGTGAGCCGCTTAAAG...GTCTAG"
[1]  "5008"  "5064"  "ATGTGGCTAAATACGTTAACAAAA...GTCTAG"
[1]  "17"    "136"   "ATGACGCAGAAGTTAACACTTTCG...AAATGA"
```

```
[1]  "230"   "331"   "ATGAGGAGAAGTGGCTTAATATGC...TTTTAA"
[1]  "284"   "331"   "ATGAGTCACATTTTGTTCATGGTA...TTTTAA"
[1]  "302"   "331"   "ATGGTAGAGATTCTCTTGTTGACA...TTTTAA"
[1]  "848"   "964"   "ATGTCTAAAGGTAAAAAACGTTCT...TTTTAA"
[1]  "1001"  "2284"  "ATGTCTAATATTCAAACTGGCGCC...TCGTGA"
[1]  "1031"  "2284"  "ATGCCGCATGACCTTTCCCATCTT...TCGTGA"
[1]  "1130"  "2284"  "ATGGACGCCGTTGGCGCTCTCCGT...TCGTGA"
[1]  "1256"  "2284"  "ATGAAGGATGGTGTTAATGCCACT...TCGTGA"
[1]  "1421"  "2284"  "ATGCCTGACCGTACCGAGGCTAAC...TCGTGA"
[1]  "1550"  "2284"  "ATGACGACTTCTACCACATCTATT...TCGTGA"
[1]  "1580"  "2284"  "ATGGGTCTGCAAGCTGCTTATGCT...TCGTGA"
[1]  "1637"  "2284"  "ATGCAGCGTTACCATGATGTTATT...TCGTGA"
[1]  "1715"  "2284"  "ATGCGCTCTAATCTCTGGGCATCT...TCGTGA"
[1]  "1850"  "2284"  "ATGTTTACTCTTGCGCTTGTTCGT...TCGTGA"
[1]  "1991"  "2284"  "ATGAAGGATGTTTTCCGTTCTGGT...TCGTGA"
[1]  "2543"  "2731"  "ATGCTGGTAATGGTGGTTTTCTTC...CTTTGA"
[1]  "2552"  "2731"  "ATGGTGGTTTTCTTCATTGCATTC...CTTTGA"
[1]  "2639"  "2731"  "ATGCCGACCCTAAATTTTTTGCCT...CTTTGA"
[1]  "2732"  "2776"  "ATGGTCGCCATGATGGTGGTTATT...GTGTGA"
[1]  "2741"  "2776"  "ATGATGGTGGTTATTATACCGTCA...GTGTGA"
[1]  "2744"  "2776"  "ATGGTGGTTATTATACCGTCAAGG...GTGTGA"
[1]  "2816"  "2878"  "ATGTTGGTTTCATGGTTTGGTCTA...CGCTGA"
[1]  "4349"  "4405"  "ATGGTTTTTAGAGAACGAGAAGAC...TGCTGA"
[1]  "51"    "221"   "ATGAGTCGAAAAATTATCTTGATA...AACTGA"
[1]  "390"   "848"   "ATGAGTCAAGTTACTGAACAATCC...ATGTAA"
[1]  "681"   "848"   "ATGGAAGGCGCTGAATTTACGGAA...ATGTAA"
[1]  "1038"  "1196"  "ATGACCTTTCCCATCTTGGCTTCC...CTGTAG"
[1]  "1212"  "1259"  "ATGTCCCTCATCGTCACGTTTATG...TCATGA"
[1]  "1263"  "1388"  "ATGGTGTTAATGCCACTCCTCTCC...ATTTGA"
[1]  "1272"  "1388"  "ATGCCACTCCTCTCCCGACTGTTA...ATTTGA"
[1]  "1317"  "1388"  "ATGCCGCTTTTCTTGGCACGATTA...ATTTGA"
[1]  "1449"  "1553"  "ATGAGCTTAATCAAGATGATGCTC...AAATGA"
[1]  "1464"  "1553"  "ATGATGCTCGTTATGGTTTCCGTT...AAATGA"
[1]  "1467"  "1553"  "ATGCTCGTTATGGTTTCCGTTGCT...AAATGA"
[1]  "1476"  "1553"  "ATGGTTTCCGTTGCTGCCATCTCA...AAATGA"
[1]  "1599"  "1775"  "ATGCTAATTTGCATACTGACCAAG...CGTTAG"
[1]  "1650"  "1775"  "ATGATGTTATTTCTTCATTTGGAG...CGTTAG"
[1]  "1653"  "1775"  "ATGTTATTTCTTCATTTGGAGGTA...CGTTAG"
[1]  "1686"  "1775"  "ATGACGCTGACAACCGTCCTTTAC...CGTTAG"
[1]  "1743"  "1775"  "ATGATGTTGATGGAACTGACCAAA...CGTTAG"
[1]  "1746"  "1775"  "ATGTTGATGGAACTGACCAAACGT...CGTTAG"
[1]  "1842"  "1928"  "ATGGCACTATGTTTACTCTTGCGC...CTTTGA"
[1]  "1962"  "1994"  "ATGGCAACTTGCCGCCGCGTGAAA...CTATGA"
[1]  "1998"  "2234"  "ATGTTTTCCGTTCTGGTGATTCGT...ATGTGA"
[1]  "2061"  "2234"  "ATGCGCCTTCGTATGTTTCTCCTG...ATGTGA"
```

```
[1]  "2073"  "2234"  "ATGTTTCTCCTGCTTATCACCTTC...ATGTGA"
[1]  "2166"  "2234"  "ATGATTATGACCAGTGTTTCCAGT...ATGTGA"
[1]  "2172"  "2234"  "ATGACCAGTGTTTCCAGTCCGTTC...ATGTGA"
[1]  "2856"  "2891"  "ATGCCGCGGATTGGTTTCGCTGAA...TATTAA"
[1]  "2931"  "3917"  "ATGTTTGGTGCTATTGCTGGCGGT...AAATAA"
[1]  "2982"  "3917"  "ATGTCTAAATTGTTTGGAGGCGGT...AAATAA"
[1]  "3069"  "3917"  "ATGGGTGATGCTGGTATTAAATCT...AAATAA"
[1]  "3156"  "3917"  "ATGGCTAAAGCTGGTAAAGGACTT...AAATAA"
[1]  "3357"  "3917"  "ATGGTTGACGCCGGATTTGAGAAT...AAATAA"
[1]  "3399"  "3917"  "ATGCAACTGGACAATCAGAAAGAG...AAATAA"
[1]  "3432"  "3917"  "ATGCAAAATGAGACTCAAAAAGAG...AAATAA"
[1]  "3522"  "3917"  "ATGCTTGCTTATCAACAGAAGGAG...AAATAA"
[1]  "3570"  "3917"  "ATGGAAAACACCAATCTTTCCAAG...AAATAA"
[1]  "3615"  "3917"  "ATGCGCCAAATGCTTACTCAAGCT...AAATAA"
[1]  "3624"  "3917"  "ATGCTTACTCAAGCTCAAACGGCT...AAATAA"
[1]  "3681"  "3917"  "ATGACTCGCAAGGTTAGTGCTGAG...AAATAA"
```

A related algorithmic problem consists of finding the longest open reading frame of a given DNA sequence fragment. The longest open reading frame often determines the correct reading frame for eukaryotes, where translation usually takes place in one reading frame only. Again, the problem has to be solved on the reverse complement of the sequence as well if the DNA is double stranded.

The previous algorithm for extracting all open reading frames can be extended to find the longest open reading frame, by keeping the position of the start and stop codon of the longest open reading frame found so far. In the following description, the start codon of the longest open reading frame found so far is $S[i', \ldots, i'+2]$, and the corresponding stop codon is $S[j', \ldots, j'+2]$.

---

function longest_open_reading_frame(S)
    $i' \leftarrow j' \leftarrow 0$
    $n \leftarrow \text{length}(S)$
    for $i \leftarrow 1, 2, 3$ **do**
        while $i + 2 \leqslant n$ **do**
            if $S[i, \ldots, i+2] = \text{AGT}$ **then**
                $j \leftarrow i + 3$
                while $j + 2 \leqslant n$ **and** $S[j, \ldots, j+2] \notin \{\text{TAA}, \text{TAG}, \text{TGA}\}$ **do**
                    $j \leftarrow j + 3$
                if $j + 2 \leqslant n$ **then**
                    if $j + 2 - i + 1 > j' + 2 - i' + 1$ **then**
                        $i' \leftarrow i$
                        $j' \leftarrow j$
            $i \leftarrow i + 3$
    return $(i', j' + 2)$

The previous algorithm for finding the longest open reading frame of a given DNA sequence fragment can be implemented in Perl in a straightforward way, as shown in the following Perl script.

```perl
sub longest_open_reading_frame {
  my $seq = shift;
  my ($ii,$jj) = (0,0);
  for my $r (0,1,2) {
    for (my $i = $r; $i <= length($seq)-3; $i += 3) {
      if (substr($seq,$i,3) eq "ATG") {
        my $j = $i+3;
        while ($j <= length($seq)-3 &&
            substr($seq,$j,3) ne "TAA" &&
            substr($seq,$j,3) ne "TAG" &&
            substr($seq,$j,3) ne "TGA") {
          $j += 3;
        }
        if ($j <= length($seq)-3) {
          my $len = $j+2-$i+1;
          if ($j+2-$i+1 > $jj+2-$ii+1) {
            $ii = $i;
            $jj = $j;
          }
        }
      }
    }
  }
  return [$ii,$jj+2];
}
```

The algorithm for finding the longest open reading frame of a given DNA sequence fragment can also be easily implemented in R, as shown in the following R script.

```r
longest.open.reading.frame <- function (seq) {
  ii <- jj <- 0
  for (i in 1:3) {
    while (i+2 <= nchar(seq)) {
      if (substr(seq,i,i+2) == "ATG") {
        j <- i + 3
        while (j+2 <= nchar(seq) &&
            substr(seq,j,j+2) != "TAA" &&
            substr(seq,j,j+2) != "TAG" &&
            substr(seq,j,j+2) != "TGA") {
          j <- j + 3
        }
```

```
        if (j+2 <= nchar(seq)) {
            if (j+2-i+1 > jj+2-ii+1) {
                ii <- i
                jj <- j
            }
        }
    }
    i <- i + 3
    }
}
c(ii,jj+2)
}
```

The longest of the 104 open reading frames with at least 30 nucleotides in the DNA sequence of *Bacteriophage φ-X174* has indeed $2{,}284 - 1{,}001 + 1 = 1{,}284$ nucleotides.

```
> longest.open.reading.frame(seq)
[1]  1001  2284
```

The actual reading frame it belongs to can be obtained by integer division. Open reading frame $S[i, \ldots, j]$ comes from reading frame $((i - 1) \bmod 3) + 1$.

```
> ((1001-1) %% 3) + 1
[1] 2
```

Bibliographic Notes

Most of the research in combinatorial pattern matching is reflected in the various editions of the Annual Symposium on Combinatorial Pattern Matching (Apostolico et al. 1992; 1993; Crochemore and Gusfield 1994; Galil and Ukkonen 1995; Hirschberg and Myers 1996; Apostolico and Hein 1997; Farach-Colton 1998; Crochemore and Paterson 1999; Giancarlo and Sankoff 2000; Amir and Landau 2001; Apostolico and Takeda 2002; Baeza-Yates et al. 2003; Sahinalp et al. 2004; Apostolico et al. 2005; Lewenstein and Valiente 2006; Ma and Zhang 2007; Ferragina and Landau 2008). There are also several books on specific aspects of combinatorial pattern matching, focused on algorithms on sequences (Stephen 1998; Crochemore and Rytter 1994; Navarro and Raffinot 2002; Crochemore and Rytter 2003; Smyth 2003; Crochemore et al. 2007).

There are several books on algorithms in computational biology which also address combinatorial pattern matching, including (Waterman 1995; Gusfield 1997; Pevzner 2000; Valiente 2002; Dwyer 2003; Jones and Pevzner 2004; Deonier et al. 2005; Kasahara and Morishita 2006). A brief introduction to bioinformatics was written by Cohen (2004). Systems biology is a quite recent discipline, although there are already a couple of textbooks (Alon 2006; Palsson 2006), and a brief introduction to systems biology was written by Kitano (2002a;b).

The algorithmic techniques used in this book are rather simple, in order to make life easier for the biologist reader. A basic understanding of algorithms and computing will be more than sufficient to follow this book, with the most advanced algorithmic technique used in the book being perhaps dynamic programming, and the presentation of the algorithms is iterative rather than recursive. The use of dynamic programming in computational biology was reviewed by Giegerich (2000) and Eddy (2004b).

Alternative Perl and R implementations for some of the algorithms presented in this book can be found within the BioPerl project (Stajich et al. 2002; Birney et al. 2009) and in the Bioconductor project (Gentleman et al. 2005; Hahne et al. 2008), respectively. See the appendices for further bibliographic notes on Perl and R.

The DNA sequence of *Bacteriophage φ-X174*, the first complete genome to be sequenced, was determined by Sanger et al. (1977). The first complete mitochondrial genome sequence of an extinct species was reported by Haddrath and Baker (2001). See also (Sanger and Dowding 1996; Green et al. 2008).

Part I

Sequence Pattern Matching

Chapter 2

Sequences

Sequences are fundamental mathematical objects that count among the most common combinatorial structures in computer science and computational biology. Basic notions underlying combinatorial algorithms on sequences, such as counting, generation, and traversal algorithms, as well as appropriate data structures for the representation of sequences, are the subject of this introductory chapter.

2.1 Sequences in Mathematics

The notion of sequence most often found in discrete mathematics is that of a (finite or infinite) ordered list of elements. The same element can appear multiple times at different positions in the sequence. A sequence thus defines an ordered multiset, that is, an ordered set of elements, each belonging to the multiset with a certain multiplicity.

Some applications of sequences in mathematics involve *labeled* sequences, where the elements have additional attributes such as, for instance, their multiplicity in the sequence.

Example 2.1
The following three sequences (shown with the elements separated by spaces, for clarity) are identical as multisets of elements, but they are all different sequences.

```
A B C C D D D E E E E E F F F F F F F F
A B C D E F C D E F D E F E F E F F F F
F F F F F F F F E E E E E D D D C C B A
```

Actually, they all define the same labeled sequence: an ordered multiset with elements A and B, element C twice, three occurrences of element D, five occurrences of element E, and eight occurrences of element F.

```
(A,1) (B,1) (C,2) (D,3) (E,5) (F,8)
```

2.1.1   Counting Labeled Sequences

Determining the number of possible labeled sequences is a trivial exercise in mathematics. Here, *counting* refers to determining the number of possible sequences that have certain properties, while *generation* is the process of obtaining the actual sequences with these properties such as, for instance, all labeled sequences.

Assume the elements are drawn from the alphabet $\{A, B\}$. There are $2^1 = 2$ ways to make a sequence of length 1 with elements from this alphabet:

```
A
B
```

Each of these two sequences can be extended in two different ways to make a sequence of length 2 and, thus, there are $2 \cdot 2 = 2^2 = 4$ possible sequences of length 2 with elements from that alphabet:

```
A A
A B
B A
B B
```

Each of these four sequences can now be extended in two different ways to make a sequence of length 3 and, thus, there are $2 \cdot 4 = 2^3 = 8$ possible sequences of length 3 with elements from that alphabet:

```
A A A
A A B
A B A
A B B
B A A
B A B
B B A
B B B
```

In general, there are 2^n possible sequences of length n with elements from that alphabet, and there are m^n possible sequences of length n with elements from an alphabet of size m, as shown in the following R script for sequence length $1 \leqslant n \leqslant 12$ and alphabet size $1 \leqslant m \leqslant 6$.

```
> outer(1:12,1:6,function(n,m)m^n)
      [,1] [,2]  [,3]   [,4]    [,5]    [,6]
[1,]     1    2     3      4       5       6
[2,]     1    4     9     16      25      36
[3,]     1    8    27     64     125     216
[4,]     1   16    81    256     625    1296
[5,]     1   32   243   1024    3125    7776
[6,]     1   64   729   4096   15625   46656
[7,]     1  128  2187  16384   78125  279936
```

[8,]	1	256	6561	65536	390625	1679616
[9,]	1	512	19683	262144	1953125	10077696
[10,]	1	1024	59049	1048576	9765625	60466176
[11,]	1	2048	177147	4194304	48828125	362797056
[12,]	1	4096	531441	16777216	244140625	2176782336

Assume now the elements are drawn from the alphabet $\{A, B, C\}$. There are $\binom{3+1-1}{3-1} = \binom{3}{2} = 3$ ways to make a labeled sequence of length 1 with elements from this alphabet:

```
(A,1)
(B,1)
(C,1)
```

There are $\binom{3+2-1}{3-1} = \binom{4}{2} = 6$ ways to make a labeled sequence of length 2 with elements from that alphabet:

```
(A,1) (B,1)
(A,1) (C,1)
(A,2)
(B,1) (C,1)
(B,2)
(C,2)
```

Also, there are $\binom{3+3-1}{3-1} = \binom{5}{2} = 10$ ways to make a labeled sequence of length 3 with elements from that alphabet:

```
(A,1) (B,1) (C,1)
(A,1) (B,2)
(A,1) (C,2)
(A,2) (B,1)
(A,2) (C,1)
(A,3)
(B,1) (C,2)
(B,2) (C,1)
(B,3)
(C,3)
```

In general, there are $\binom{3+n-1}{3-1} = (n+2)(n+1)/2$ possible sequences of length n with elements from that alphabet, and there are $\binom{m+n-1}{m-1}$ possible labeled sequences of length n with elements from an alphabet of size m, as shown in the following R script for labeled sequence length $1 \leqslant n \leqslant 20$ and alphabet size $1 \leqslant m \leqslant 8$.

```
> outer(1:20,1:8,function(n,m)choose(m+n-1,m-1))
```

	[,1]	[,2]	[,3]	[,4]	[,5]	[,6]	[,7]	[,8]
[1,]	1	2	3	4	5	6	7	8
[2,]	1	3	6	10	15	21	28	36

[3,]	1	4	10	20	35	56	84	120
[4,]	1	5	15	35	70	126	210	330
[5,]	1	6	21	56	126	252	462	792
[6,]	1	7	28	84	210	462	924	1716
[7,]	1	8	36	120	330	792	1716	3432
[8,]	1	9	45	165	495	1287	3003	6435
[9,]	1	10	55	220	715	2002	5005	11440
[10,]	1	11	66	286	1001	3003	8008	19448
[11,]	1	12	78	364	1365	4368	12376	31824
[12,]	1	13	91	455	1820	6188	18564	50388
[13,]	1	14	105	560	2380	8568	27132	77520
[14,]	1	15	120	680	3060	11628	38760	116280
[15,]	1	16	136	816	3876	15504	54264	170544
[16,]	1	17	153	969	4845	20349	74613	245157
[17,]	1	18	171	1140	5985	26334	100947	346104
[18,]	1	19	190	1330	7315	33649	134596	480700
[19,]	1	20	210	1540	8855	42504	177100	657800
[20,]	1	21	231	1771	10626	53130	230230	888030

2.2 Sequences in Computer Science

The notion of sequence most often found in computer science is also that of an ordered list of elements, where it is often called a *string* of characters or symbols drawn from an underlying set or alphabet. The alphabet itself is usually ordered, thus allowing the definition of an ordering among sequences, called the dictionary or *lexicographical* order.

A sequence $(x_1, x_2, \ldots, x_k)$ precedes a sequence $(x'_1, x'_2, \ldots, x'_\ell)$ in lexicographical order if $x_1 < x'_1$, or $x_1 = x'_1$ and $(x_2, \ldots, x_k)$ precedes $(x'_2, \ldots, x'_\ell)$ in lexicographical order, where the empty sequence precedes any non-empty sequence.

Example 2.2
The $2 + 4 + 8 + 16 = 30$ sequences of length 1 through 4 over the alphabet $\{A, B\}$ are shown in lexicographical order.

```
A
AA
AAA
AAAA
AAAB
AAB
```

```
AABA
AABB
AB
ABA
ABAA
ABAB
ABB
ABBA
ABBB
B
BA
BAA
BAAA
BAAB
BAB
BABA
BABB
BB
BBA
BBAA
BBAB
BBB
BBBA
BBBB
```

In labeled sequences, both the alphabet and the attributes associated with the elements of the sequences are usually ordered, allowing also the definition of a lexicographical ordering among labeled sequences. A labeled sequence $((x_1, n_1), (x_2, n_2), \ldots, (x_k, n_k))$ with $x_1 < x_2 < \cdots < x_n$ precedes another labeled sequence $((x_1', n_1'), (x_2', n_2'), \ldots, (x_\ell', n_\ell'))$ with $x_1' < x_2' < \cdots < x_\ell'$ in lexicographical order if $x_1 < x_1'$, or $x_1 = x_1'$ and $n_1 < n_1'$, or $x_1 = x_1'$ and $n_1 = n_1'$ and $(x_2, \ldots, x_k)$ precedes $(x_2', \ldots, x_\ell')$ in lexicographical order, where the empty labeled sequence precedes any non-empty labeled sequence.

Example 2.3

The $2 + 3 + 4 + 5 = 14$ labeled sequences of length 1 through 4 over the alphabet $\{A, B\}$ are shown in lexicographical order.

```
(A,1)
(A,1)  (B,1)
(A,1)  (B,2)
(A,1)  (B,3)
(A,2)
(A,2)  (B,1)
```

```
(A,2) (B,2)
(A,3)
(A,3) (B,1)
(A,4)
(B,1)
(B,2)
(B,3)
(B,4)
```

A first assessment of the similarities and differences between two sequences can be made by means of the symmetric difference of the corresponding labeled sequences, that is, the number of elements in which the two sequences differ. While the length of the symmetric difference of the labeled sequences is equal to the difference in length of the two sequences, different weights can be given to each element of the alphabet in a particular application. Distance measures over sequences will be discussed in the next two chapters.

Example 2.4
Consider the following two sequences over the alphabet $\{A, B\}$.

```
AABAAABBAAAABBB
ABBAABBBAAABBBBAAAABBBBB
```

The corresponding labeled sequences, of length 15 and 24, are as follows.

```
(A,9)  (B,6)
(A,10) (B,14)
```

Their symmetric difference is thus the following labeled sequence, of length 9.

```
(A,1) (B,8)
```

2.2.1 Traversing Labeled Sequences

Most algorithms on sequences require a systematic method of accessing the elements of a sequence, and combinatorial pattern matching algorithms are no exception. The most common method for accessing the elements of a sequence is by traversing the ordered list of elements, from first to last.

The following Perl script illustrates the traversal of a sequence represented as a character string.

```perl
my $seq = "AABAAABBAAAABBB";

for (my $i = 0; $i < length($seq); $i++) {
   print substr($seq,$i,1),"\n";
}
```

The following R script illustrates the traversal of a sequence represented as a vector of characters.

```
> seq <- "AABAAABBAAAABBB"
> str(seq)
 chr "AABAAABBAAAABBB"
> nchar(seq)
[1] 15
> for (i in 1:nchar(seq)) print(substr(seq,i,i))
[1] "A"
[1] "A"
[1] "B"
[1] "A"
[1] "A"
[1] "A"
[1] "B"
[1] "B"
[1] "A"
[1] "A"
[1] "A"
[1] "A"
[1] "B"
[1] "B"
[1] "B"
```

The labeled sequence corresponding to a given sequence can be obtained by traversing the sequence, counting the number of occurrences of each element in the sequence.

The following Perl script illustrates the traversal of a sequence $seq, starting with position 0, to obtain the corresponding labeled sequence %seq, represented as a hash of integers indexed by the alphabet.

```
sub seq_to_labeled_seq {
  my $seq = shift;
  my %seq;
  for (my $i = 0; $i < length($seq); $i++) {
    my $elem = substr($seq,$i,1);
    $seq{$elem}++;
  }
  return \%seq;
}
```

The following R script illustrates the traversal of a sequence to obtain the corresponding labeled sequence, represented as a list of character vectors.

```
> seq.to.labeled.seq <- function (seq) {
  alphabet <- sort(unique(strsplit(seq,"")[[1]]))
```

```
  lab <- matrix(0,nrow=length(alphabet),ncol=1)
  dimnames(lab) <- list(alphabet,"count")
  for (i in 1:nchar(seq))
    lab[substr(seq,i,i),] <-
      lab[substr(seq,i,i),] + 1
  lab
}

> seq <- "AABAAABBAAAABBB"
> seq.to.labeled.seq(seq)
  count
A      9
B      6
```

The most common method for accessing the elements of a labeled sequence is by traversing the ordered list of labeled elements, from first to last. The following Perl script illustrates the traversal of a labeled sequence represented as a hash of integers indexed by the alphabet.

```
my $seq = "AABAAABBAAAABBB";
my %seq = %{ seq_to_labeled_seq($seq) };

for my $elem (sort keys %seq) {
  print "($elem,$seq{$elem})\n";
}
```

The following R script illustrates the traversal of a labeled sequence represented as a list of character vectors.

```
> seq <- "AABAAABBAAAABBB"
> lab <- seq.to.labeled.seq(seq)
> for (elem in row.names(lab))
    print(paste("(",elem,",",lab[elem,],")",sep=""))
[1] "(A,9)"
[1] "(B,6)"
```

The symmetric difference of two sequences can be obtained by traversing each of the corresponding labeled sequences in turn, computing the absolute difference of the number of occurrences of each element in the two sequences. In the following Perl script, the multiplicities in the second sequence are subtracted from the multiplicities in the first sequence, keeping the absolute value of the result.

```
sub symmetric_difference {
  my $seq1 = shift;
  my $seq2 = shift;

  my %seq1 = %{ seq_to_labeled_seq($seq1) };
```

```
  my %seq2 = %{ seq_to_labeled_seq($seq2) };

  for my $elem (keys %seq2) {
     $seq1{$elem} = abs($seq1{$elem}-$seq2{$elem});
  }

  return \%seq1;
}
```

In the following R script, the multiplicities of the elements in the second sequence are subtracted from their multiplicities in the first sequence, keeping again the absolute value of the result.

```
> symmetric.difference <- function (seq1,seq2) {
     lab1 <- seq.to.labeled.seq(seq1)
     lab2 <- seq.to.labeled.seq(seq2)
     for (elem in row.names(lab2))
        lab1[elem,] <- abs(lab1[elem,] - lab2[elem,])
     lab1
}

> seq1 <- "AABAAABBAAAABBB"
> seq2 <- "ABBAABBBAAABBBBAAAABBBBB"
> symmetric.difference(seq1,seq2)
  count
A     1
B     8
```

2.3 Sequences in Computational Biology

One of the notions of sequence most often found in computational biology is that of a genomic sequence, that is, a sequence over the alphabet of deoxyribonucleic acid (DNA) or ribonucleic acid (RNA) nucleotides. The primary structure of DNA can be represented as a sequence over the alphabet of nucleotides: the purines A (adenine) and G (guanine), and the pyrimidines C (cytosine) and T (thymine). The primary structure of RNA can also be represented as a sequence over the alphabet of nucleotides, where the pyrimidine T (thymine) is replaced by U (uracil).

Along a genomic sequence, the nucleotides are held together by a backbone of alternating phosphate and sugar residues. The sugar in DNA is 2-deoxyribose, a pentose (five-carbon) sugar, whereas it is ribose, also a pentose sugar, in RNA. The carbon atoms are numbered 1, 2, 3, 4, 5 in a nucleotide and

$1', 2', 3', 4', 5'$ in a sugar, following the conventions of organic chemistry. The phosphates form covalent bonds between the $3'$ carbon of one sugar and the $5'$ carbon of the next sugar along the backbone, thus defining a direction of the DNA sequence from the unbound $5'$ carbon to the unbound $3'$ carbon.

In vivo, DNA consists of two strands held together by hydrogen bonds between *complementary* nucleotides, which fold in space in the shape of a double helix. Adenine and thymine are complementary, and the AT base pair has two hydrogen bonds. Guanine and cytosine are also complementary, and the GC base pair has three hydrogen bonds instead. Each of the two ends of a double helix has the $3'$ end of one DNA strand and the $5'$ end of the other one and, thus, two DNA sequences are complementary if one is the reverse complement of the other, that is, if one sequence can be obtained from the other by replacing each nucleotide by its complement (A by T, C by G, G by C, T by A) and then reversing the resulting sequence.

Example 2.5
The DNA sequences AAAGGAGGTGGTCCA and TGGACCACCTCCTTT, which are common to most organisms, are complementary.

```
5'-AAAGGAGGTGGTCCA-3'
3'-TGGACCACCTCCTTT-5'
```

Unlike DNA, however, RNA often consists of a single strand, which folds back on itself in space by hydrogen bonds between short stretches of complementary nucleotides. The most usual ones are the *canonical* or Watson-Crick nucleotide pairs: adenine with uracil, and guanine with cytosine.

Example 2.6
The large subunit ribosomal RNA sequence

```
5'-GGGUGCUCAGUACGAGAGGAACCGCACCC-3'
```

folds back on itself, forming the sarcin/ricin loop, a highly conserved form of RNA secondary structure across different species.

$$
\begin{array}{llccccccc}
 & & & & ^{C}{}^{A}{}^{G}{}^{U}{}^{A} & & \\
 & & & _{U} & & {}_{C} & \\
5' & \text{GGGUGC} & & & & & G \\
 & |\ |\ |\ |\ |\ | & & & & & A \\
3' & \text{CCCACG} & & & & & {}^{G} \\
 & & & {}_{C}{}_{C} & {}_{A}{}_{A}{}_{G}{}_{G}{}^{A} \\
\end{array}
$$

The secondary structure of an RNA molecule is the spatial conformation adopted by the molecule, and it is usually divided into various forms of loops of unpaired nucleotides joined by helical stems of paired nucleotides within the RNA molecule.

Sequences also arise as a mathematical model of protein structure. In fact, the primary structure of a protein can be represented as a sequence over the alphabet of amino acids or *residues*: the *hydrophobic* or water-insoluble A (alanine), C (cysteine), I (isoleucine), L (leucine), M (methionine), F (phenylalanine), and V (valine), and the *hydrophilic* or water-soluble R (arginine), N (asparagine), D (aspartate), E (glutamate), Q (glutamine), G (glycine), H (histidine), K (lysine), P (proline), S (serine), T (threonine), W (tryptophan), and Y (tyrosine).

Example 2.7
The following protein sequence, 2bop, corresponds to a DNA binding protein from a bovine virus, the *Bovine papillomavirus*.

```
SCFALISGTANQVKCYRFRVKKNHRHRYENCTTTWFTVADNGAERQGQAQILI
TFGSPSQRQDFLKHVPLPPGMNISGFTASLDF
```

Protein sequences also fold in space, driven by various interactions among the residues and adopting particular spatial conformations that will ultimately determine their biological function.

2.3.1   Reverse Complementing DNA Sequences

Recall that DNA consists of two strands of complementary nucleotides that fold in space in the shape of a double helix, each of whose two ends has the $3'$ end of one DNA strand and the $5'$ end of the other DNA strand. Reading one such sequence in the $5'$ to $3'$ direction, the complementary sequence will be read in the reversed $3'$ to $5'$ direction, and corresponding nucleotides in the two sequences will be complementary.

Example 2.8
The following DNA sequence is also shown reversed, complemented, and reverse complemented.

```
5'-AAAGGAGGTGGTCCA-3'
3'-ACCTGGTGGAGGAAA-5'
5'-TTTCCTCCACCAGGT-3'
3'-TGGACCACCTCCTTT-5'
```

The reverse of a DNA sequence can be obtained by traversing the DNA sequence from the $3'$ to the $5'$ end, and the reverse complement can be obtained by replacing the nucleotides by their complementary nucleotides during the traversal. In the following description, a DNA sequence S of length n is reversed by putting, for each i with $1 \leqslant i \leqslant n$, the i-th nucleotide in position $n - i + 1$ of the reverse sequence, and it is reverse complemented by also replacing each nucleotide with the complementary one.

function reverse_complement(S)
    $n \leftarrow$ length(S)
    for $i \leftarrow 1$ **to** n **do**
        if $S[i] = $ A **then**
            $R[n - i + 1] \leftarrow$ T
        else if $S[i] = $ C **then**
            $R[n - i + 1] \leftarrow$ G
        else if $S[i] = $ G **then**
            $R[n - i + 1] \leftarrow$ C
        else if $S[i] = $ T **then**
            $R[n - i + 1] \leftarrow$ A
    return R

The previous algorithm for computing the reverse complement of a DNA sequence can be easily implemented in Perl by taking advantage of the array *reverse* method, which, when used in scalar context instead of array context, concatenates the elements of the array and returns a string with all elements in the opposite order, and also by taking advantage of the transliteration operator *tr* of Perl, as shown in the following Perl script.

```perl
sub reverse_complement {
  my $seq = shift;
  $seq = reverse $seq;
  $seq =~ tr/ACGT/TGCA/;
  return $seq;
}
```

The previous algorithm for computing the reverse complement of a DNA sequence can also be easily implemented in R by taking advantage of the *chartr* (character translation) function, as illustrated by the following R script.

```r
> reverse.complement <- function (seq) {
    rev <- paste(rev(unlist(strsplit(seq,split=""))),
      sep="",collapse="")
    chartr("ACGT","TGCA",rev)
}

> seq <- "AAAGGAGGTGGTCCA"
> reverse.complement(seq)
[1] "TGGACCACCTCCTTT"
```

2.3.2 Counting RNA Sequences

RNA sequences are labeled over the alphabet $\{A, C, G, U\}$ and, thus, there are 4^n possible RNA sequences of length n, as shown in the following R script for sequence length $1 \leqslant n \leqslant 12$.

```
> t(sapply(1:12,function(n)c(n,4^n)))
      [,1]      [,2]
 [1,]    1         4
 [2,]    2        16
 [3,]    3        64
 [4,]    4       256
 [5,]    5      1024
 [6,]    6      4096
 [7,]    7     16384
 [8,]    8     65536
 [9,]    9    262144
[10,]   10   1048576
[11,]   11   4194304
[12,]   12  16777216
```

A more interesting problem consists of counting the number $R(n)$ of possible RNA secondary structures of length n. While there is only one possible RNA secondary structure of length 0 (the empty sequence) and only one possible RNA secondary structure of length 1 or 2, there are two possible RNA secondary structures of length 3,

$$\text{X X X} \qquad \overset{\frown}{\text{X X X}}$$

where X stands for an A, C, G, U base along the RNA sequence, in 5′ to 3′ order. Further, there are four possible RNA secondary structures of length 4,

$$\text{X X X X} \qquad \overset{\frown}{\text{X X X X}} \qquad \overset{\frown}{\text{X X X X}} \qquad \overset{\frown}{\text{X X X X}}$$

and eight possible RNA secondary structures of length 5,

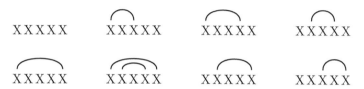

In general, in a sequence of length $n + 1$, the base at position $n + 1$ is either not paired or it is paired with the base at position j, where $1 \leqslant j \leqslant n - 1$. In the latter case, the bases at positions 1 through $j - 1$ can form any of the $R(j-1)$ possible secondary structures of length $j-1$, and the bases in positions $j+1$ through n can also form any of the $R(n-j)$ possible secondary structures of length $n-j$. Therefore, $R(0) = R(1) = R(2) = 1$ and, for $n \geqslant 2$,

$$R(n+1) = R(n) + \sum_{j=1}^{n-1} R(j-1)R(n-j)$$

This gives an algorithm for counting the number $R(n)$ of RNA secondary structures of length n. The computation of $R(n+1)$ requires the values of $R(j-1)$ and $R(n-j)$ for each $j = 1, \ldots, n-1$; that is, it requires each of the values $R(0), R(1), \ldots, R(n-1)$. In the following description, these values are computed in that order and stored in a vector, so that they are already available whenever needed during the computation of $R(n)$.

function count(n)
  $R[0] \leftarrow R[1] \leftarrow R[2] \leftarrow 1$
  for $i = 2$ **to** $n-1$ **do**
    $R[i+1] \leftarrow R[i]$
    for $j = 1$ **to** $i-1$ **do**
      $R[i+1] \leftarrow R[i+1] + R[j-1] \cdot R[i-j]$
  return $R[n]$

The previous algorithm for counting the number of RNA secondary structures of a given length can be implemented in Perl in a straightforward way, with the values $R(0), R(1), \ldots, R(n)$ stored in positions $0, 1, \ldots, n$ of an array R, as shown in the following Perl script.

```perl
sub count {
  my $n = shift;
  my @R;
  $R[0] = $R[1] = $R[2] = 1;
  for (my $i = 2; $i < $n; $i++) {
    $R[$i+1] = $R[$i];
    for (my $j = 1; $j < $i; $j++) {
      $R[$i+1] += $R[$j-1] * $R[$i-$j]
    }
  }
  return $R[$n];
}
```

The previous algorithm for counting the number of RNA secondary structures of a given length can also be easily implemented in R. The values $R(0), R(1), \ldots, R(n)$ are stored in positions $1, 2, \ldots, n+1$ of vector R, however, because unlike Perl array indexes that start at 0, R vector indexes do start at 1. This is all illustrated by the following R script.

```
> count <- function (n) {
    R <- rep(0,n)
    R[1:3] <- 1
    if (n > 2) {
      for (i in 2:(n-1)) {
        R[i+2] <- R[i+1]
        for (j in 1:(i-1)) {
          R[i+2] <- R[i+2] + R[j] * R[i-j+1]
        }
      }
    }
    R[n+1]
  }
> t(sapply(0:12,function(n)c(n,count(n))))
      [,1] [,2]
 [1,]    0    1
 [2,]    1    1
 [3,]    2    1
 [4,]    3    2
 [5,]    4    4
 [6,]    5    8
 [7,]    6   17
 [8,]    7   37
 [9,]    8   82
[10,]    9  185
[11,]   10  423
[12,]   11  978
[13,]   12 2283
```

Example 2.9
The following large subunit ribosomal RNA sequence, of length 29,

`GGGUGCUCAGUACGAGAGGAACCGCACCC`

has 8,622,571,758 possible secondary structures, only 789,564 of which are indeed possible for this RNA sequence. The remaining 8,621,782,194 secondary structures involve base pairs other than AU or CG.

2.3.3   Generating DNA Sequences

All the DNA sequences with n nucleotides can be generated by taking each of the DNA sequences with $n-1$ nucleotides in turn and then extending them with one more nucleotide.

Example 2.10

The four DNA sequences of length 1,

A	C	G	T

can each be extended in four different ways to give sequences of length 2.

AA	AC	AG	AT
CA	CC	CG	CT
GA	GC	GG	GT
TA	TC	TG	TT

Each of these DNA sequences of length 2 can in turn be extended in four different ways to give sequences of length 3.

AAA	AAC	AAG	AAT
ACA	ACC	ACG	ACT
AGA	AGC	AGG	AGT
ATA	ATC	ATG	ATT
CAA	CAC	CAG	CAT
CCA	CCC	CCG	CCT
CGA	CGC	CGG	CGT
CTA	CTC	CTG	CTT
GAA	GAC	GAG	GAT
GCA	GCC	GCG	GCT
GGA	GGC	GGG	GGT
GTA	GTC	GTG	GTT
TAA	TAC	TAG	TAT
TCA	TCC	TCG	TCT
TGA	TGC	TGG	TGT
TTA	TTC	TTG	TTT

Thus, the DNA sequences of length 1 are just the elements of the alphabet $\Sigma = \{A, C, G, T\}$, and the DNA sequences of length $n > 1$ are the result of extending the DNA sequences of length $n - 1$ with an element of Σ. This gives an algorithm for generating all DNA sequences of length $n \geqslant 1$. In the following description, *concat* refers not only to the concatenation of two sequences but also to the concatenation of a sequence of elements from an alphabet and a single element from that alphabet.

```
function words(n, Σ)
    if n = 1 then
        L ← Σ
    else
        S ← words(n − 1, Σ)
        L ← ∅
```

> **for** each word w of S **do**
>> **for** each element s of Σ **do**
>>> $w' \leftarrow \text{concat}(w, s)$
>>> $L \leftarrow L \cup \{w'\}$
> **return** L

The representation of sequences in BioPerl does not include any method to generate all DNA sequences of a certain length. However, the previous algorithm can be easily implemented in Perl by representing sequences as strings and sets of sequences as arrays of strings, while using the string concatenation operator to extend each sequence of length $n - 1$ with one nucleotide more to obtain a sequence of length n, as illustrated by the following Perl script.

```perl
sub words {
  my $n = shift;
  my $alphabet = shift;
  my @n = split "", $alphabet;
  if ($n == 1) {
    return \@n;
  } else {
    my $short = words($n - 1, $alphabet);
    my @long;
    for my $seq (@$short) {
      for my $n (@n) {
        push @long, $seq . $n;
      }
    }
    return \@long;
  }
}

my $k = 3; my $alphabet = "ACGT";
my $words = words($k, $alphabet);
my $count = 0;
map {
  $count++;
  if ($count % 4) {
    print "$_ "
  } else {
    print "$_\n"
  }
} @$words;
```

Running the previous Perl script produces the following output, which consists of the 64 DNA sequences with 3 nucleotides.

```
AAA  AAC  AAG  AAT
```

```
ACA ACC ACG ACT
AGA AGC AGG AGT
ATA ATC ATG ATT
CAA CAC CAG CAT
CCA CCC CCG CCT
CGA CGC CGG CGT
CTA CTC CTG CTT
GAA GAC GAG GAT
GCA GCC GCG GCT
GGA GGC GGG GGT
GTA GTC GTG GTT
TAA TAC TAG TAT
TCA TCC TCG TCT
TGA TGC TGG TGT
TTA TTC TTG TTT
```

The representation of sequences in R, on the other hand, includes a function words to obtain all the sequences of length n, as illustrated by the following R script.

```
> library(seqinr)
> options(width="32")
> words(length=3,alphabet=s2c("ACGT"))
 [1] "AAA" "AAC" "AAG" "AAT"
 [5] "ACA" "ACC" "ACG" "ACT"
 [9] "AGA" "AGC" "AGG" "AGT"
[13] "ATA" "ATC" "ATG" "ATT"
[17] "CAA" "CAC" "CAG" "CAT"
[21] "CCA" "CCC" "CCG" "CCT"
[25] "CGA" "CGC" "CGG" "CGT"
[29] "CTA" "CTC" "CTG" "CTT"
[33] "GAA" "GAC" "GAG" "GAT"
[37] "GCA" "GCC" "GCG" "GCT"
[41] "GGA" "GGC" "GGG" "GGT"
[45] "GTA" "GTC" "GTG" "GTT"
[49] "TAA" "TAC" "TAG" "TAT"
[53] "TCA" "TCC" "TCG" "TCT"
[57] "TGA" "TGC" "TGG" "TGT"
[61] "TTA" "TTC" "TTG" "TTT"
```

2.3.4 Representing Sequences in Perl

There are many ways in which sequences can be represented in Perl, let alone the simple representation of a sequence as a character string, and, as a matter of fact, many different Perl modules implementing various types of sequences are available for download from CPAN, the Comprehensive Perl Archive Network, at http://www.cpan.org/. Among them, let us focus on the BioPerl sequence representation, which is essentially an object-oriented representation of sequences together with a collection of sequence features.

A sequence is represented in BioPerl as a `Bio::Seq` object, which consists of the actual character string of the sequence along with a collection of optional sequence features, such as an accession number, the species name, and an indication of whether it is a DNA, an RNA, or a protein sequence.

For instance, the DNA sequence AAAGGAGGTGGTCCA can be obtained by just creating a `Bio::Seq` object for the sequence, as shown in the following Perl script.

```
use Bio::Seq;
my $seq = Bio::Seq->new(-seq => "AAAGGAGGTGGTCCA");
```

The character string is recognized then as a DNA sequence, although this information can also be made explicit when creating the `Bio::Seq` object for the sequence.

```
use Bio::Seq;
my $seq = Bio::Seq->new(-seq => "AAAGGAGGTGGTCCA",
  -alphabet => "dna");
```

However, a `Bio::Seq` object can also be obtained from the character string representing the sequence with the help of the `Bio::SeqIO` module in BioPerl, as shown in the following Perl script, where the sequence is stored in a file in the popular FASTA format.

```
use Bio::SeqIO;
my $seqio = Bio::SeqIO->new(-file => "seq.fas",
  -format => "fasta");
my $seq = $seqio->next_seq;
```

The `Bio::SeqIO` module in BioPerl can also be used to retrieve sequences from genomic databases, as shown in the following Perl script, where the complete genome sequence (4,639,675 nucleotides) of the bacterium *Escherichia coli* K-12, strain MG1655, is retrieved from the GenBank database.

```
use Bio::DB::GenBank;
my $db = Bio::DB::GenBank->new;
my $seq = $db->get_Seq_by_gi("48994873");
```

The representation of sequences in BioPerl includes additional methods for performing various operations on sequences; for instance, to access the identifier of a sequence,

```
my $id = $seq->id;
```

to obtain the length of a sequence,

```
my $len = $seq->length;
```

to get the accession number or unique biological identifier for a sequence,

```
my $acc = $seq->accession_number;
```

to access the description of a sequence,

```
my $desc = $seq->desc;
```

to obtain the subsequence of a DNA, RNA, or protein sequence contained between an initial and a final position, as a character string,

```
use Bio::Seq;
my $s = "GGGUGCUCAGUACGAGAGGAACCGCACCC";
my $seq = Bio::Seq->new(-seq => $s);
my $prefix = $seq->subseq(1,12);
my $suffix = $seq->subseq(9,$seq->length);
```

to truncate a DNA, RNA, or protein sequence from an initial to a final position into a sequence instead of just a character string,

```
my $s = "SCFALISGTANQVKCYRFRVKKNHRHRYENCTTTWFTVADNGAE
RQGQAQILITFGSPSQRQDFLKHVPLPPGMNISGFTASLDF";
my $seq = Bio::Seq->new(-seq => $s);
my $t = $seq->trunc(4,9);
```

and to obtain the reverse complement of a DNA or RNA sequence.

```
my $seq = Bio::Seq->new(-seq => "AAAGGAGGTGGTCCA");
my $rev_com = $seq->revcom;
```

A fragment of a DNA sequence can also be translated into the corresponding protein coding sequence, according to the mapping of triplets of nucleotides (codons) to amino acids that underlies the genetic code. For example, the DNA sequence fragment AAAGGAGGTGGTCCA, with the five codons AAA, GGA, GGT, GGT, CCA, translates into the protein sequence fragment KGGGP, with the five amino acids K (lysine), G (glycine), G, G, P (proline).

```
my $seq = Bio::Seq->new(-seq => "AAAGGAGGTGGTCCA");
my $prot_seq = $seq->translate;
```

2.3.5 Representing Sequences in R

There are also many ways in which sequences can be represented in R, let alone the simple representation of a sequence as a vector of characters and, as a matter of fact, many different R contributed packages implementing various types of sequences are available for download from CRAN, the Comprehensive R Archive Network, at http://cran.r-project.org/. Among them, let us focus on the seqinr (Sequences in R) sequence representation, which is essentially a vector-based representation of sequences together with a collection of retrieval and analysis functions.

A sequence is represented in the R package `seqinr` as a vector of characters, and there is a function `c2s` (character vector to string) to convert a sequence to a character string, as well as a function `s2c` (string to character vector) to convert a character string to a sequence.

```
> library(seqinr)
> s <- "TGCTTCTGACTATAATAG"
> options(width=54)
> s2c(s)
 [1] "T" "G" "C" "T" "T" "C" "T" "G" "A" "C" "T" "A"
[13] "T" "A" "A" "T" "A" "G"
> c2s(s2c(s))
[1] "TGCTTCTGACTATAATAG"
```

Further, a sequence can also be obtained from the character string representing the sequence with the help of the `read.fasta` function of the R package `seqinr`, as shown in the following R script, where the sequence is stored in a file in the popular FASTA format.

```
> fas <- read.fasta(file="seq.fas",forceDNAtolower=
    FALSE)
> getSequence(fas[[1]])
 [1] "T" "G" "C" "T" "T" "C" "T" "G" "A" "C" "T" "A"
[13] "T" "A" "A" "T" "A" "G"
```

The R package `seqinr` can also be used to retrieve sequences from genomic databases, as shown in the following R script, where the complete genome sequence (4,639,675 nucleotides) of the bacterium *Escherichia coli* K-12, strain MG1655, is retrieved from the GenBank database.

```
> library(seqinr)
> choosebank("genbank")
> query("eco","AC=U00096")
> seq <- getSequence(eco$req[[1]])
> closebank()
> length(seq)
[1] 4639675
```

The representation of sequences in R package `seqinr` includes additional functions for performing various operations on sequences; for instance, to access the accession number or unique biological identifier for a sequence,

```
> getName(eco$req[[1]])
[1] "U00096"
```

to obtain the length of a sequence,

```
> getLength(eco$req[[1]])
[1] 4639675
```

to obtain the subsequence of a DNA, RNA, or protein sequence contained between an initial and a final position,

```
> getSequence(getFrag(fas[[1]],1,12))
 [1] "T" "G" "C" "T" "T" "C" "T" "G" "A" "C" "T" "A"
> getSequence(getFrag(fas[[1]],9,length(fas[[1]])))
 [1] "A" "C" "T" "A" "T" "A" "A" "T" "A" "G"
```

and to translate a fragment of DNA sequence into the corresponding protein coding sequence, according to the mapping of triplets of nucleotides (codons) to amino acids that underlies the genetic code.

```
> translate(s2c("AAAGGAGGTGGTCCA"))
[1] "K" "G" "G" "G" "P"
```

Bibliographic Notes

The problem of counting the number of possible RNA secondary structures was first studied by Waterman (1978). See also (Schmitt and Waterman 1994).

The fact that the size of the symmetric difference of two sets or multisets is a metric is a well-known result. For a proof in the case of sets, see (Restle 1959). See also (Deza and Deza 2006, ch. 1).

The FASTA format was developed by Pearson and Lipman (1988) for representing sequences in a sequence alignment program. The GenBank database of genetic sequences is described in (Benson et al. 2008).

The complete genome sequence of *Escherichia coli* K-12 was first published in (Blattner et al. 1997).

The representation of sequences in BioPerl (Stajich et al. 2002) is described in more detail in (Birney et al. 2009). The representation of sequences in the R package seqinr is described in more detail in (Charif and Lobry 2007).

Chapter 3

Simple Pattern Matching in Sequences

Combinatorial pattern matching is the search for exact or approximate occurrences of a given pattern within a given text. When it comes to biological sequences, both the pattern and the text are sequences and the pattern matching problem becomes one of finding the occurrences of a sequence within another sequence. For instance, scanning a protein sequence for the presence of a known pattern can help annotate both the protein and the corresponding genome, and finding a sequence within another sequence can help in assessing their similarities and differences. This will be the subject of the next chapter.

A related pattern matching problem consists in finding the patterns themselves that occur within a given sequence. For instance, finding all occurrences of short words within a sequence is useful for analyzing the sequence and also for computing distances between two sequences. This is the subject of this chapter.

3.1   Finding Words in Sequences

A short word or pattern can appear many times within a genomic sequence. An example of a very short, one-letter pattern in molecular biology is the guanine cytosine content of a fragment of DNA, accounting for how often the patterns G and C occur within a given DNA sequence. Further examples of patterns in DNA sequences are the TATA box found in the promoter region of most eukaryotic genes (an occurrence of the TATAAAA pattern within the sequence) and the Pribnow box found in the promoter region of most prokaryotic genes (an occurrence of the TATAAT pattern within the DNA sequence).

3.1.1   Word Composition of Sequences

A nucleic acid or amino acid sequence can be seen as composed of a number of possibly overlapping k-mers or words of length k, for a certain $k \geqslant 1$. The k-mer composition of a sequence is given by the frequency with which each possible k-mer occurs within the sequence. The 1-mer composition is related to the GC content of a DNA sequence, and the 2-mer, 3-mer, and 4-mer compositions are also known as the di-nucleotide, tri-nucleotide, and tetra-nucleotide compositions of a DNA sequence.

Example 3.1

The fragment of DNA sequence

TTGATTACCTTATTTGATCATTACACATTGTACGCTTGTGTCAAAATATCACATGTGCCT

has the following 1-mer composition,

word	frequency
A	16
C	12
G	8
T	24

the following di-nucleotide composition,

word	freq	word	freq	word	freq	word	freq
AA	3	AC	5	AG	0	AT	8
CA	6	CC	2	CG	1	CT	3
GA	2	GC	2	GG	0	GT	4
TA	5	TC	3	TG	7	TT	8

and the following 3-mer composition.

word	freq	word	freq	word	freq	word	freq
AAA	2	AAC	0	AAG	0	AAT	1
ACA	3	ACC	1	ACG	1	ACT	0
AGA	0	AGC	0	AGG	0	AGT	0
ATA	1	ATC	2	ATG	1	ATT	4
CAA	1	CAC	2	CAG	0	CAT	3
CCA	0	CCC	0	CCG	0	CCT	2
CGA	0	CGC	1	CGG	0	CGT	0
CTA	0	CTC	0	CTG	0	CTT	2
GAA	0	GAC	0	GAG	0	GAT	2
GCA	0	GCC	1	GCG	0	GCT	1
GGA	0	GGC	0	GGG	0	GGT	0
GTA	1	GTC	1	GTG	2	GTT	0
TAA	0	TAC	3	TAG	0	TAT	2
TCA	3	TCC	0	TCG	0	TCT	0
TGA	2	TGC	1	TGG	0	TGT	4
TTA	3	TTC	0	TTG	4	TTT	1

The k-mer composition of a sequence can be computed by first obtaining all the sequences of length k over the alphabet Σ and then counting the number of occurrences of each possible k-mer within the given sequence. In the following description, the sequences of length k over the alphabet Σ are obtained using the *words* algorithm, and $S[i..i + k - 1]$ denotes the word of length k starting at position i of sequence S.

> **function** word_composition(S, k, Σ)
> $L \leftarrow$ words(k, Σ)
>   **for** each word w of L **do**
>   freq[w] $\leftarrow 0$
> $n \leftarrow$ length(S)
>   **for** $i \leftarrow 1$ **to** $n - k + 1$ **do**
> $w \leftarrow S[i..i + k - 1]$
>   freq[w] $\leftarrow$ freq[w] $+ 1$
>   **return** freq

The representation of sequences in BioPerl does not include any method to compute the k-mer composition of a sequence. However, the previous algorithm can be easily implemented in Perl by using a hash *freq* of word frequencies indexed by k-mers, as shown in the following Perl script.

```perl
sub word_composition {
  my $seq = shift;
  my $k = shift;
  my $alphabet = shift;

  my $words = words($k,$alphabet);

  my %freq;
  for my $word (@$words) {
    $freq{$word} = 0;
  }

  for my $i (0 .. length($seq)-$k) {
    my $word = substr($seq,$i,$k);
    $freq{$word}++;
  }

  return \%freq;
}
```

Running the Perl script

```perl
my $seq = "TTGATTACCTTATTTGATCATTACACATTGTACGCTTGTGTC
AAAATATCACATGTGCCT";
```

```perl
my $k = 3;
my $alphabet = "ACGT";
my $words = words($k,$alphabet);
my %freq = %{ word_composition($seq,$k,$alphabet) };
my $count = 0;
map {
  $count++;
  print "$_ ",$freq{$_};
  if ($count % 4) { print "  " } else { print "\n" }
} @$words;
```

produces the following output, which consists of the 3-mer composition of DNA sequence TTGATTACCTTATTTGATCATTACACATTGTACGCTTG TGTCAAAATATCACATGTGCCT.

AAA	2	AAC	0	AAG	0	AAT	1
ACA	3	ACC	1	ACG	1	ACT	0
AGA	0	AGC	0	AGG	0	AGT	0
ATA	1	ATC	2	ATG	1	ATT	4
CAA	1	CAC	2	CAG	0	CAT	3
CCA	0	CCC	0	CCG	0	CCT	2
CGA	0	CGC	1	CGG	0	CGT	0
CTA	0	CTC	0	CTG	0	CTT	2
GAA	0	GAC	0	GAG	0	GAT	2
GCA	0	GCC	1	GCG	0	GCT	1
GGA	0	GGC	0	GGG	0	GGT	0
GTA	1	GTC	1	GTG	2	GTT	0
TAA	0	TAC	3	TAG	0	TAT	2
TCA	3	TCC	0	TCG	0	TCT	0
TGA	2	TGC	1	TGG	0	TGT	4
TTA	3	TTC	0	TTG	4	TTT	1

The previous algorithm can also be used to obtain the k-mer composition of an amino acid sequence, as illustrated by the following Perl script.

```perl
my $seq = "MKPVTLYDVAEYAGVSYQTVSRVVNQASHVSAKTREKVEAAM
AELNYIPN";
my $k = 1;
my $alphabet = "ACDEFGHIKLMNPQRSTVWY";
my $words = words($k,$alphabet);
my %freq = %{ word_composition($seq,$k,$alphabet) };
my $count = 0;
map {
  $count++;
  print "$_ ",$freq{$_};
  if ($count % 10) { print "  " } else { print "\n" }
} @$words;
```

Running the previous script produces the following output, which consists of the amino acid composition of sequence MKPVTLYDVAEYAGVSYQTV SRVVNQASHVSAKTREKVEAAMAELNYIPN.

A	7	C	0	D	1	E	4	F	0	G	1	H	1	I	1	K	3	L	2
M	2	N	3	P	2	Q	2	R	2	S	4	T	3	V	8	W	0	Y	4

The representation of sequences in R, on the other hand, includes a function count to obtain the k-mer composition of a sequence, as illustrated by the following R script.

```
> library(seqinr)
> seq <- "TTGATTACCTTATTTGATCATTACACATTGTACGCTTGTGTCA
AAATATCACATGTGCCT"
> options(width="32")
> count(s2c(seq),word=3,alphabet=s2c("ACGT"))

AAA AAC AAG AAT ACA ACC ACG ACT
 2   0   0   1   3   1   1   0
AGA AGC AGG AGT ATA ATC ATG ATT
 0   0   0   0   1   2   1   4
CAA CAC CAG CAT CCA CCC CCG CCT
 1   2   0   3   0   0   0   2
CGA CGC CGG CGT CTA CTC CTG CTT
 0   1   0   0   0   0   0   2
GAA GAC GAG GAT GCA GCC GCG GCT
 0   0   0   2   0   1   0   1
GGA GGC GGG GGT GTA GTC GTG GTT
 0   0   0   0   1   1   2   0
TAA TAC TAG TAT TCA TCC TCG TCT
 0   3   0   2   3   0   0   0
TGA TGC TGG TGT TTA TTC TTG TTT
 2   1   0   4   3   0   4   1
```

Again, the R function *count* can also be used to obtain the k-mer composition of an amino acid sequence, as shown in the following R script.

```
> library(seqinr)
> seq <- "MKPVTLYDVAEYAGVSYQTVSRVVNQASHVSAKTREKVEAAMA
ELNYIPN"
> count(s2c(seq),word=1,alphabet=s2c("
   ACDEFGHIKLMNPQRSTVWY"))

A C D E F G H I K L M N P Q R S T V W Y
7 0 1 4 0 1 1 1 3 2 2 3 2 2 4 3 8 0 4
> options(width="48")
> count(s2c(seq),word=2,alphabet=s2c("
   ACDEFGHIKLMNPQRSTVWY"))
```

AA	AC	AD	AE	AF	AG	AH	AI	AK	AL	AM	AN	AP	AQ	AR	AS
1	0	0	2	0	1	0	0	1	0	1	0	0	0	0	1

AT	AV	AW	AY	CA	CC	CD	CE	CF	CG	CH	CI	CK	CL	CM	CN
0	0	0	0	0	0	0	0	0	0	0	0	0	0	0	0

CP	CQ	CR	CS	CT	CV	CW	CY	DA	DC	DD	DE	DF	DG	DH	DI
0	0	0	0	0	0	0	0	0	0	0	0	0	0	0	0

DK	DL	DM	DN	DP	DQ	DR	DS	DT	DV	DW	DY	EA	EC	ED	EE
0	0	0	0	0	0	0	0	0	1	0	0	1	0	0	0

EF	EG	EH	EI	EK	EL	EM	EN	EP	EQ	ER	ES	ET	EV	EW	EY
0	0	0	0	1	1	0	0	0	0	0	0	0	0	0	1

FA	FC	FD	FE	FF	FG	FH	FI	FK	FL	FM	FN	FP	FQ	FR	FS
0	0	0	0	0	0	0	0	0	0	0	0	0	0	0	0

FT	FV	FW	FY	GA	GC	GD	GE	GF	GG	GH	GI	GK	GL	GM	GN
0	0	0	0	0	0	0	0	0	0	0	0	0	0	0	0

GP	GQ	GR	GS	GT	GV	GW	GY	HA	HC	HD	HE	HF	HG	HH	HI
0	0	0	0	0	1	0	0	0	0	0	0	0	0	0	0

HK	HL	HM	HN	HP	HQ	HR	HS	HT	HV	HW	HY	IA	IC	ID	IE
0	0	0	0	0	0	0	0	0	1	0	0	0	0	0	0

IF	IG	IH	II	IK	IL	IM	IN	IP	IQ	IR	IS	IT	IV	IW	IY
0	0	0	0	0	0	0	0	1	0	0	0	0	0	0	0

KA	KC	KD	KE	KF	KG	KH	KI	KK	KL	KM	KN	KP	KQ	KR	KS
0	0	0	0	0	0	0	0	0	0	0	0	1	0	0	0

KT	KV	KW	KY	LA	LC	LD	LE	LF	LG	LH	LI	LK	LL	LM	LN
1	1	0	0	0	0	0	0	0	0	0	0	0	0	0	1

LP	LQ	LR	LS	LT	LV	LW	LY	MA	MC	MD	ME	MF	MG	MH	MI
0	0	0	0	0	0	0	1	1	0	0	0	0	0	0	0

MK	ML	MM	MN	MP	MQ	MR	MS	MT	MV	MW	MY	NA	NC	ND	NE
1	0	0	0	0	0	0	0	0	0	0	0	0	0	0	0

NF	NG	NH	NI	NK	NL	NM	NN	NP	NQ	NR	NS	NT	NV	NW	NY
0	0	0	0	0	0	0	0	0	1	0	0	0	0	0	1

PA	PC	PD	PE	PF	PG	PH	PI	PK	PL	PM	PN	PP	PQ	PR	PS
0	0	0	0	0	0	0	0	0	0	0	1	0	0	0	0

PT	PV	PW	PY	QA	QC	QD	QE	QF	QG	QH	QI	QK	QL	QM	QN
0	1	0	0	1	0	0	0	0	0	0	0	0	0	0	0

QP	QQ	QR	QS	QT	QV	QW	QY	RA	RC	RD	RE	RF	RG	RH	RI
0	0	0	0	1	0	0	0	0	0	0	1	0	0	0	0

RK	RL	RM	RN	RP	RQ	RR	RS	RT	RV	RW	RY	SA	SC	SD	SE
0	0	0	0	0	0	0	0	0	1	0	0	1	0	0	0

SF	SG	SH	SI	SK	SL	SM	SN	SP	SQ	SR	SS	ST	SV	SW	SY
0	0	1	0	0	0	0	0	0	0	1	0	0	0	0	1

TA	TC	TD	TE	TF	TG	TH	TI	TK	TL	TM	TN	TP	TQ	TR	TS
0	0	0	0	0	0	0	0	0	1	0	0	0	0	1	0

TT	TV	TW	TY	VA	VC	VD	VE	VF	VG	VH	VI	VK	VL	VM	VN
0	1	0	0	1	0	0	0	0	0	0	0	0	0	0	1

VP	VQ	VR	VS	VT	VV	VW	VY	WA	WC	WD	WE	WF	WG	WH	WI
0	0	0	3	1	1	0	0	0	0	0	0	0	0	0	0
WK	WL	WM	WN	WP	WQ	WR	WS	WT	WV	WW	WY	YA	YC	YD	YE
0	0	0	0	0	0	0	0	0	0	0	0	1	0	1	0
YF	YG	YH	YI	YK	YL	YM	YN	YP	YQ	YR	YS	YT	YV	YW	YY
0	0	0	1	0	0	0	0	0	1	0	0	0	0	0	0

3.1.2 Alignment Free Comparison of Sequences

The similarities and differences between two biological sequences can be assessed by computing a distance measure between the two sequences. The alignment free distance is based on the word composition of the sequences. While similar DNA sequences have similar GC content, the k-mer frequencies for larger values of k reveal similarities and differences between two sequences.

Example 3.2
The fragments of DNA sequence

$$\text{CCCCAATATGGGCGCGACCCCCCGGAATCTCTATTCACCAGCTT} \quad (1)$$
$$\text{CCCCAATATGGGCGCGACCCCCCGGAATCTGTCTCCGCCAGCCT} \quad (2)$$
$$\text{CCCCAATATGGGCGCTACTTTCACAATAACCCACTAGACAGCCT} \quad (3)$$

have the following 1-mer frequencies,

word	A	C	G	T
(1)	9	18	8	9
(2)	7	20	10	7
(3)	13	16	6	9

and they have the following 2-mer frequencies.

word	AA	AC	AG	AT	CA	CC	CG	CT	GA	GC	GG	GT	TA	TC	TG	TT
(1)	2	2	1	4	3	9	3	3	2	3	3	0	2	3	1	2
(2)	2	1	1	3	2	11	4	3	2	4	3	1	1	3	2	0
(3)	3	5	2	3	5	6	1	4	1	3	2	0	4	1	1	2

The extent to which the k-mer frequencies of two sequences differ can be measured in a number of ways, such as by computing their covariance or correlation. The alignment free distance between two sequences is given by the linear correlation coefficient of their k-mer frequencies.

Example 3.3
The alignment free distance between each pair of the DNA sequences from the previous example is as follows.

sequences		$k = 1$	$k = 2$	$k = 3$	$k = 4$
(1)	(2)	0.9453431	0.9205409	0.8260229	0.7631025
(1)	(3)	0.8081352	0.6153795	0.5388881	0.3561884
(2)	(3)	0.6148987	0.4210917	0.3995599	0.3561884

The first sequence is thus more similar to the second sequence than to the third sequence, no matter the word length k used to assess similarities.

Given the k-mer composition of two biological sequences, their alignment free distance can be obtained by computing the linear correlation coefficient of the k-mer frequencies, that is, by dividing the covariance of the k-mer frequencies by the product of their standard deviations.

> **function** alignment_free_distance(S_1, S_2, k, Σ)
> $\quad F_1 \leftarrow$ word_composition(S_1, k, Σ)
> $\quad F_2 \leftarrow$ word_composition(S_2, k, Σ)
> $\quad cov \leftarrow$ covariance(F_1, F_2)
> $\quad sd_1 \leftarrow$ standard_deviation(F_1)
> $\quad sd_2 \leftarrow$ standard_deviation(F_2)
> $\quad$ **return** $cov/(sd_1 sd_2)$

The representation of sequences in BioPerl does not include any method to compute the linear correlation coefficient of the k-mer frequencies of two sequences. However, the previous algorithm can be easily implemented in Perl by first implementing methods to compute the mean and standard deviation of an array of values and the covariance of two arrays of values.

The mean of an array of values can be obtained by dividing the sum of the values by the number of values, as shown in the following Perl script.

```perl
sub mean {
  my $a = shift;
  my $res;
  foreach (@$a) { $res += $_ }
  $res /= @$a;
}
```

The standard deviation of an array of values, on the other hand, can be obtained by computing first the mean of the squared values and the square of the mean of the values and then taking the square root of their difference, as illustrated by the following Perl script.

```perl
sub sd {
  my $a = shift;
  my $mean = mean($a);
  return sqrt( mean( [map $_ ** 2, @$a] ) -
    ($mean ** 2) );
}
```

The covariance of two arrays of values of the same length can be obtained by computing first the sum of the product of their values and then dividing it by the number of values and subtracting the product of the mean of the values of each of the two arrays, as show in the following Perl script.

```perl
sub cov {
  my $a1 = shift;
  my $a2 = shift;
  my $res;
  foreach (0 .. @$a1 - 1) {
    $res += $a1->[$_] * $a2->[$_];
  }
  $res /= @$a1;
  $res -= mean($a1) * mean($a2);
}
```

Finally, the linear correlation coefficient of two arrays of values of the same length can be obtained by just dividing their covariance by the product of their standard deviations, as shown in the following Perl script.

```perl
sub cor {
  my $a1 = shift;
  my $a2 = shift;
  return cov($a1,$a2) / (sd($a1) * sd($a2));
}
```

The representation of sequences in R, on the other hand, does include a function **cor** to compute the linear correlation coefficient of two vectors of values of the same length. The computation of the alignment free distance between two sequences can thus be implemented in R by first obtaining their k-mer frequencies with the **count** function and then computing their correlation, as illustrated by the following R script.

```r
> library(seqinr)

> alignment.free.distance <- function (seq1,seq2,k,
    sigma) {
  freq1 <- count(seq1,word=k,alphabet=s2c(sigma))
  freq2 <- count(seq2,word=k,alphabet=s2c(sigma))
  cor(freq1,freq2)
}

> seq1 <- s2c("
    CCCCAATATGGGCGCGACCCCCCGGAATCTCTATTCACCAGCTT")
> seq2 <- s2c("
    CCCCAATATGGGCGCGACCCCCCGGAATCTGTCTCCGCCAGCCT")
> seq3 <- s2c("
    CCCCAATATGGGCGCTACTTTCACAATAACCCACTAGACAGCCT")
```

```
> alignment.free.distance(seq1,seq2,1,"ACGT")
[1] 0.9453431
> alignment.free.distance(seq1,seq3,1,"ACGT")
[1] 0.8081352
> alignment.free.distance(seq2,seq3,1,"ACGT")
[1] 0.6148987
>
> alignment.free.distance(seq1,seq2,2,"ACGT")
[1] 0.9205409
> alignment.free.distance(seq1,seq3,2,"ACGT")
[1] 0.6153795
> alignment.free.distance(seq2,seq3,2,"ACGT")
[1] 0.4210917
>
> alignment.free.distance(seq1,seq2,3,"ACGT")
[1] 0.8260229
> alignment.free.distance(seq1,seq3,3,"ACGT")
[1] 0.5388881
> alignment.free.distance(seq2,seq3,3,"ACGT")
[1] 0.3995599

> alignment.free.distance(seq1,seq2,4,"ACGT")
[1] 0.7631025
> alignment.free.distance(seq1,seq3,4,"ACGT")
[1] 0.3561884
> alignment.free.distance(seq2,seq3,4,"ACGT")
[1] 0.3561884
```

Bibliographic Notes

There are several approaches to the alignment free comparison of sequences, reviewed, for instance, by Vinga and Almeida (2003) and further assessed in (Ferragina et al. 2007).

Chapter 4

General Pattern Matching in Sequences

Combinatorial pattern matching is the search for exact or approximate occurrences of a given pattern within a given text. When it comes to biological sequences, both the pattern and the text are sequences and the pattern matching problem becomes one of finding the occurrences of a sequence within another sequence. For instance, scanning a protein sequence for the presence of a known pattern can help annotate both the protein and the corresponding genome, and finding a sequence within another sequence can help in assessing their similarities and differences. This is the subject of this chapter.

4.1 Finding Subsequences

There are several ways in which a sequence can be contained in another sequence. A sequence can be a prefix of a longer sequence, or it can be a suffix of the longer sequence. Also, a sequence can be contained deeper within another sequence: it can be a suffix of a prefix or, equivalently, a prefix of a suffix of the other sequence.

Example 4.1
The sequence TATTTGATCATT is contained in the following DNA sequence.

```
TTGATTACCTTATTTGATCATTACACATTGTACGCTTGTG
```

It is a suffix of a prefix of the longer sequence, as shown in the following alignment.

```
            TATTTGATCATT
TTGATTACCTTATTTGATCATT
TTGATTACCTTATTTGATCATTACACATTGTACGCTTGTG
```

It is also a prefix of a suffix of the longer sequence, as illustrated by the following alignment.

```
            TATTTGATCATT
```

```
                TATTTGATCATTACACATTGTACGCTTGTG
TTGATTACCTTATTTGATCATTACACATTGTACGCTTGTG
```

A sequence can also be contained many times in another sequence, and the occurrences of the shorter sequence in the longer sequence may even overlap.

Example 4.2

There are four occurrences of the sequence ATT (top) but only two occurrences of the sequence ATTAC (middle) in the following longer DNA sequence (bottom).

```
  ATT         ATT         ATT     ATT
   ATTAC               ATTAC
TTGATTACCTTATTTGATCATTACACATTGTACGCTTGTG
```

There are also two overlapping occurrences of the short sequence ACA within the longer sequence.

```
                     ACA
                      ACA
TTGATTACCTTATTTGATCATTACACATTGTACGCTTGTG
```

The occurrences of a given sequence within another given sequence can be found by traversing the longer sequence, starting at each possible initial position in turn where the shorter sequence could occur. In the following description, the starting positions of the occurrences of a *pattern* sequence P of length m in a *text* sequence T of length n, with $m \leqslant n$, are collected in a list L.

```
function occurrences(P, T)
    m ← length(P)
    n ← length(T)
    L ← ∅
    for i ← 1 to n − m + 1 do
        if P[1, 2, . . . , m] = T[i, i + 1, . . . , i + m − 1] then
            L ← L ∪ {i}
    return L
```

The previous algorithm for finding all occurrences of a sequence P in another sequence T can be implemented in Perl in a straightforward way, as shown in the following Perl script.

```
sub occurrences {
  my $p = shift;
  my $t = shift;
```

```perl
  my $m = length $p;
  my $n = length $t;
  my @L;
  for (my $i = 0; $i < $n-$m+1; $i++) {
    if ($p eq substr($t,$i,$m)) {
      push @L, $i;
    }
  }
  return \@L;
}
```

Notice, however, that the Perl regular expression matching method *m* can also be used to find all occurrences of a sequence in another sequence.

```perl
my @L = $t =~ m/$p/g;
```

While the *m* method gives back the actual sequences as occurrences, the starting position of each such occurrence can be obtained with the *index* method, as illustrated by the following Perl script.

```perl
my @L;
my $i = index($t, $p, 0);
while ($i != -1) {
  push @L, $i;
  $i = index($t, $p, $i+1);
}
```

The previous algorithm for finding all occurrences of a sequence P in another sequence T can also be implemented in R in a straightforward way, as shown in the following R script.

```r
> occurrences <- function (p,t) {
  L <- c()
  for (i in 1:nchar(t)-nchar(p)+1)
    if (p == substr(t,i,i+nchar(p)-1))
      L <- c(L,i)
  L
}

> p <- "ATT"
> t <- "TTGATTACCTTATTTGATCATTACACATTGTACGCTTGTG"
> occurrences(p,t)
[1]   4 12 20 27

> sapply(occurrences(p,t),
    function(i)substr(t,i,i+nchar(p)-1))
[1]  "ATT" "ATT" "ATT" "ATT"
```

The algorithm for finding all occurrences of a sequence within a longer sequence by traversing part of the latter once for each possible starting position of the shorter sequence has the disadvantage that it can be rather slow in practice for large sequences, because it cannot always be established whether or not $P[1, 2, \ldots, m]$ matches $T[i, i+1, \ldots, i+m-1]$ until m elementary comparison operations have been made and this test is repeated $n - m + 1$ times altogether. Finding all occurrences of a sequence in another sequence is indeed a classic pattern matching problem, for which faster algorithms are known. Some of them are based on storing all the suffixes of a sequence in a compact representation called a *suffix array*.

4.1.1 Suffix Arrays

The suffix array of a sequence is a permutation of all starting positions of the suffixes of the sequence, lexicographically sorted. Despite its simplicity, the suffix array of a sequence is very useful for finding subsequences, because all the occurrences of a sequence within another sequence appear together in the suffix array, as prefixes of suffixes of the longer sequence.

Example 4.3
The DNA sequence

```
TTGATTACCTTATTTGATCATTACACATTGTACGCTTGTG
```

has the following suffixes, starting at positions 1 through 40.

```
 [1]  TTGATTACCTTATTTGATCATTACACATTGTACGCTTGTG
 [2]  TGATTACCTTATTTGATCATTACACATTGTACGCTTGTG
 [3]  GATTACCTTATTTGATCATTACACATTGTACGCTTGTG
 [4]  ATTACCTTATTTGATCATTACACATTGTACGCTTGTG
 [5]  TTACCTTATTTGATCATTACACATTGTACGCTTGTG
 [6]  TACCTTATTTGATCATTACACATTGTACGCTTGTG
 [7]  ACCTTATTTGATCATTACACATTGTACGCTTGTG
 [8]  CCTTATTTGATCATTACACATTGTACGCTTGTG
 [9]  CTTATTTGATCATTACACATTGTACGCTTGTG
[10]  TTATTTGATCATTACACATTGTACGCTTGTG
[11]  TATTTGATCATTACACATTGTACGCTTGTG
[12]  ATTTGATCATTACACATTGTACGCTTGTG
[13]  TTTGATCATTACACATTGTACGCTTGTG
[14]  TTGATCATTACACATTGTACGCTTGTG
[15]  TGATCATTACACATTGTACGCTTGTG
[16]  GATCATTACACATTGTACGCTTGTG
[17]  ATCATTACACATTGTACGCTTGTG
[18]  TCATTACACATTGTACGCTTGTG
[19]  CATTACACATTGTACGCTTGTG
[20]  ATTACACATTGTACGCTTGTG
```

```
[21]  TTACACATTGTACGCTTGTG
[22]  TACACATTGTACGCTTGTG
[23]  ACACATTGTACGCTTGTG
[24]  CACATTGTACGCTTGTG
[25]  ACATTGTACGCTTGTG
[26]  CATTGTACGCTTGTG
[27]  ATTGTACGCTTGTG
[28]  TTGTACGCTTGTG
[29]  TGTACGCTTGTG
[30]  GTACGCTTGTG
[31]  TACGCTTGTG
[32]  ACGCTTGTG
[33]  CGCTTGTG
[34]  GCTTGTG
[35]  CTTGTG
[36]  TTGTG
[37]  TGTG
[38]  GTG
[39]  TG
[40]  G
```

In lexicographical order, these suffixes define the suffix array of the sequence.

```
[23]  ACACATTGTACGCTTGTG
[25]  ACATTGTACGCTTGTG
 [7]  ACCTTATTTGATCATTACACATTGTACGCTTGTG
[32]  ACGCTTGTG
[17]  ATCATTACACATTGTACGCTTGTG
[20]  ATTACACATTGTACGCTTGTG
 [4]  ATTACCTTATTTGATCATTACACATTGTACGCTTGTG
[27]  ATTGTACGCTTGTG
[12]  ATTTGATCATTACACATTGTACGCTTGTG
[24]  CACATTGTACGCTTGTG
[19]  CATTACACATTGTACGCTTGTG
[26]  CATTGTACGCTTGTG
 [8]  CCTTATTTGATCATTACACATTGTACGCTTGTG
[33]  CGCTTGTG
 [9]  CTTATTTGATCATTACACATTGTACGCTTGTG
[35]  CTTGTG
[40]  G
[16]  GATCATTACACATTGTACGCTTGTG
 [3]  GATTACCTTATTTGATCATTACACATTGTACGCTTGTG
[34]  GCTTGTG
[30]  GTACGCTTGTG
[38]  GTG
[22]  TACACATTGTACGCTTGTG
```

```
 [6]  TACCTTATTTGATCATTACACATTGTACGCTTGTG
[31]  TACGCTTGTG
[11]  TATTTGATCATTACACATTGTACGCTTGTG
[18]  TCATTACACATTGTACGCTTGTG
[39]  TG
[15]  TGATCATTACACATTGTACGCTTGTG
 [2]  TGATTACCTTATTTGATCATTACACATTGTACGCTTGTG
[29]  TGTACGCTTGTG
[37]  TGTG
[21]  TTACACATTGTACGCTTGTG
 [5]  TTACCTTATTTGATCATTACACATTGTACGCTTGTG
[10]  TTATTTGATCATTACACATTGTACGCTTGTG
[14]  TTGATCATTACACATTGTACGCTTGTG
 [1]  TTGATTACCTTATTTGATCATTACACATTGTACGCTTGTG
[28]  TTGTACGCTTGTG
[36]  TTGTG
[13]  TTTGATCATTACACATTGTACGCTTGTG
```

The actual suffix array of the sequence is just the array of starting positions.

```
23  25   7  32  17  20   4  27  12  24  19  26   8  33   9  35
40  16   3  34  30  38  22   6  31  11  18  39  15   2  29  37
21   5  10  14   1  28  36  13
```

The suffix array of a sequence can be obtained by various methods. The simplest way to obtain the suffix array of a sequence consists of sorting an array of positions by comparing the corresponding suffixes of the sequence. In the following description, the suffix array for a sequence S of length n is built by sorting an array A of positions 1 through n, where the comparison of array entries $A[i]$ and $A[j]$ is based on the comparison of suffixes $S[A[i], \ldots, n]$ and $S[A[j], \ldots, n]$.

> **procedure** suffix_array(S, A)
> $\quad n \leftarrow \text{length}(S)$
> $\quad$ **for** $i \leftarrow 1$ **to** n **do**
> $\quad\quad A[i] \leftarrow i$
> $\quad$ sort A by comparing $S[A[i], \ldots, n]$ with $S[A[j], \ldots, n]$

The previous algorithm for building the suffix array of a sequence can be implemented in Perl in a straightforward way. The suffix array @sa of a sequence seq of length n will be a permutation of array positions $0, 1, \ldots, n-1$, however, because Perl arrays do not start with position 1 but, rather, with position 0. This is all shown in the following Perl script.

```
sub suffix_array {
```

```
  my $s = shift;
  my @sa = (0 .. length($s)-1);
  @sa = sort {substr($s,$a) cmp substr($s,$b)} @sa;
  return \@sa;
}
```

The previous algorithm for building the suffix array of a sequence can also be implemented in R in a straightforward way, as shown in the following R script.

```
> suffix.array <- function (seq) {
  sa <- apply(matrix(1:nchar(seq)),1,function(i)
      substr(seq,i,nchar(seq)))
  order(sa)
}

> seq <- "TTGATTACCTTATTTGATCATTACACATTGTACGCTTGTG"
> options("width"=40)
> suffix.array(seq)
 [1] 23 25  7 32 17 20  4 27 12 24 19 26
[13]  8 33  9 35 40 16  3 34 30 38 22  6
[25] 31 11 18 39 15  2 29 37 21  5 10 14
[37]  1 28 36 13
```

The suffixes of the sequence appear indeed in lexicographical order within the suffix array.

```
> t(sapply(suffix.array(seq),function(i)c(i,substr(
    seq,i,nchar(seq)))))
       [,1] [,2]
 [1,] "23" "ACACATTGTACGCTTGTG"
 [2,] "25" "ACATTGTACGCTTGTG"
 [3,] "7"  "ACCTTATTTGATCATTACACATTGTACGCTTGTG"
 [4,] "32" "ACGCTTGTG"
 [5,] "17" "ATCATTACACATTGTACGCTTGTG"
 [6,] "20" "ATTACACATTGTACGCTTGTG"
 [7,] "4"  "ATTACCTTATTTGATCATTACACATTGTACGCTTGTG"
 [8,] "27" "ATTGTACGCTTGTG"
 [9,] "12" "ATTTGATCATTACACATTGTACGCTTGTG"
[10,] "24" "CACATTGTACGCTTGTG"
[11,] "19" "CATTACACATTGTACGCTTGTG"
[12,] "26" "CATTGTACGCTTGTG"
[13,] "8"  "CCTTATTTGATCATTACACATTGTACGCTTGTG"
[14,] "33" "CGCTTGTG"
[15,] "9"  "CTTATTTGATCATTACACATTGTACGCTTGTG"
[16,] "35" "CTTGTG"
[17,] "40" "G"
```

```
[18,]  "16"  "GATCATTACACATTGTACGCTTGTG"
[19,]  "3"   "GATTACCTTATTTGATCATTACACATTGTACGCTTGTG"
[20,]  "34"  "GCTTGTG"
[21,]  "30"  "GTACGCTTGTG"
[22,]  "38"  "GTG"
[23,]  "22"  "TACACATTGTACGCTTGTG"
[24,]  "6"   "TACCTTATTTGATCATTACACATTGTACGCTTGTG"
[25,]  "31"  "TACGCTTGTG"
[26,]  "11"  "TATTTGATCATTACACATTGTACGCTTGTG"
[27,]  "18"  "TCATTACACATTGTACGCTTGTG"
[28,]  "39"  "TG"
[29,]  "15"  "TGATCATTACACATTGTACGCTTGTG"
[30,]  "2"   "TGATTACCTTATTTGATCATTACACATTGTACGCTTGTG"
[31,]  "29"  "TGTACGCTTGTG"
[32,]  "37"  "TGTG"
[33,]  "21"  "TTACACATTGTACGCTTGTG"
[34,]  "5"   "TTACCTTATTTGATCATTACACATTGTACGCTTGTG"
[35,]  "10"  "TTATTTGATCATTACACATTGTACGCTTGTG"
[36,]  "14"  "TTGATCATTACACATTGTACGCTTGTG"
[37,]  "1"   "TTGATTACCTTATTTGATCATTACACATTGTACGCTTGTG"
[38,]  "28"  "TTGTACGCTTGTG"
[39,]  "36"  "TTGTG"
[40,]  "13"  "TTTGATCATTACACATTGTACGCTTGTG"
```

Given a pattern sequence and the suffix array of another sequence, an occurrence of the pattern in the sequence can be obtained by traversing the suffix array and comparing the pattern sequence with the suffixes of the sequence that start at the positions stored in the suffix array. Since the occurrences of the pattern sequence within the other sequence appear together, as prefixes of suffixes, in the suffix array of the longer sequence, however, an occurrence of the pattern sequence can be found much more quickly by performing a *binary search* on the suffix array of the longer sequence.

In the following description, an occurrence of a pattern sequence P of length m in another sequence T of length n, with $m \leqslant n$, is obtained by binary search on the suffix array $A[1, \ldots, n]$ of T. Starting with the index $i = (1 + n)/2$, the pattern P is compared to the first m characters of the suffix of T starting at position $A[i]$, and, as a result, either the search continues in $A[1, \ldots, i-1]$ because $P < T[A[i], \ldots, A[i] + m - 1]$ in lexicographical order, it continues in $A[i+1, \ldots, n]$ because $P > T[A[i], \ldots, A[i] + m - 1]$ in lexicographical order, or it finishes because an occurrence of P as a prefix of the suffix of T starting at position $A[i]$ was found. The portion of the suffix array being searched is always $A[\ell, \ldots, r]$.

Once an occurrence, at position $A[i]$, of pattern sequence P in sequence T has been found, the remaining occurrences (if any) are obtained by extending the interval (i, j) of suffix array indices as much as possible, where initially

$i = j$, profiting from the fact that all occurrences appear together in A. The lower index i of the interval is decreased by one for each further occurrence $T[A[i-1], \ldots, A[i-1]+m-1]$ towards the beginning of the suffix array, and the upper index $j = i$ is increased by one for each further occurrence $T[A[j+1], \ldots, A[j+1]+m-1]$ towards the end of the suffix array. The resulting interval (i, j) comprises the starting positions in suffix array A of all the occurrences of sequence P in sequence T, with $T[A[i], \ldots, A[i]+m-1]$ as the first occurrence and with $T[A[j], \ldots, A[j]+m-1]$ as the last occurrence.

```
function occurrences(P, T, A)
    m ← length(P)
    n ← length(T)
    ℓ ← 1
    r ← n
    while ℓ ⩽ r do
        i ← (ℓ + r)/2
        if P < T[A[i], ..., A[i] + m − 1] then
            r ← i − 1
        else if P > T[A[i], ..., A[i] + m − 1] then
            ℓ ← i + 1
        else
            j ← i
            while i > 1 and P = T[A[i − 1], ..., A[i − 1] + m − 1] do
                i ← i − 1
            while j < n and P = T[A[j + 1], ..., A[j + 1] + m − 1] do
                j ← j + 1
            return (i, j)
    return (−1, −1)
```

The previous algorithm for finding all occurrences of a sequence in another sequence using the suffix array of the latter sequence can be implemented in Perl in a straightforward way, as shown in the following Perl script. Recall that the suffix array of a sequence of length n will be a permutation of array positions $0, 1, \ldots, n-1$, because Perl arrays do not start with position 1 but, rather, with position 0.

```perl
sub occurrences {
    my $p = shift;
    my $t = shift;
    my $sa = shift;
    my ($l, $r) = (0, $#sa);
    while ($l <= $r) {
        my $i = int(($l + $r)/2);
        my $c = $p cmp substr($t, $sa[$i], length $p);
        if ($c < 0) {
```

```
      $r = $i - 1;
   } elsif ($c > 0) {
      $l = $i + 1;
   } else {
      my $j = $i;
      while ($i > 0 &&
          $p eq substr($t, $sa[$i-1], length $p)) {
         $i--;
      }
      while ($j < $#sa-1 &&
          $p eq substr($t, $sa[$j+1], length $p)) {
         $j++;
      }
      return ($i, $j);
   }
}
return (-1, -1);
}
```

The previous algorithm for finding all occurrences of a sequence in another sequence using the suffix array of the latter sequence can also be implemented in R in a straightforward way, as shown in the following R script.

```
> occurrences <- function (p,t,sa) {
  ell <- 1
  r <- nchar(t)
  while (ell <= r) {
    i <- as.integer((ell + r) / 2)
    s <- substr(t,sa[i],sa[i]+nchar(p)-1)
    if (p < s) {
      r <- i - 1
    } else if (p > s) {
      ell <- i + 1
    } else {
      j <- i
      while (i > 1 &&
          p==substr(t,sa[i-1],sa[i-1]+nchar(p)-1)) {
        i <- i - 1
      }
      while (j < nchar(t) &&
          p==substr(t,sa[j+1],sa[j+1]+nchar(p)-1)) {
        j <- j + 1
      }
      return(c(i,j))
    }
  }
}
```

```
  return(c(-1,-1))
}


> t <- "TTGATTACCTTATTTGATCATTACACATTGTACGCTTGTG"
> sa <- suffix.array(t)

> occurrences("ATTAC",t,sa)
[1] 6 7
> occurrences("ATT",t,sa)
[1] 6 9
> substr(t,sa[6],nchar(t))
[1] "ATTACACATTGTACGCTTGTG"
> substr(t,sa[7],nchar(t))
[1] "ATTACCTTATTTGATCATTACACATTGTACGCTTGTG"
> substr(t,sa[8],nchar(t))
[1] "ATTGTACGCTTGTG"
> substr(t,sa[9],nchar(t))
[1] "ATTTGATCATTACACATTGTACGCTTGTG"
```

The occurrences of a sequence in another sequence can be found even more quickly when the suffixes of the sequence share long prefixes, because when comparing $P[1,\ldots,m]$ with $T[A[i],\ldots,A[i]+m-1]$ in the previous algorithm, the length of the longest common prefix between them can be obtained as a by-product of their comparison, and it can be used in later iterations to shorten the suffixes that remain to be compared.

In fact, when searching for the occurrences of sequence P between positions ℓ and r of the suffix array A of sequence T, it suffices to keep track of the length $llcp$ of the longest common prefix between P and the suffix $T[A[\ell],\ldots,n]$ as well as the length $rlcp$ of the longest common prefix between P and the suffix $T[A[r],\ldots,n]$, because the shortest of these two longest common prefixes will be a prefix common to all the suffixes starting at positions $A[\ell]$ through $A[r]$ of sequence T. That is, if $P[1,\ldots,llcp] = T[A[\ell],\ldots,A[\ell]+llcp-1]$ and $P[1,\ldots,rlcp] = T[A[r],\ldots,A[r]+rlcp-1]$, then $P[1,\ldots,h] = T[A[k],\ldots,A[k]+h-1]$ for all $\ell \leqslant k \leqslant r$, where h is the smallest of $llcp$ and $rlcp$.

The length of the longest common prefix between $P[i,\ldots,m]$ and $T[j,\ldots,n]$ can be obtained by traversing the two sequences and counting the number c of common elements in their prefixes until $P[i+c] \neq T[j+c]$.

```
function lcp(P[i,...,m], T[j,...,n])
    c ← 0
    while i + c ⩽ m and j + c ⩽ n and P[i+c] = T[j+c] do
        c ← c + 1
    return c
```

Now, in the following description, the occurrences of sequence $P[1, \ldots, m]$ in sequence $T[1, \ldots, n]$, where $m \leqslant n$, are obtained by binary search on the portion $A[\ell, \ldots, r]$ of the suffix array of T, while keeping track of the length $llcp$ of the longest common prefix between P and the suffix $T[A[\ell], \ldots, n]$ as well as the length $rlcp$ of the longest common prefix between P and the suffix $T[A[r], \ldots, n]$. Then the lexicographical comparison of $P[1, \ldots, m]$ with $T[A[i], \ldots, A[i] + m - 1]$ is replaced by the computation of the length c of the longest common prefix between $P[h + 1, \ldots, m]$ and $T[A[i] + h, \ldots, n]$, where h is the smallest of $llcp$ and $rlcp$, and $P[1, \ldots, m] = T[A[i], \ldots, A[i] + m - 1]$ if $h + c = m$. Otherwise, the positions of the $h + c$ common elements are skipped, and $P[h + c + 1]$ and $T[A[i] + h + c]$ are compared next.

```
function occurrences(P, T, A)
    m ← length(P)
    n ← length(T)
    ℓ ← 1
    r ← n
    llcp ← rlcp ← 0
    while ℓ ≤ r do
        i ← (ℓ + r)/2
        h ← min{llcp, rlcp}
        c ← lcp(P[h + 1, ..., m], T[A[i] + h, ..., n])
        if h + c = m then
            j ← i
            while i > 1 and P = T[A[i − 1], ..., A[i − 1] + m − 1] do
                i ← i − 1
            while j < n and P = T[A[j + 1], ..., A[j + 1] + m − 1] do
                j ← j + 1
            return (i, j)
        else if P[h + c + 1] < T[A[i] + h + c] then
            r ← i − 1
            rlcp ← h + c
        else
            ℓ ← i + 1
            llcp ← h + c
    return (−1, −1)
```

The algorithm for computing the length of the longest common prefix between $P[i, \ldots, m]$ and $T[j, \ldots, n]$ can be implemented in Perl in a straightforward way, as shown in the following Perl script.

```
sub lcp {
    my ($p, $i, $t, $j) = @_;
    my $c = 0;
    while ($i+$c < length $p && $j+$c < length $t &&
```

```
          substr($p,$i+$c,1) eq substr($t,$j+$c,1)) {
     $c++;
   }
   return $c;
}
```

The improved algorithm for finding the occurrences of sequence $P[1,\ldots,m]$ in sequence $T[1,\ldots,n]$ can also be implemented in Perl in a straightforward way, as shown in the following Perl script. Recall that the suffix array of the sequence $T[1,\ldots,n]$ will be a permutation of array positions $0,1,\ldots,n-1$, because Perl arrays do not start with position 1 but, rather, with position 0.

```
sub occurrences {
  my $p = shift;
  my $t = shift;
  my $sa = shift;
  my ($l, $r) = (0, $#sa);
  my ($llcp, $rlcp) = (0, 0);
  while ($l <= $r) {
    my $i = int(($l + $r)/2);
    my $h = $llcp; $h = $rlcp if $rlcp < $h;
    my $c = lcp($p,$h,$t,$sa[$i]+$h);
    if ($h+$c == length $p) {
      my $j = $i;
      while ($i > 0 &&
          $p eq substr($t, $sa[$i-1], length $p)) {
        $i--;
      }
      while ($j < $#sa-1 &&
          $p eq substr($t, $sa[$j+1], length $p)) {
        $j++;
      }
      return ($i, $j);
    } elsif (substr($p,$h+$c,1) lt
        substr($t,$sa[$i]+$h+$c,1)) {
      $r = $i - 1;
      $rlcp = $h + $c;
    } else {
      $l = $i + 1;
      $llcp = $h + $c;
    }
  }
  return (-1, -1);
}
```

The algorithm for computing the length of the longest common prefix between $P[i,\ldots,m]$ and $T[j,\ldots,n]$ and the improved algorithm for finding the occurrences of sequence $P[1, \ldots, m]$ in sequence $T[1, \ldots, n]$ can both be implemented in R in a straightforward way, as shown in the following R script.

```
> lcp <- function (seq1,pos1,seq2,pos2) {
  n1 <- nchar(seq1)
  n2 <- nchar(seq2)
  c <- 0
  while (pos1+c <= n1 && pos2+c <= n2 &&
      substr(seq1,pos1+c,pos1+c) ==
      substr(seq2,pos2+c,pos2+c)) {
    c <- c + 1
  }
  c
}
> occurrences <- function (p,t,sa) {
  ell <- 1
  r <- nchar(t)
  llcp <- 0
  rlcp <- 0
  while (ell <= r) {
    i <- as.integer((ell + r) / 2)
    h <- min(llcp,rlcp)
    c <- lcp(p,h+1,t,sa[i]+h)
    if (h+c == nchar(p)) {
      j <- i
      while (i > 1 &&
          p==substr(t,sa[i-1],sa[i-1]+nchar(p)-1)) {
        i <- i - 1
      }
      while (j < nchar(t) &&
          p==substr(t,sa[j+1],sa[j+1]+nchar(p)-1)) {
        j <- j + 1
      }
      return(c(i,j))
    } else if (substr(p,h+c+1,h+c+1) <
        substr(t,sa[i]+h+c,sa[i]+h+c)) {
      r <- i - 1
      rlcp <- h + c
    } else {
      ell <- i + 1
      llcp <- h + c
    }
  }
  return(c(-1,-1))
}
```

```
> t <- "TTGATTACCTTATTTGATCATTACACATTGTACGCTTGTG"
> sa <- suffix.array(t)

> occurrences("ATTAC",t,sa)
[1] 6 7
> occurrences("ATT",t,sa)
[1] 6 9
> substr(t,sa[6],nchar(t))
[1] "ATTACACATTGTACGCTTGTG"
> substr(t,sa[7],nchar(t))
[1] "ATTACCTTATTTGATCATTACACATTGTACGCTTGTG"
> substr(t,sa[8],nchar(t))
[1] "ATTGTACGCTTGTG"
> substr(t,sa[9],nchar(t))
[1] "ATTTGATCATTACACATTGTACGCTTGTG"
```

4.2   Finding Common Subsequences

Subsequences shared by two sequences reveal information common to the two sequences. As there are several ways in which a sequence can be contained in another sequence, common subsequences can be common prefixes, suffixes, suffixes of prefixes, or prefixes of suffixes. Further, in order to reveal the most of their shared information, it is interesting to find common subsequences of largest size between two given sequences.

Example 4.4
The following fragments of DNA sequence

```
TGCTTCTGACTATAATAG
GCTTCCGGCTCGTATAATGTGTGG
```

contain the Pribnow box TATAAT as their longest common subsequence, as shown in the two alignments below.

```
        TATAAT                          TATAAT
TGCTTCTGACTATAATAG           GCTTCCGGCTCGTATAATGTGTGG
```

The common occurrences of a pattern as a subsequence of two given sequences can be found by traversing the two sequences, starting at each possible initial position in turn where the pattern sequence could occur. In the following description, the starting positions of the occurrences of a *pattern* sequence P of length m in each of the two *text* sequences T_1 of length n_1 and T_2 of length n_2, with $m \leqslant \min\{n_1, n_2\}$, are collected in a list L.

function common_occurrences(P, T_1, T_2)
    $m \leftarrow$ length(P)
    $n_1 \leftarrow$ length(T_1)
    $n_2 \leftarrow$ length(T_2)
    $L \leftarrow \emptyset$
    for $i \leftarrow 1$ **to** $n_1 - m + 1$ **do**
        if $P[1, \ldots, m] = T_1[i, \ldots, i + m - 1]$ **then**
            for $j \leftarrow 1$ **to** $n_2 - m + 1$ **do**
                if $P[1, \ldots, m] = T_2[j \ldots, j + m - 1]$ **then**
                    $L \leftarrow L \cup \{(i, j)\}$
    return L

The previous algorithm for finding all occurrences of a sequence P common to two sequences T_1 and T_2 can be implemented in Perl in a straightforward way, as shown in the following Perl script.

```perl
sub common_occurrences {
  my $p = shift;
  my $t1 = shift;
  my $t2 = shift;
  my $m = length $p;
  my $n1 = length $t1;
  my $n2 = length $t2;
  my @L;
  for (my $i = 0; $i < $n1-$m+1; $i++) {
    if ($p eq substr($t1,$i,$m)) {
      for (my $j = 0; $j < $n2-$m+1; $j++) {
        if ($p eq substr($t2,$j,$m)) {
          push @L, $i, $j;
        }
      }
    }
  }
  return \@L;
}
```

The previous algorithm for finding all occurrences of a sequence P common to two sequences T_1 and T_2 can also be implemented in R in a straightforward way, as shown in the following R script.

```r
> common.occurrences <- function (p,t1,t2) {
  L <- c()
  for (i in 1:nchar(t1)-nchar(p)+1)
```

```
      if (p == substr(t1,i,i+nchar(p)-1))
        for (j in 1:nchar(t2)-nchar(p)+1)
          if (p == substr(t2,j,j+nchar(p)-1))
            L <- c(L,i,j)
  L
}

> p <- "TATAAT"
> t1 <- "TGCTTCTGACTATAATAG"
> t2 <- "GCTTCCGGCTCGTATAATGTGTGG"
> common.occurrences(p,t1,t2)
[1]  11 13

> L <- common.occurrences(p,t1,t2)
> sapply(split(L,1:2)[[1]],
    function(i)substr(t1,i,i+nchar(p)-1))
[1] "TATAAT"
> sapply(split(L,1:2)[[2]],
    function(j)substr(t2,j,j+nchar(p)-1))
[1] "TATAAT"
```

On the other hand, the longest common subsequences of two given sequences can be found by traversing the two sequences, starting at each possible initial position of each sequence in turn where a common subsequence could occur. In the following description, given two sequences S_1 and S_2, the subsequences $S_1[i, \ldots, j]$, of length $j - i + 1$, and $S_2[k, \ldots, k + j - i]$, also of length $k + j - i - k + 1 = j - i + 1$, are compared to each other, and the longest common subsequences, of length ℓ, are collected in a list L.

---

function longest_common_subsequences(S_1, S_2)
    $L \leftarrow \emptyset$
    $\ell \leftarrow 0$
    for $i \leftarrow 1$ **to** $length(S_1)$ **do**
        for $j \leftarrow i$ **to** $length(S_1)$ **do**
            $X \leftarrow S_1[i, \ldots, j]$
            for $k \leftarrow 1$ **to** $length(S_2) - j + i$ **do**
                $Y \leftarrow S_2[k, \ldots, k + j - i]$
            if $X = Y$ **then**
                if $length(X) = \ell$ **then**
                    $L \leftarrow L \cup \{X\}$
                else if $length(X) > \ell$ **then**
                    $L \leftarrow \{X\}$
                    $\ell \leftarrow length(X)$
    return L

The previous algorithm for finding the longest common subsequences of two given sequences S_1 and S_2 can be implemented in Perl in a straightforward way, as shown in the following Perl script. Repetitions in the list L of longest common subsequences are discarded with the help of a hash %count indexed by sequences.

```perl
sub longest_common_subsequences {
  my $s1 = shift;
  my $s2 = shift;
  my $n1 = length($s1);
  my $n2 = length($s2);
  my @L;
  my $l = 0;
  for (my $i = 0; $i < $n1; $i++) {
    for (my $j = $i; $j < $n1; $j++) {
      my $seq = substr($s1,$i,$j-$i+1);
      for (my $k = 0; $k < $n2-$j+$i; $k++) {
        if ($seq eq substr($s2,$k,$j-$i+1)) {
          if (length($seq) == $l) {
            push @L, $seq;
          } elsif (length($seq) > $l) {
            @L = $seq;
            $l = length($seq);
          }
        }
      }
    }
  }
  my %count;
  for my $seq (@L) { $count{$seq}++; }
  @L = sort keys %count;
  return \@L;
}
```

The previous algorithm for finding the longest common subsequences of two given sequences S_1 and S_2 can also be implemented in R in a straightforward way, with any repetitions in the list L of longest common subsequences being discarded by the unique function, as shown in the following R script.

```r
> longest.common.subsequences <- function (s1,s2) {
  L <- c()
  ell <- 0
  for (i in 1:nchar(s1)) {
    for (j in i:nchar(s1)) {
      seq <- substr(s1,i,j)
      for (k in 1:(nchar(s2)-j+i)) {
```

```
          if (seq == substr(s2,k,k+j-i)) {
              if (nchar(seq) == ell) {
                L <- c(L,seq)
              } else if (nchar(seq) > ell) {
                L <- c(seq)
                ell <- nchar(seq)
              }
          }
        }
      }
    }
  }
  unique(L)
}


> s1 <- "TGCTTCTGACTATAATAG"
> s2 <- "GCTTCCGGCTCGTATAATGTGTGG"
> longest.common.subsequences(s1,s2)
[1] "TATAAT"
```

Now, the algorithm for finding the longest common subsequences of two sequences S_1 and S_2 by traversing part of the sequences once for each possible starting position of a common subsequence has the disadvantage that it can be slow in practice for large sequences, because a subsequence $S_1[i, \ldots, j]$ is compared to a subsequence $S_2[k, \ldots, k+j-i]$ even if the comparison of a prefix of the former to a prefix of the latter has already failed and, thus, they cannot be common subsequences. Finding the longest common subsequences of two sequences is indeed another classic pattern matching problem, for which faster algorithms are known. Some of them are based on dynamic programming, while others are based on storing all the suffixes of the two sequences in a compact representation called a *generalized suffix array*.

The longest common subsequences of two sequences can be obtained by finding the longest common suffixes between each pair of prefixes of the sequences, keeping all the longest ones. In general, the length $LCS(S_1[1, \ldots, i], S_2[1, \ldots, j])$ of a longest common suffix between two prefixes $S_1[1, \ldots, i]$ and $S_2[1, \ldots, j]$ of the sequences S_1 and S_2 is given by the recurrence

$$LCS(S_1[1, \ldots, i], S_2[1, \ldots, j]) = \begin{cases} LCS(S_1[1, \ldots, i-1], S_2[1, \ldots, j-1]) + 1 \\ \quad \text{if } S_1[i] = S_2[j] \\ 0 \quad \text{otherwise} \end{cases}$$

Computation of this recurrence by dynamic programming involves the use of a dynamic programming table to store each $LCS(S_1[1, \ldots, i], S_2[1, \ldots, j])$, for $1 \leqslant i \leqslant n_1$ and $1 \leqslant j \leqslant n_2$, where n_1 is the length of S_1 and n_2 is the length of S_2.

Example 4.5

The longest common subsequences of the two sequences TGCTTCTGACT
ATAATAG and GCTTCCGGCTCGTATAATGTGTGG have length 6, as
shown in entry $(16, 18)$ of the following dynamic programming table. Prefix TGCTTCTGACTATAAT of the first sequence and prefix GCTTCCG
GCTCGTATAAT of the second sequence share TATAAT as a common suffix.

		G 1	C 2	T 3	T 4	C 5	C 6	G 7	G 8	C 9	T 10	C 11	G 12	T 13	A 14	T 15	A 16	A 17	T 18	G 19	T 20	G 21	T 22	G 23	G 24
T	1	0	0	1	1	0	0	0	0	0	1	0	0	1	0	1	0	0	1	0	1	0	1	0	0
G	2	1	0	0	0	0	0	1	1	0	0	0	1	0	0	0	0	0	0	2	0	2	0	2	1
C	3	0	2	0	0	1	1	0	0	2	0	1	0	0	0	0	0	0	0	0	0	0	0	0	0
T	4	0	0	3	1	0	0	0	0	0	3	0	0	1	0	1	0	0	1	0	1	0	1	0	0
T	5	0	0	1	4	0	0	0	0	0	1	0	0	1	0	1	0	0	1	0	1	0	1	0	0
C	6	0	1	0	0	5	1	0	0	1	0	2	0	0	0	0	0	0	0	0	0	0	0	0	0
T	7	0	0	2	1	0	0	0	0	0	2	0	0	1	0	1	0	0	1	0	1	0	1	0	0
G	8	1	0	0	0	0	0	1	1	0	0	0	1	0	0	0	0	0	0	2	0	2	0	2	1
A	9	0	0	0	0	0	0	0	0	0	0	0	0	0	1	0	1	1	0	0	0	0	0	0	0
C	10	0	1	0	0	1	1	0	0	1	0	1	0	0	0	0	0	0	0	0	0	0	0	0	0
T	11	0	0	2	1	0	0	0	0	0	2	0	0	1	0	1	0	0	1	0	1	0	1	0	0
A	12	0	0	0	0	0	0	0	0	0	0	0	0	0	2	0	2	1	0	0	0	0	0	0	0
T	13	0	0	1	1	0	0	0	0	0	1	0	0	1	0	3	0	0	2	0	1	0	1	0	0
A	14	0	0	0	0	0	0	0	0	0	0	0	0	0	2	0	4	1	0	0	0	0	0	0	0
A	15	0	0	0	0	0	0	0	0	0	0	0	0	0	1	0	1	5	0	0	0	0	0	0	0
T	16	0	0	1	1	0	0	0	0	0	1	0	0	1	0	2	0	0	**6**	0	1	0	1	0	0
A	17	0	0	0	0	0	0	0	0	0	0	0	0	0	2	0	3	1	0	0	0	0	0	0	0
G	18	1	0	0	0	0	0	1	1	0	0	0	1	0	0	0	0	0	0	1	0	1	0	1	1

In the following description, the dynamic programming table LCS is filled
in for each $1 \leqslant i \leqslant n_1$ and $1 \leqslant j \leqslant n_2$ while keeping track of the length ℓ of a
longest common suffix between all pairs of prefixes considered so far, and all
common suffixes of prefixes of length $i - (i - \ell + 1) + 1 = \ell$ are collected in a
list L.

```
function longest_common_subsequences(S₁, S₂)
    L ← ∅
    ℓ ← 0
    for i ← 1 to length(S₁) do
        for j ← i to length(S₂) do
            LCS[i, j] ← 0
            if S₁[i] = S₂[j] then
                if i = 1 or j = 1 then
                    LCS[i, j] ← 1
                else
                    LCS[i, j] ← LCS[i − 1, j − 1] + 1
```

$$\textbf{if } LCS[i,j] > \ell \textbf{ then}$$
$$\ell \leftarrow LCS[i,j]$$
$$L \leftarrow \emptyset$$
$$\textbf{if } LCS[i,j] = \ell \textbf{ then}$$
$$L \leftarrow L \cup \{S_1[i - \ell + 1, \ldots, i]\}$$
return L

The previous dynamic programming algorithm for finding the longest common subsequences of two given sequences S_1 and S_2 can be implemented in Perl in a straightforward way, as shown in the following Perl script.

```perl
sub longest_common_subsequences {
  my $s1 = shift;
  my $s2 = shift;
  my @L;
  my $l = 0;
  my @LCS;
  for (my $i = 0; $i < length $s1; $i++ ) {
    for (my $j = 0; $j < length $s2; $j++ ) {
      $LCS[$i][$j] = 0;
      if (substr($s1,$i,1) eq substr($s2,$j,1)) {
        if ($i == 0 || $j == 0) {
          $LCS[$i][$j] = 1;
        } else {
          $LCS[$i][$j] = $LCS[$i-1][$j-1] + 1;
        }
        if ($LCS[$i][$j] > $l) {
          $l = $LCS[$i][$j];
          @L = ();
        }
        if ($LCS[$i][$j] == $l) {
          push @L, substr($s1,$i-$l+1,$l);
        }
      }
    }
  }
  return @L;
}
```

The previous dynamic programming algorithm for finding the longest common subsequences of two given sequences S_1 and S_2 can also be implemented in R in a straightforward way, as illustrated by the following R script.

```r
> longest.common.subsequences <- function (s1,s2) {
  L <- c()
  ell <- 0
```

```
LCS <- matrix(0,nrow=nchar(s1),ncol=nchar(s2))
for (i in 1:nchar(s1)) {
  for (j in 1:nchar(s2)) {
    if (substr(s1,i,i) == substr(s2,j,j)) {
      if (i == 1 || j == 1) {
        LCS[i,j] <- 1
      } else {
        LCS[i,j] <- LCS[i-1,j-1] + 1
      }
      if (LCS[i,j] > ell) {
        ell <- LCS[i,j]
        L <- c()
      }
      if (LCS[i,j] == ell) {
        L <- c(L,substr(s1,i-ell+1,i))
      }
    }
  }
}
L
}

> s1 <- "TGCTTCTGACTATAATAG"
> s2 <- "GCTTCCGGCTCGTATAATGTGTGG"
> longest.common.subsequences(s1,s2)
[1] "TATAAT"
```

4.2.1 Generalized Suffix Arrays

The generalized suffix array of two sequences is a permutation of all starting positions of the suffixes of the sequences, lexicographically sorted, where the starting position of each suffix is tagged as coming from the first or the second sequence. Despite its simplicity, the generalized suffix array of two sequences is very useful for finding common subsequences, because all the occurrences of a sequence within both of the sequences appear together in the generalized suffix array, as prefixes of suffixes of the sequences.

As in the case of the suffix array of a sequence, the generalized suffix array of two sequences can also be obtained by various methods. The simplest way to obtain the generalized suffix array of two sequences consists in first tagging the sequences and then sorting an array of positions by comparing the corresponding suffixes of the tagged sequences. In the following description, the generalized suffix array for a sequence S_1 of length n_1 and a sequence S_2 of length n_2 is built by sorting an array A of positions 1 through n_1 tagged by the sequence identifier 1 and positions 1 through n_2 tagged by the sequence identifier 2, where the comparison of an array entry $A[i]$ tagged by, say, 1 and an array entry $A[j]$ tagged by, say, 2 is based on the comparison of suffixes $S_1[A[i], \ldots, n_1]$ and $S_2[A[j], \ldots, n_2]$.

---

procedure generalized_suffix_array(S_1, S_2, A)
  $n_1 \leftarrow \text{length}(S_1)$
  $n_2 \leftarrow \text{length}(S_2)$
  for $i \leftarrow 1$ **to** n_1 **do**
    $A[i] \leftarrow (i, 1)$
  for $i \leftarrow 1$ **to** n_2 **do**
    $A[n_1 + i] \leftarrow (i, 2)$
  sort A by comparing $S_1[A[i], \ldots, n_1]$ with $S_2[A[j], \ldots, n_2]$

The previous algorithm for building the generalized suffix array of two sequences can be implemented in Perl in a straightforward way. The generalized suffix array @gsa of a sequence %seq{1} of length n_1 and a sequence %seq{2} of length n_2 will be a permutation of array positions $0, 1, \ldots, n_1 - 1$ and a permutation of array positions $0, 1, \ldots, n_2 - 1$, however, because Perl arrays do not start with position 1 but, rather, with position 0. Also, storing the two sequences in a hash makes it easier to sort their suffixes, as shown in the following Perl script.

```
sub generalized_suffix_array {
  my $seq = shift;
  my %seq = %{$seq};

  my @gsa;
  for my $id (keys %seq) {
    for my $i (0 .. length($seq{$id})-1) {
      push @gsa, [$i, $id];
    }
  }

  @gsa = sort {substr($seq{@{$a}[1]}, @{$a}[0]) cmp
    substr($seq{@{$b}[1]}, @{$b}[0])} @gsa;

  return \@gsa;
}
```

The previous algorithm for building the generalized suffix array of two sequences can also be easily implemented in R. The suffixes sa1 and sa2 of the two sequences are sorted together in lexicographical order and tagged with the identifier tag of the sequence they come from. Then the positions in lexicographical order are transformed back to positions in the respective sequences. This is all shown in the following R script.

```
> generalized.suffix.array <- function (s1,s2) {
  sa1 <- apply(matrix(1:nchar(s1)),1,
    function (i) substr(s1,i,nchar(s1)))
  sa2 <- apply(matrix(1:nchar(s2)),1,
    function (i) substr(s2,i,nchar(s2)))
  arr <- order(c(sa1,sa2))
  tag <- sapply(arr,function (i)
    if (i <= nchar(s1)) { 1 } else { 2 } )
  arr <- sapply(arr,function (i)
    if (i <= nchar(s1)) { i } else { i-nchar(s1) } )
  cbind(arr,tag)
}

> s1 <- "TGCTTCTGACTATAATAG"
> s2 <- "GCTTCCGGCTCGTATAATGTGTGG"
> options("width"=40)
> gsa <- generalized.suffix.array(s1,s2)
> gsa[,1]
 [1] 14 16  9 17 12 14 15 17  5  6 11 10
[13]  9  6  2  3 18 24  8  8  1  2 23  7
[25] 12 21 19 13 15 16 11 13  4 10  5  7
[37]  1 22 20 18  3  4
> gsa[,2]
 [1] 1 2 1 1 1 2 1 2 2 2 2 1 2 1 2 1 1 2
[19] 1 2 2 1 2 2 2 2 1 2 1 1 2 2 2 1 1
[37] 1 2 2 2 2 1
```

The suffixes of the two sequences appear indeed in lexicographical order within the generalized suffix array.

```
> options("width"=60)
> cbind(gsa,apply(gsa,1,function (i)
    if (i[2]==1) { substr(s1,i[1],nchar(s1)) }
    else { substr(s2,i[1],nchar(s2)) } ))
       arr  tag
 [1,] "14" "1" "AATAG"
 [2,] "16" "2" "AATGTGTGG"
 [3,] "9"  "1" "ACTATAATAG"
 [4,] "17" "1" "AG"
 [5,] "12" "1" "ATAATAG"
 [6,] "14" "2" "ATAATGTGTGG"
 [7,] "15" "1" "ATAG"
 [8,] "17" "2" "ATGTGTGG"
 [9,] "5"  "2" "CCGGCTCGTATAATGTGTGG"
[10,] "6"  "2" "CGGCTCGTATAATGTGTGG"
[11,] "11" "2" "CGTATAATGTGTGG"
```

```
[12,]  "10"  "1"  "CTATAATAG"
[13,]  "9"   "2"  "CTCGTATAATGTGTGG"
[14,]  "6"   "1"  "CTGACTATAATAG"
[15,]  "2"   "2"  "CTTCCGGCTCGTATAATGTGTGG"
[16,]  "3"   "1"  "CTTCTGACTATAATAG"
[17,]  "18"  "1"  "G"
[18,]  "24"  "2"  "G"
[19,]  "8"   "1"  "GACTATAATAG"
[20,]  "8"   "2"  "GCTCGTATAATGTGTGG"
[21,]  "1"   "2"  "GCTTCCGGCTCGTATAATGTGTGG"
[22,]  "2"   "1"  "GCTTCTGACTATAATAG"
[23,]  "23"  "2"  "GG"
[24,]  "7"   "2"  "GGCTCGTATAATGTGTGG"
[25,]  "12"  "2"  "GTATAATGTGTGG"
[26,]  "21"  "2"  "GTGG"
[27,]  "19"  "2"  "GTGTGG"
[28,]  "13"  "1"  "TAATAG"
[29,]  "15"  "2"  "TAATGTGTGG"
[30,]  "16"  "1"  "TAG"
[31,]  "11"  "1"  "TATAATAG"
[32,]  "13"  "2"  "TATAATGTGTGG"
[33,]  "4"   "2"  "TCCGGCTCGTATAATGTGTGG"
[34,]  "10"  "2"  "TCGTATAATGTGTGG"
[35,]  "5"   "1"  "TCTGACTATAATAG"
[36,]  "7"   "1"  "TGACTATAATAG"
[37,]  "1"   "1"  "TGCTTCTGACTATAATAG"
[38,]  "22"  "2"  "TGG"
[39,]  "20"  "2"  "TGTGG"
[40,]  "18"  "2"  "TGTGTGG"
[41,]  "3"   "2"  "TTCCGGCTCGTATAATGTGTGG"
[42,]  "4"   "1"  "TTCTGACTATAATAG"
```

Now, the common occurrences of a pattern as a subsequence of two given sequences can also be found by binary search on the generalized suffix array of the two sequences. In the following description, a common occurrence of a pattern sequence P of length m in a sequence S_1 of length n_1, with $m \leqslant n_1$, and a sequence S_2 of length n_2, with $m \leqslant n_2$, is obtained by binary search on the generalized suffix array $A[1, \ldots, n_1 + n_2]$ of S_1 and S_2. Starting with the index $i = (1 + n_1 + n_2)/2$, the pattern P is compared to the first m characters of the suffix of S_t starting at position k, where $(k, t) = A[i]$, and, as a result, either the search continues in $A[1, \ldots, i - 1]$ because $P < S_t[k, \ldots, k + m - 1]$ in lexicographical order, or it continues in $A[i + 1, \ldots, n]$ because $P > S_t[k, \ldots, k + m - 1]$ in lexicographical order, or it finishes because an occurrence of P as a prefix of the suffix of S_t starting at position k was found. Again, the portion of the generalized suffix array being searched is always $A[\ell, \ldots, r]$.

Once a common occurrence, at position k, of pattern sequence P in sequence S_t has been found, where $(k, t) = A[i]$, the remaining occurrences (if any) are obtained by extending the interval (i, i) of generalized suffix array indices as much as possible, profiting from the fact that all occurrences appear together in A. The lower index i of the interval is decreased by one for each further occurrence towards the beginning of the generalized suffix array, and the upper index $j = i$ is increased by one for each further occurrence towards the end of the generalized suffix array. The resulting interval (i, j) comprises the starting positions in A of all the common occurrences of sequence P in sequences S_1 and S_2.

```
function common_occurrences(P, S1, S2, A)
    m ← length(P)
    n1 ← length(S1)
    n2 ← length(S2)
    ℓ ← 1
    r ← n1 + n2
    while ℓ ⩽ r do
        i ← (ℓ + r)/2
        (k, t) ← A[i]
        if P < St[k, ..., k + m − 1] then
            r ← i − 1
        else if P > St[k, ..., k + m − 1] then
            ℓ ← i + 1
        else
            j ← i
            while i > 1, (k, t) ← A[i − 1] and P = St[k, ..., k + m − 1] do
                i ← i − 1
            while j < n, (k, t) ← A[j + 1] and P = St[k, ..., k + m − 1] do
                j ← j + 1
            return (i, j)
    return (−1, −1)
```

The previous algorithm for finding all occurrences of a sequence P common to two sequences S_1 and S_2 using the generalized suffix array of S_1 and S_2 can be implemented in Perl in a straightforward way, as shown in the following Perl script.

```
sub common_occurrences {
    my $p = shift;
    my $seq = shift;
    my %seq = %{$seq};
    my $gsa = shift;
    my @gsa = @{$gsa};
```

```perl
my ($l, $r) = (0, $#gsa);
while ($l <= $r) {
  my $i = int(($l + $r)/2);
  my $c = $p cmp substr($seq{$gsa[$i][1]}, $gsa[$i
    ][0], length $p);
  if ($c < 0) {
    $r = $i - 1;
  } elsif ($c > 0) {
    $l = $i + 1;
  } else {
    my $j = $i;
    while ($i > 0 &&
        $p eq substr($seq{$gsa[$i-1][1]}, $gsa[$i
          -1][0], length $p)) {
      $i--;
    }
    while ($j < $#gsa &&
        $p eq substr($seq{$gsa[$j+1][1]}, $gsa[$j
          +1][0], length $p)) {
      $j++;
    }
    return ($i, $j);
  }
}
return (-1, -1);
}
```

The algorithm for finding all occurrences of a sequence P common to two sequences S_1 and S_2 using their generalized suffix array can also be implemented in R in a straightforward way, as shown in the following R script.

```r
> common.occurrences <- function (p,seq,gsa) {
  ell <- 1
  r <- n <- nchar(seq[1]) + nchar(seq[2])
  while (ell <= r) {
    i <- as.integer((ell + r) / 2)
    s <- substr(seq[gsa[i,2]],
      gsa[i,1],gsa[i,1]+nchar(p)-1)
    if (p < s) {
      r <- i - 1
    } else if (p > s) {
      ell <- i + 1
    } else {
      j <- i
      while (i > 1 && p == s) {
        s <- substr(seq[gsa[i-1,2]],
```

```
            gsa[i-1,1],gsa[i-1,1]+nchar(p)-1)
        if (p == s) i <- i - 1
      }
      s <- p
      while (j < n && p == s) {
        s <- substr(seq[gsa[j+1,2]],
          gsa[j+1,1],gsa[j+1,1]+nchar(p)-1)
        if (p == s) j <- j + 1
      }
      return(c(i,j))
    }
  }
  return(c(-1,-1))
}

> p <- "TATAAT"
> s1 <- "TGCTTCTGACTATAATAG"
> s2 <- "GCTTCCGGCTCGTATAATGTGTGG"
> gsa <- generalized.suffix.array(s1,s2)
> common.occurrences(p,c(s1,s2),gsa)
[1] 31 32
> gsa[31:32,]
     arr tag
[1,]  11   1
[2,]  13   2
> substr(s1,gsa[31,1],gsa[31,1]+nchar(p)-1)
[1] "TATAAT"
> substr(s2,gsa[32,1],gsa[32,1]+nchar(p)-1)
[1] "TATAAT"
```

Now, as in the case of suffix arrays, the occurrences of a sequence common to two sequences can be found even more quickly when the suffixes of the sequences share long prefixes, because when comparing $P[1,\ldots,m]$ with $S_t[k,\ldots,k+m-1]$ in the previous algorithm, the length of the longest common prefix between them can be obtained as a by-product of their comparison, and it can be used in later iterations to shorten the suffixes that remain to be compared.

In fact, when searching for the occurrences of sequence P between positions ℓ and r of the generalized suffix array A of sequences S_1 and S_2, it suffices to keep track of the length $llcp$ of the longest common prefix between P and the suffix $S_t[k,\ldots,n_t]$, where $A[\ell] = (k,t)$, as well as the length $rlcp$ of the longest common prefix between P and the suffix $S_t[k\ldots,n_t]$, where $A[r] = (k,t)$, because the shortest of these two longest common prefixes will be a prefix common to all the suffixes starting at positions $A[\ell]$ through $A[r]$ of sequences S_1 or S_2.

In the following description, a common occurrence of a pattern sequence P of length m in a sequence S_1 of length n_1, with $m \leqslant n_1$, and a sequence S_2 of length n_2, with $m \leqslant n_2$, is obtained by binary search on the portion $A[\ell, \ldots, r]$ of the generalized suffix array of S_1 and S_2, while keeping track of the length *llcp* of the longest common prefix between P and the suffix $S_t[k, \ldots, n_t]$, where $A[\ell] = (k, t)$, as well as the length *rlcp* of the longest common prefix between P and the suffix $S_t[k, \ldots, n_t]$, where $A[r] = (k, t)$. Then the lexicographical comparison of $P[1, \ldots, m]$ with $S_t[k, \ldots, k + m - 1]$ is replaced by the computation of the length c of the longest common prefix between $P[h + 1, \ldots, m]$ and $S_t[k + h, \ldots, n_t]$, where h is the smallest of *llcp* and *rlcp*, and $P[1, \ldots, m] = S_t[k, \ldots, k + m - 1]$ if $h + c = m$. Otherwise, the positions of the $h + c$ common elements are skipped, and $P[h + c + 1]$ and $S_t[k + h + c]$ are compared next.

```
function common_occurrences(P, S₁, S₂, A)
    m ← length(P)
    n₁ ← length(S₁)
    n₂ ← length(S₂)
    ℓ ← 1
    r ← n ← n₁ + n₂
    llcp ← rlcp ← 0
    while ℓ ≤ r do
        i ← (ℓ + r)/2
        (k, t) ← A[i]
        h ← min{llcp, rclp}
        c ← lcp(P[h + 1, ..., m], Sₜ[k + h, ..., nₜ])
        if h + c = m then
            j ← i
            while i > 1, (k, t) ← A[i − 1] and P = Sₜ[k, ..., k + m − 1] do
                i ← i − 1
            while j < n, (k, t) ← A[j + 1] and P = Sₜ[k, ..., k + m − 1] do
                j ← j + 1
            return (i, j)
        else if P[h + c + 1] < Sₜ[k + h + c] then
            r ← i − 1
            rlcp ← h + c
        else
            ℓ ← i + 1
            llcp ← h + c
    return (−1, −1)
```

The improved algorithm for finding the occurrences of a sequence P common to two sequences S_1 and S_2 using the generalized suffix array of S_1 and S_2 can be implemented in Perl in a straightforward way, as shown in the following Perl script.

```perl
sub common_occurrences {
  my $p = shift;
  my $seq = shift;
  my %seq = %{$seq};
  my $gsa = shift;
  my @gsa = @{$gsa};
  my ($l, $r) = (0, $#gsa);
  my ($llcp, $rlcp) = (0, 0);
  while ($l <= $r) {
    my $i = int(($l + $r)/2);
    my $h = $llcp; $h = $rlcp if $rlcp < $h;
    my $c = lcp($p,$h,$seq{$gsa[$i][1]},$gsa[$i][0]+
        $h);
    if ($h+$c == length $p) {
      my $j = $i;
      while ($i > 0 &&
          $p eq substr($seq{$gsa[$i-1][1]}, $gsa[$i
          -1][0], length $p)) {
        $i--;
      }
      while ($j < $#gsa-1 &&
          $p eq substr($seq{$gsa[$j+1][1]}, $gsa[$j
          +1][0], length $p)) {
        $j++;
      }
      return ($i, $j);
    } elsif (substr($p,$h+$c,1) lt
        substr($seq{$gsa[$i][1]},$gsa[$i][0]+$h+$c,1)
        ) {
      $r = $i - 1;
      $rlcp = $h + $c;
    } else {
      $l = $i + 1;
      $llcp = $h + $c;
    }
  }
  return (-1, -1);
}
```

The improved algorithm for finding the occurrences of a sequence P common to two sequences S_1 and S_2 using the generalized suffix array of S_1 and S_2 can also be implemented in R in a straightforward way, where the two sequences are input as a vector seq in order to be able to access them according to the second column of the generalized suffix array, as seq[gsa[i,2]], for instance. This is all shown in the following R script.

```
> common.occurrences <- function (p,seq,gsa) {
  ell <- 1
  r <- n <- nchar(seq[1]) + nchar(seq[2])
  llcp <- 0
  rlcp <- 0
  while (ell <= r) {
    i <- as.integer((ell + r) / 2)
    s <- substr(seq[gsa[i,2]],gsa[i,1],
      gsa[i,1]+nchar(p)-1)
    h <- min(llcp,rlcp)
    c <- lcp(p,h+1,s,1+h)
    if (h+c == nchar(p)) {
      j <- i
      while (i > 1 && p == s) {
        s <- substr(seq[gsa[i-1,2]],gsa[i-1,1],
          gsa[i-1,1]+nchar(p)-1)
        if (p == s) i <- i - 1
      }
      s <- p
      while (j < n && p == s) {
        s <- substr(seq[gsa[j+1,2]],gsa[j+1,1],
          gsa[j+1,1]+nchar(p)-1)
        if (p == s) j <- j + 1
      }
      return(c(i,j))
    } else {
      s <- substr(seq[gsa[i,2]],gsa[i,1]+h+c,
        gsa[i,1]+h+c)
      if (substr(p,h+c+1,h+c+1) < s) {
        r <- i - 1
        rlcp <- h + c
      } else {
        ell <- i + 1
        llcp <- h + c
      }
    }
  }
  return(c(-1,-1))
}

> p <- "TATAAT"
> s1 <- "TGCTTCTGACTATAATAG"
> s2 <- "GCTTCCGGCTCGTATAATGTGTGG"
```

```
> gsa <- generalized.suffix.array(s1,s2)
> common.occurrences(p,c(s1,s2),gsa)
[1] 31 32
> gsa[31:32,]
     arr tag
[1,]  11   1
[2,]  13   2
> substr(s1,gsa[31,1],gsa[31,1]+nchar(p)-1)
[1] "TATAAT"
> substr(s2,gsa[32,1],gsa[32,1]+nchar(p)-1)
[1] "TATAAT"
```

On the other hand, the longest common subsequences of two given sequences can be found by traversing the generalized suffix array of the two sequences and computing the longest common prefix of each pair of consecutive entries if they correspond to different sequences. In the following description, the length p of the longest common prefix between suffixes F_1 and F_2 in each pair of consecutive entries of the generalized suffix array A is computed, and the largest length ℓ of the longest common prefixes found so far is kept together with a list L of all such longest common subsequences.

function longest_common_subsequences(S_1, S_2, A)
    $L \leftarrow \emptyset$
    $\ell \leftarrow 0$
    $(k,t) \leftarrow A[1]$
    $F_2 \leftarrow S_t[k,\ldots,n_t]$
    for $i \leftarrow 2$ **to** $n_1 + n_2$ **do**
        $F_1 \leftarrow F_2$
        $(k,t) \leftarrow A[i]$
        $F_2 \leftarrow S_t[k,\ldots,n_t]$
        $(k',t') \leftarrow A[i-1]$
        if $t \neq t'$ **then**
            $p \leftarrow lcp(F_1, F_2)$
            if $p > \ell$ **then**
                $\ell \leftarrow p$
                $L \leftarrow \emptyset$
            if $p = \ell$ **then**
                $L \leftarrow L \cup \{S_t[1,\ldots,p]\}$
    return L

The previous algorithm for finding the longest common occurrences of two sequences using their generalized suffix array can be implemented in Perl in a straightforward way, as shown in the following Perl script.

```
sub longest_common_subsequences {
  my $seq = shift;
```

```
my %seq = %{$seq};
my $gsa = shift;
my @gsa = @{$gsa};
my @L;
my $l = 0;
my $f2 = substr($seq{$gsa[0][1]}, $gsa[0][0],
    length($seq{$gsa[0][1]}));
for (my $i = 1; $i < length($seq{1})+length($seq
    {2}); $i++) {
  my $f1 = $f2;
  $f2 = substr($seq{$gsa[$i][1]}, $gsa[$i][0],
      length($seq{$gsa[$i][1]}));
  if ($gsa[$i][1] != $gsa[$i-1][1]) {
    my $p = lcp($f1,0,$f2,0);
    if ($p > $l) {
      $l = $p;
      @L = ();
    }
    if ($p == $l) {
      push @L,
          substr($seq{$gsa[$i][1]},$gsa[$i][0],$p);
    }
  }
}
return \@L;
}
```

The previous algorithm for finding the longest common occurrences of two sequences using their generalized suffix array can also be implemented in R in a straightforward way, as shown in the following R script. Recall that the two sequences are input as a vector **seq** in order to be able to access them according to the second column of the generalized suffix array **gsa**.

```
> longest.common.subsequences <- function (seq,gsa) {
L <- c()
ell <- 0
suf2 <- substr(seq[gsa[1,2]],gsa[1,1],
    nchar(seq[gsa[1,2]]))
for (i in 2:(nrow(gsa))) {
  suf1 <- suf2
  suf2 <- substr(seq[gsa[i,2]],gsa[i,1],
      nchar(seq[gsa[i,2]]))
  if (gsa[i-1,2] != gsa[i,2]) {
    pref <- substr(suf1,1,lcp(suf1,1,suf2,1))
    if (nchar(pref) > ell) {
      ell <- nchar(pref)
```

```
        L <- c()
     }
     if (nchar(pref) == ell) {
       L <- c(L,pref)
     }
   }
 }
 L
}


> s1 <- "TGCTTCTGACTATAATAG"
> s2 <- "GCTTCCGGCTCGTATAATGTGTGG"
> gsa <- generalized.suffix.array(s1,s2)
> longest.common.subsequences(c(s1,s2),gsa)
[1] "TATAAT"
```

4.3   Comparing Sequences

The similarities and differences between two sequences can be assessed by computing a distance measure between the two sequences. The *edit distance* is based on the elementary edit operations of inserting an element in a sequence, deleting an element from a sequence, and substituting another element for an element in a sequence. The *alignment* of two sequences is an explicit description of the correspondence between the elements of the two sequences, together with the positions at which element insertions and deletions take place in each sequence.

4.3.1   Edit Distance-Based Comparison of Sequences

The *edit distance* is based on the insertion, deletion, and substitution of elements in the two sequences under comparison. The number and type of edit operations needed to transform one sequence into the other reveal similarities and differences between two sequences. There are several types of edit distance between sequences, depending on the type of edit operations allowed.

The *Hamming distance* between two sequences of the same length is defined as the number of positions at which the two sequences differ. This is the same as the smallest number of element substitutions needed to transform one sequence into the other, meaning that insertions and deletions are forbidden and only substitutions are allowed in the Hamming distance.

Example 4.6

At least 12 element substitutions are needed to transform the DNA sequence TGCTTCTGACTATAATAG into GCTTCCGGCTCGTATAAT, as shown in the following alignment. Therefore, the Hamming distance between the two sequences is 12.

```
TGCTTCTGACTATAATAG
||| | | |||| || |
GCTTCCGGCTCGTATAAT
```

The Hamming distance between two sequences of the same length can be computed by traversing the sequences and counting the number of positions at which they differ. In the following description, the Hamming distance d between two sequences S_1 and S_2 is set to -1 when they have different lengths; otherwise, it is obtained as the number of sequence positions i such that $S_1[i] \neq S_2[i]$.

> **function** hamming_distance(S_1, S_2)
> $\quad n \leftarrow length(S_1)$
> $\quad$ **if** $n \neq length(S_2)$ **then**
> $\quad\quad d \leftarrow -1$
> $\quad$ **else**
> $\quad\quad d \leftarrow 0$
> $\quad\quad$ **for** $i \leftarrow 1$ **to** n **do**
> $\quad\quad\quad$ **if** $S_1[i] \neq S_2[i]$ **then**
> $\quad\quad\quad\quad d \leftarrow d + 1$
> $\quad$ **return** d

The previous algorithm for computing the Hamming distance between two sequences of the same length can be implemented in Perl in a straightforward way, as shown in the following Perl script.

```perl
sub hamming_distance {
  my $s1 = shift;
  my $s2 = shift;
  my $d = -1;
  if (length $s1 == length $s2) {
    $d = 0;
    for (my $i = 0; $i < length $s1; $i++) {
      $d++ if substr($s1,$i,1) ne substr($s2,$i,1);
    }
  }
  return $d;
}
```

The algorithm for computing the Hamming distance between two sequences of the same length can also be easily implemented in R, for instance, by first splitting the sequences into vectors with the `strsplit` function and then using the `sum` function to count the number of positions at which the two vectors differ. This is all shown in the following R script.

```
> hamming.distance <- function (s1,s2) {
  if (nchar(s1) != nchar(s2))
    -1
  else
    sum(strsplit(s1,"")[[1]] != strsplit(s2,"")[[1]])
}

> s1 <- "TGCTTCTGACTATAATAG"
> s2 <- "GCTTCCGGCTCGTATAAT"
> hamming.distance(s1,s1)
[1] 0
> hamming.distance(s1,s2)
[1] 12
```

The *Levenshtein distance* between two sequences (not necessarily of the same length) is defined as the smallest number of element insertions and deletions needed to transform one sequence into the other. Unlike the Hamming distance, in which only element substitutions are allowed, in the Levenshtein distance only element insertions and deletions are allowed.

Example 4.7
The DNA sequence GCTTCCGGCTCGTATAATGTGTGG can be transformed into TGCTTCTGACTATAATAG by 4 insertions and 10 deletions, as shown in the following alignment, and this is the least possible number of insertions and deletions to transform one of these sequences into the other. Therefore, the Levenshtein distance between them is 14.

```
-GCTTCC-GG-CTCGTATAAT-GTGTGG
 |||||   |  ||   |||||  |
TGCTTC-TG-ACT---ATAATAG-----
```

Given two sequences S_1 and S_2, assume a prefix $S_1[1, \ldots, i-1]$ can be transformed into a prefix $S_2[1, \ldots, j]$ by x insertions and deletions, a prefix $S_1[1, \ldots, i]$ can be transformed into a prefix $S_2[1, \ldots, j-1]$ by y insertions and deletions, and $S_1[1, \ldots, i-1]$ can be transformed into $S_2[1, \ldots, j-1]$ by z insertions and deletions. Then $S_1[1, \ldots, i]$ can also be transformed into $S_2[1, \ldots, j]$ by z insertions and deletions if $S_1[i] = S_2[j]$ or else by the x edit operations plus the insertion of element $S_1[i]$ or the y edit operations plus the deletion of element $S_2[j]$. In general, the Levenshtein distance d between two sequences S_1 and S_2 is given by the recurrence

$$d(S_1[1,\ldots,i], S_2[1,\ldots,j]) =$$
$$= \min \begin{cases} d(S_1[1,\ldots,i-1], S_2[1,\ldots,j]) + 1, \\ d(S_1[1,\ldots,i], S_2[1,\ldots,j-1]) + 1, \\ d(S_1[1,\ldots,i-1], S_2[1,\ldots,j-1]) & \text{if } S_1[i] = S_2[j] \end{cases}$$

where $d(S_1[1,\ldots,i], S_2[1,\ldots,j]) = 0$ if both $i = 0$ and $j = 0$, $d(S_1[1,\ldots,i], S_2[1,\ldots,j]) = i$ if $i \neq 0$ and $j = 0$, and $d(S_1[1,\ldots,i], S_2[1,\ldots,j]) = j$ if $i = 0$ and $j \neq 0$.

Computation of this recurrence by dynamic programming involves the use of a dynamic programming table to store each $d(S_1[1,\ldots,i], S_2[1,\ldots,j])$, for $1 \leqslant i \leqslant n_1$ and $1 \leqslant j \leqslant n_2$, where n_1 is the length of S_1 and n_2 is the length of S_2.

Example 4.8

The Levenshtein distance between the DNA sequences GCTTCCGGCTCG TATAATGTGTGG and TGCTTCTGACTATAATAG is 14, as shown in entry $(24, 18)$ of the following dynamic programming table.

			T	G	C	T	T	C	T	G	A	C	T	A	T	A	A	T	A	G
		0	1	2	3	4	5	6	7	8	9	10	11	12	13	14	15	16	17	18
	0	0	1	2	3	4	5	6	7	8	9	10	11	12	13	14	15	16	17	18
G	1	1	2	1	2	3	4	5	6	7	8	9	10	11	12	13	14	15	16	17
C	2	2	3	2	1	2	3	4	5	6	7	8	9	10	11	12	13	14	15	16
T	3	3	2	3	2	1	2	3	4	5	6	7	8	9	10	11	12	13	14	15
T	4	4	3	4	3	2	1	2	3	4	5	6	7	8	9	10	11	12	13	14
C	5	5	4	5	4	3	2	1	2	3	4	5	6	7	8	9	10	11	12	13
C	6	6	5	6	5	4	3	2	3	4	5	4	5	6	7	8	9	10	11	12
G	7	7	6	5	6	5	4	3	4	3	4	5	6	7	8	9	10	11	12	11
G	8	8	7	6	7	6	5	4	5	4	5	6	7	8	9	10	11	12	13	12
C	9	9	8	7	6	7	6	5	6	5	6	5	6	7	8	9	10	11	12	13
T	10	10	9	8	7	6	7	6	5	6	7	6	5	6	7	8	9	10	11	12
C	11	11	10	9	8	7	8	7	6	7	8	7	6	7	8	9	10	11	12	13
G	12	12	11	10	9	8	9	8	7	6	7	8	7	8	9	10	11	12	13	12
T	13	13	12	11	10	9	8	9	8	7	8	9	8	9	8	9	10	11	12	13
A	14	14	13	12	11	10	9	10	9	8	7	8	9	8	9	8	9	10	11	12
T	15	15	14	13	12	11	10	11	10	9	8	9	8	9	8	9	10	9	10	11
A	16	16	15	14	13	12	11	12	11	10	9	10	9	8	9	8	9	10	9	10
A	17	17	16	15	14	13	12	13	12	11	10	11	10	9	10	9	8	9	10	11
T	18	18	17	16	15	14	13	14	13	12	11	12	11	10	9	10	9	8	9	10
G	19	19	18	17	16	15	14	15	14	13	12	13	12	11	10	11	10	9	10	9
T	20	20	19	18	17	16	15	16	15	14	13	14	13	12	11	12	11	10	11	10
G	21	21	20	19	18	17	16	17	16	15	14	15	14	13	12	13	12	11	12	11
T	22	22	21	20	19	18	17	18	17	16	15	16	15	14	13	14	13	12	13	12
G	23	23	22	21	20	19	18	19	18	17	16	17	16	15	14	15	14	13	14	13
G	24	24	23	22	21	20	19	20	19	18	17	18	17	16	15	16	15	14	15	14

In the following description, the dynamic programming table D is filled in for each $0 \leqslant i \leqslant n_1$ and $0 \leqslant j \leqslant n_2$, and the Levenshtein distance between S_1 and S_2 is stored in entry $D[n_1, n_2]$.

> **function** levenshtein_distance(S_1, S_2)
> $\quad n_1 \leftarrow \text{length}(S_1)$
> $\quad n_2 \leftarrow \text{length}(S_2)$
> $\quad D[0,0] \leftarrow 0$
> $\quad$ **for** $i \leftarrow 1$ **to** n_1 **do**
> $\quad\quad D[i,0] \leftarrow i$
> $\quad$ **for** $j \leftarrow 1$ **to** n_2 **do**
> $\quad\quad D[0,j] \leftarrow j$
> $\quad$ **for** $i \leftarrow 1$ **to** n_1 **do**
> $\quad\quad$ **for** $j \leftarrow 1$ **to** n_2 **do**
> $\quad\quad\quad$ **if** $S_1[i] = S_2[j]$ **then**
> $\quad\quad\quad\quad D[i,j] \leftarrow D[i-1, j-1]$
> $\quad\quad\quad$ **else**
> $\quad\quad\quad\quad D[i,j] \leftarrow \min(D[i-1,j]+1, D[i,j-1]+1)$
> $\quad$ **return** $D[n_1, n_2]$

The previous algorithm for computing the Levenshtein distance between two sequences can be implemented in Perl in a straightforward way, as shown in the following Perl script.

```perl
sub levenshtein_distance {
  my $s1 = shift;
  my $s2 = shift;
  my ($n1, $n2) = (length $s1, length $s2);

  my @d;
  for ( my $i = 0; $i <= $n1; $i++ ) {
    $d[$i][0] = $i;
  }
  for ( my $j = 0; $j <= $n2; $j++ ) {
    $d[0][$j] = $j;
  }

  my @t1 = split //, $s1;
  my @t2 = split //, $s2;

  for ( my $i = 1; $i <= $n1; $i++ ) {
    for ( my $j = 1; $j <= $n2; $j++ ) {
      if ($t1[$i-1] eq $t2[$j-1]) {
        $d[$i][$j] = $d[$i-1][$j-1];
```

```
    } else {
      $d[$i][$j] = $d[$i-1][$j] + 1;
      if ($d[$i][$j-1] + 1 < $d[$i][$j]) {
        $d[$i][$j] = $d[$i][$j-1] + 1;
      }
    }
  }
}

return \@d;
}
```

The previous algorithm for computing the Levenshtein distance between two sequences can also be implemented in R in a straightforward way. Unlike Perl array indices that start at 0, however, R array indexes do start at 1 and, thus, the values $d(S_1[1, \ldots, i], S_2[1, \ldots, j])$ are stored in positions $i + 1, j + 1$ of array D. This is all illustrated by the following R script.

```
> levenshtein.distance <- function(s1,s2) {
    t1 <- strsplit(s1,split="")[[1]]
    t2 <- strsplit(s2,split="")[[1]]
    n1 <- length(t1)
    n2 <- length(t2)
    d <- array(0,dim=c(n1+1,n2+1))
    d[,1] <- 0:n1
    d[1,] <- 0:n2
    for (i in 2:(n1+1))
      for (j in 2:(n2+1))
        if (t1[i-1] == t2[j-1])
          d[i,j] <- d[i-1,j-1]
        else
          d[i,j] <- min(d[i-1,j]+1, d[i,j-1]+1)
    d
}

> s1 <- "GCTTCCGGCTCGTATAATGTGTGG"
> s2 <- "TGCTTCTGACTATAATAG"
> d <- levenshtein.distance(s1,s1)
> d[nchar(s1)+1,nchar(s1)+1]
[1] 0
> d <- levenshtein.distance(s1,s2)
> d[nchar(s1)+1,nchar(s2)+1]
[1] 14
```

In general, the *edit distance* between two sequences (not necessarily of the same length) is defined as the smallest number of insertions, deletions, and substitutions needed to transform one sequence into the other. The edit distance thus combines the Hamming distance, in which only element substitutions are allowed, with the Levenshtein distance, in which only element insertions and deletions are allowed.

Example 4.9

The DNA sequence GCTTCCGGCTCGTATAATGTGTGG can be transformed into TGCTTCTGACTATAATAG by 1 insertion, 7 deletions, and 3 substitutions, as shown in the following alignment, and this is the least possible number of insertions, deletions, and substitutions to transform one of these sequences into the other. Therefore, the edit distance between them is 11.

```
-GCTTCCGGCTCGTATAATGTGTGG
 |||||*|*||    |||||* |
TGCTTCTGACT---ATAATA-G---
```

Given two sequences S_1 and S_2, assume a prefix $S_1[1,\ldots,i-1]$ can be transformed into a prefix $S_2[1,\ldots,j]$ by x insertions, deletions, and substitutions, a prefix $S_1[1,\ldots,i]$ can be transformed into a prefix $S_2[1,\ldots,j-1]$ by y insertions, deletions, and substitutions, and $S_1[1,\ldots,i-1]$ can be transformed into $S_2[1,\ldots,j-1]$ by z insertions, deletions, and substitutions. Then $S_1[1,\ldots,i]$ can also be transformed into $S_2[1,\ldots,j]$ by z edit operations if $S_1[i] = S_2[j]$ or else by the x edit operations plus the insertion of element $S_1[i]$, the y edit operations plus the deletion of element $S_2[j]$, or the z edit operations plus the substitution of element $S_2[j]$ for element $S_1[i]$. In general, the edit distance d between two sequences S_1 and S_2 is given by the recurrence

$$d(S_1[1,\ldots,i], S_2[1,\ldots,j]) =$$
$$= \min \begin{cases} d(S_1[1,\ldots,i-1], S_2[1,\ldots,j]) + 1, \\ d(S_1[1,\ldots,i], S_2[1,\ldots,j-1]) + 1, \\ d(S_1[1,\ldots,i-1], S_2[1,\ldots,j-1]) \quad \text{if } S_1[i] = S_2[j] \\ d(S_1[1,\ldots,i-1], S_2[1,\ldots,j-1]) + 1 \text{ if } S_1[i] \neq S_2[j] \end{cases}$$

where $d(S_1[1,\ldots,i], S_2[1,\ldots,j]) = 0$ if both $i = 0$ and $j = 0$, $d(S_1[1,\ldots,i], S_2[1,\ldots,j]) = i$ if $i \neq 0$ and $j = 0$, and $d(S_1[1,\ldots,i], S_2[1,\ldots,j]) = j$ if $i = 0$ and $j \neq 0$.

Computation of this recurrence by dynamic programming involves the use of a dynamic programming table to store each $d(S_1[1,\ldots,i], S_2[1,\ldots,j])$, for $1 \leqslant i \leqslant n_1$ and $1 \leqslant j \leqslant n_2$, where n_1 is the length of S_1 and n_2 is the length of S_2.

Example 4.10

The edit distance between the DNA sequences GCTTCCGGCTCGTATAAT

GTGTGG and TGCTTCTGACTATAATAG is 11, as shown in entry $(24, 18)$ of the following dynamic programming table.

			T	G	C	T	T	C	T	G	A	C	T	A	T	A	A	T	A	G
		0	1	2	3	4	5	6	7	8	9	10	11	12	13	14	15	16	17	18
	0	0	1	2	3	4	5	6	7	8	9	10	11	12	13	14	15	16	17	18
G	1	1	1	1	2	3	4	5	6	7	8	9	10	11	12	13	14	15	16	17
C	2	2	2	2	1	2	3	4	5	6	7	8	9	10	11	12	13	14	15	16
T	3	3	2	3	2	1	2	3	4	5	6	7	8	9	10	11	12	13	14	15
T	4	4	3	3	3	2	1	2	3	4	5	6	7	8	9	10	11	12	13	14
C	5	5	4	4	3	3	2	1	2	3	4	5	6	7	8	9	10	11	12	13
C	6	6	5	5	4	4	3	2	2	3	4	4	5	6	7	8	9	10	11	12
G	7	7	6	5	5	5	4	3	3	2	3	4	5	6	7	8	9	10	11	11
G	8	8	7	6	6	6	5	4	4	3	3	4	5	6	7	8	9	10	11	11
C	9	9	8	7	6	7	6	5	5	4	4	3	4	5	6	7	8	9	10	11
T	10	10	9	8	7	6	7	6	5	5	5	4	3	4	5	6	7	8	9	10
C	11	11	10	9	8	7	7	7	6	6	6	5	4	4	5	6	7	8	9	10
G	12	12	11	10	9	8	8	8	7	6	7	6	5	5	5	6	7	8	9	9
T	13	13	12	11	10	9	8	9	8	7	7	7	6	6	5	6	7	7	8	9
A	14	14	13	12	11	10	9	9	9	8	7	8	7	6	6	5	6	7	7	8
T	15	15	14	13	12	11	10	10	9	9	8	8	8	7	6	6	6	6	7	8
A	16	16	15	14	13	12	11	11	10	10	9	9	9	8	7	6	6	7	6	7
A	17	17	16	15	14	13	12	12	11	11	10	10	10	9	8	7	6	7	7	7
T	18	18	17	16	15	14	13	13	12	12	11	11	10	10	9	8	7	6	7	8
G	19	19	18	17	16	15	14	14	13	12	12	12	11	11	10	9	8	7	7	7
T	20	20	19	18	17	16	15	15	14	13	13	13	12	12	11	10	9	8	8	8
G	21	21	20	19	18	17	16	16	15	14	14	14	13	13	12	11	10	9	9	8
T	22	22	21	20	19	18	17	17	16	15	15	15	14	14	13	12	11	10	10	9
G	23	23	22	21	20	19	18	18	17	16	16	16	15	15	14	13	12	11	11	10
G	24	24	23	22	21	20	19	19	18	17	17	17	16	16	15	14	13	12	12	11

In the following description, the dynamic programming table D is filled in for each $0 \leqslant i \leqslant n_1$ and $0 \leqslant j \leqslant n_2$, and the edit distance between S_1 and S_2 is stored in entry $D[n_1, n_2]$.

```
function edit_distance(S₁, S₂)
    n₁ ← length(S₁)
    n₂ ← length(S₂)
    D[0, 0] ← 0
    for i ← 1 to n₁ do
        D[i, 0] ← i
    for j ← 1 to n₂ do
        D[0, j] ← j
    for i ← 1 to n₁ do
        for j ← 1 to n₂ do
            D[i, j] ← min(D[i − 1, j] + 1, D[i, j − 1] + 1)
```

$$\textbf{if } S_1[i] = S_2[j] \textbf{ then}$$
$$D[i,j] \leftarrow \min(D[i,j], D[i-1,j-1])$$
$$\textbf{else}$$
$$D[i,j] \leftarrow \min(D[i,j], D[i-1,j-1]+1)$$
$$\textbf{return } D[n_1, n_2]$$

The previous algorithm for computing the edit distance between two sequences can be implemented in Perl in a straightforward way, as shown in the following Perl script.

```perl
sub edit_distance {
  my $s1 = shift;
  my $s2 = shift;
  my ($n1, $n2) = (length $s1, length $s2);
  my @d;
  for ( my $i = 0; $i <= $n1; $i++ ) {
    $d[$i][0] = $i;
  }
  for ( my $j = 0; $j <= $n2; $j++ ) {
    $d[0][$j] = $j;
  }

  my @t1 = split //, $s1;
  my @t2 = split //, $s2;

  for ( my $i = 1; $i <= $n1; $i++ ) {
    for ( my $j = 1; $j <= $n2; $j++ ) {
      $d[$i][$j] = $d[$i-1][$j] + 1;
      if ($d[$i][$j-1] + 1 < $d[$i][$j]) {
        $d[$i][$j] = $d[$i][$j-1] + 1;
      }
      if ($t1[$i-1] eq $t2[$j-1]) {
        if ($d[$i-1][$j-1] < $d[$i][$j]) {
          $d[$i][$j] = $d[$i-1][$j-1];
        }
      } else {
        if ($d[$i-1][$j-1] + 1 < $d[$i][$j]) {
          $d[$i][$j] = $d[$i-1][$j-1] + 1;
        }
      }
    }
  }

  return \@d;
}
```

The previous algorithm for computing the edit distance between two sequences can also be implemented in R in a straightforward way. Once again, the values $d(S_1[1,\ldots,i], S_2[1,\ldots,j])$ are stored in positions $i+1, j+1$ of array d, because unlike Perl array indexes that start at 0, R array indexes do start at 1. This is all illustrated by the following R script.

```
> edit.distance <- function (s1,s2) {
    t1 <- strsplit(s1,split="")[[1]]
    t2 <- strsplit(s2,split="")[[1]]
    n1 <- length(t1)
    n2 <- length(t2)
    d <- array(0,dim=c(n1+1,n2+1))
    d[,1] <- 0:n1
    d[1,] <- 0:n2
    for (i in 2:(n1+1)) {
        for (j in 2:(n2+1)) {
            d[i,j] <- min(d[i-1,j]+1, d[i,j-1]+1)
            if (t1[i-1] == t2[j-1])
                d[i,j] <- min(d[i,j], d[i-1,j-1])
            else
                d[i,j] <- min(d[i,j], d[i-1,j-1]+1)
        }
    }
    d
}

> s1 <- "GCTTCCGGCTCGTATAATGTGTGG"
> s2 <- "TGCTTCTGACTATAATAG"
> d <- edit.distance(s1,s1)
> d[nchar(s1)+1,nchar(s1)+1]
[1] 0
> d <- edit.distance(s1,s2)
> d[nchar(s1)+1,nchar(s2)+1]
[1] 11
```

4.3.2 Alignment-Based Comparison of Sequences

An *alignment* of two sequences is an arrangement of the two sequences as rows of a matrix, with additional gaps (dashes) between the elements to make some or all of the remaining (aligned) columns contain identical elements but with no column gapped in both sequences. A dash in the first sequence of an alignment corresponds to the insertion of the opposite element into the first sequence, a dash in the second sequence of an alignment corresponds to the deletion of the opposite element from the second sequence, and two mismatched elements opposite in an alignment correspond to a substitution of the element in the second sequence for the element in the first sequence.

The Levenshtein distance between two sequences is thus given by an alignment of the two sequences with the smallest possible number of dashes (insertions or deletions) and with no mismatched elements (substitutions), while the edit distance between two sequences is given by an alignment of the two sequences with the smallest possible number of dashes (insertions or deletions) plus mismatched elements (substitutions).

Example 4.11

The DNA sequence GCTTCCGGCTCGTATAATGTGTGG can be transformed into TGCTTCTGACTATAAATAG by inserting T, T, A, A before (original) positions 1, 7, 9, 19, and deleting C, G, C, G, T, T, G, T, G, G at (original) positions 6, 8, 11, 12, 13, 20, 21, 22, 23, 24, as shown in the following alignment.

```
-GCTTCC-GG-CTCGTATAAT-GTGTGG
 |||||   |  ||    ||||| |
TGCTTC-TG-ACT---ATAATAG-----
```

Sequence GCTTCCGGCTCGTATAATGTGTGG can also be transformed into TGCTTCTGACTATAAATAG by inserting T before position 1; substituting T for C at position 6; substituting A for G at position 8; deleting C, G, T at positions 11, 12, 13; substituting A for G at position 19; deleting T at position 20; and deleting T, G, G at positions 22, 23, 24, as shown in the following alignment.

```
-GCTTCCGGCTCGTATAATGTGTGG
 |||||*|*||    |||||*  |
TGCTTCTGACT---ATAATA-G---
```

An alignment of two sequences can be obtained from the dynamic programming table, already filled in when computing the Levenshtein distance or the edit distance between the two sequences, by tracing the sequence of edit operations from the final (bottom right) position back to the initial (top left) position. In the trace back at position i of S_1 and position j of S_2,

- $D(S_1[1,\ldots,i], S_2[1,\ldots,j]) = D(S_1[1,\ldots,i], S_2[1,\ldots,j-1])+1$ indicates the insertion of a dash into S_1, and

- $D(S_1[1,\ldots,i], S_2[1,\ldots,j]) = D(S_1[1,\ldots,i-1], S_2[1,\ldots,j])+1$ indicates the insertion of a dash into S_2.

Since there is a choice of moving left or moving up (if the previous conditions are fulfilled) and also moving in diagonal, if either $S_1[i] = S_2[j]$ and $D(S_1[1,\ldots,i], S_2[1,\ldots,j]) = D(S_1[1,\ldots,i-1], S_2[1,\ldots,j-1])$ or $S_1[i] \neq S_2[j]$ and $D(S_1[1,\ldots,i], S_2[1,\ldots,j]) = D(S_1[1,\ldots,i-1], S_2[1,\ldots,j-1])+1$, several alignments may be implicit in a single dynamic programming table.

Example 4.12

The Levenshtein distance between the DNA sequences GCTTCCGGCTCG TATAATGTGTGG and TGCTTCTGACTATAATAG gives 1,430 different alignments. Each such alignment can be obtained by following a different path of shaded entries from the final (bottom right) back to the initial (top left) entry of the following dynamic programming table, inserting a dash into S_1 when moving to the left and inserting a dash into S_2 when moving up.

		0	T 1	G 2	C 3	T 4	T 5	C 6	T 7	G 8	A 9	C 10	T 11	A 12	T 13	A 14	A 15	T 16	A 17	G 18
	0	0	1	2	3	4	5	6	7	8	9	10	11	12	13	14	15	16	17	18
G	1	1	2	1	2	3	4	5	6	7	8	9	10	11	12	13	14	15	16	17
C	2	2	3	2	1	2	3	4	5	6	7	8	9	10	11	12	13	14	15	16
T	3	3	2	3	2	1	2	3	4	5	6	7	8	9	10	11	12	13	14	15
T	4	4	3	4	3	2	1	2	3	4	5	6	7	8	9	10	11	12	13	14
C	5	5	4	5	4	3	2	1	2	3	4	5	6	7	8	9	10	11	12	13
C	6	6	5	6	5	4	3	2	3	4	5	4	5	6	7	8	9	10	11	12
G	7	7	6	5	6	5	4	3	4	3	4	5	6	7	8	9	10	11	12	11
G	8	8	7	6	7	6	5	4	5	4	5	6	7	8	9	10	11	12	13	12
C	9	9	8	7	6	7	6	5	6	5	6	5	6	7	8	9	10	11	12	13
T	10	10	9	8	7	6	7	6	5	6	7	6	5	6	7	8	9	10	11	12
C	11	11	10	9	8	7	8	7	6	7	8	7	6	7	8	9	10	11	12	13
G	12	12	11	10	9	8	9	8	7	6	7	8	7	8	9	10	11	12	13	12
T	13	13	12	11	10	9	8	9	8	7	8	9	8	9	8	9	10	11	12	13
A	14	14	13	12	11	10	9	10	9	8	7	8	9	8	9	8	9	10	11	12
T	15	15	14	13	12	11	10	11	10	9	8	9	8	9	8	9	10	9	10	11
A	16	16	15	14	13	12	11	12	11	10	9	10	9	8	9	8	9	10	9	10
A	17	17	16	15	14	13	12	13	12	11	10	11	10	9	10	9	8	9	10	11
T	18	18	17	16	15	14	13	14	13	12	11	12	11	10	9	10	9	8	9	10
G	19	19	18	17	16	15	14	15	14	13	12	13	12	11	10	11	10	9	10	9
T	20	20	19	18	17	16	15	16	15	14	13	14	13	12	11	12	11	10	11	10
G	21	21	20	19	18	17	16	17	16	15	14	15	14	13	12	13	12	11	12	11
T	22	22	21	20	19	18	17	18	17	16	15	16	15	14	13	14	13	12	13	12
G	23	23	22	21	20	19	18	19	18	17	16	17	16	15	14	15	14	13	14	13
G	24	24	23	22	21	20	19	20	19	18	17	18	17	16	15	16	15	14	15	14

For instance, by moving to the left if possible, otherwise up if possible, or else in diagonal, the following alignment of the two sequences is obtained, where matches are indicated with a vertical bar.

```
-GCTTCC-GG-CTCGTATAAT-GTGTGG
 |||||   |  ||   |||||  |
TGCTTC-TG-ACT---ATAATAG-----
```

Example 4.13

The edit distance between the DNA sequences GCTTCCGGCTCGTATAAT GTGTGG and TGCTTCTGACTATAATAG gives 187 different alignments.

Each such alignment can be obtained by following a different path of shaded entries from the final (bottom right) back to the initial (top left) entry of the following dynamic programming table, inserting a dash into S_1 when moving to the left and inserting a dash into S_2 when moving up.

		T 1	**G** 2	**C** 3	**T** 4	**T** 5	**C** 6	**T** 7	**G** 8	**A** 9	**C** 10	**T** 11	**A** 12	**T** 13	**A** 14	**A** 15	**T** 16	**A** 17	**G** 18
	0	0	1	2	3	4	5	6	7	8	9	10	11	12	13	14	15	16	17	18
G 1	1	1	1	1	2	3	4	5	6	7	8	9	10	11	12	13	14	15	16	17
C 2	2	2	2	2	1	2	3	4	5	6	7	8	9	10	11	12	13	14	15	16
T 3	3	3	2	3	2	1	2	3	4	5	6	7	8	9	10	11	12	13	14	15
T 4	4	4	3	3	3	2	1	2	3	4	5	6	7	8	9	10	11	12	13	14
C 5	5	5	4	4	3	3	2	1	2	3	4	5	6	7	8	9	10	11	12	13
C 6	6	6	5	5	4	4	3	2	2	3	4	4	5	6	7	8	9	10	11	12
G 7	7	7	6	5	5	5	4	3	3	2	3	4	5	6	7	8	9	10	11	11
G 8	8	8	7	6	6	6	5	4	4	3	3	4	5	6	7	8	9	10	11	11
C 9	9	9	8	7	6	7	6	5	5	4	4	3	4	5	6	7	8	9	10	11
T 10	10	10	9	8	7	6	7	6	5	5	5	4	3	4	5	6	7	8	9	10
C 11	11	11	10	9	8	7	7	7	6	6	6	5	4	4	5	6	7	8	9	10
G 12	12	12	11	10	9	8	8	8	7	6	7	6	5	5	5	6	7	8	9	9
T 13	13	13	12	11	10	9	8	9	8	7	7	7	6	6	5	6	7	7	8	9
A 14	14	14	13	12	11	10	9	9	9	8	7	8	7	6	6	5	6	7	7	8
T 15	15	15	14	13	12	11	10	10	9	9	8	8	8	7	6	6	6	6	7	8
A 16	16	16	15	14	13	12	11	11	10	10	9	9	9	8	7	6	6	7	6	7
A 17	17	17	16	15	14	13	12	12	11	11	10	10	10	9	8	7	6	7	7	7
T 18	18	18	17	16	15	14	13	13	12	12	11	11	10	10	9	8	7	6	7	8
G 19	19	19	18	17	16	15	14	14	13	12	12	12	11	11	10	9	8	7	7	7
T 20	20	20	19	18	17	16	15	15	14	13	13	13	12	12	11	10	9	8	8	8
G 21	21	21	20	19	18	17	16	16	15	14	14	14	13	13	12	11	10	9	9	8
T 22	22	22	21	20	19	18	17	17	16	15	15	15	14	14	13	12	11	10	10	9
G 23	23	23	22	21	20	19	18	18	17	16	16	16	15	15	14	13	12	11	11	10
G 24	24	24	23	22	21	20	19	19	18	17	17	17	16	16	15	14	13	12	12	11

For instance, by moving to the left if possible, otherwise up if possible, or else in diagonal, the following alignment of the two sequences is obtained, where matches are indicated with a vertical bar and mismatches with an asterisk.

```
-GCTTCCGGCTCGTATAATGTGTGG
 |||||*|*||    |||||* |
TGCTTCTGACT---ATAATA-G---
```

Once the Levenshtein distance or the edit distance between two sequences has been computed, an alignment of the two sequences can be obtained by tracing back the dynamic programming table from the final (bottom right) to the initial (top left) entry, inserting a dash into S_1 and moving up if $D(S_1[1,\ldots,i], S_2[1,\ldots,j]) = D(S_1[1,\ldots,i], S_2[1,\ldots,j-1])+1$, otherwise inserting a dash into S_2 and moving to the left if $D(S_1[1,\ldots,i], S_2[1,\ldots,j]) =$

$D(S_1[1,\ldots,i-1], S_2[1,\ldots,j]) + 1$, or else moving up and to the left, in diagonal. Once the first row or the first column has been reached, further movements up (after reaching the first column) or to the left (after reaching the first row) may be needed in order to finish computing the alignment.

Recall that the dynamic programming table D of a sequence S_1 of length n_1 and a sequence S_2 of length n_2 has $n_1 + 1$ rows numbered $0, \ldots, n_1$ and $n_2 + 1$ columns numbered $0, \ldots, n_2$. In the following description, a centered dot $\cdot$ denotes the concatenation of single elements (dashes) and sequences.

```
function alignment(S1, S2, D)
    i ← length(S1)
    j ← length(S2)
    T1 ← T2 ← ""
    while i > 1 and j > 1 do
        if D[i,j] = D[i,j-1] + 1 then
            T1 ← "-" · T1
            T2 ← S2[j-1] · T2
            j ← j - 1
        else if D[i,j] = D[i-1,j] + 1 then
            T1 ← S1[i-1] · T1
            T2 ← "-" · T2
            i ← i - 1
        else
            T1 ← S1[i-1] · T1
            T2 ← S2[j-1] · T2
            i ← i - 1
            j ← j - 1
    while j > 1 do
        T1 ← "-" · T1
        T2 ← S2[j-1] · T2
        j ← j - 1
    while i > 1 do
        T1 ← S1[i-1] · T1
        T2 ← "-" · T2
        i ← i - 1
    return (T1, T2)
```

The previous algorithm for obtaining an alignment between two sequences can be implemented in Perl in a straightforward way, as shown in the following Perl script.

```perl
sub alignment {
    my $s1 = shift;
    my $s2 = shift;
    my $d = shift;
```

```perl
my @d = @{$d};
my ($t1, $t2) = ("", "");
my @t1 = split //, $s1;
my @t2 = split //, $s2;
my ($n1, $n2) = (length $s1, length $s2);
my ($i, $j) = ($n1, $n2);
while ($i > 0 && $j > 0) {
  if ($d[$i][$j] eq $d[$i][$j-1]+1) {
      $t1 = "-" . $t1;
      $t2 = $t2[$j-1] . $t2;
      $j--;
  } elsif ($d[$i][$j] eq $d[$i-1][$j]+1) {
      $t1 = $t1[$i-1] . $t1;
      $t2 = "-" . $t2;
      $i--;
  } else {
      $t1 = $t1[$i-1] . $t1;
      $t2 = $t2[$j-1] . $t2;
      $i--;
      $j--;
  }
}
while ($j > 0) {
  $t1 = "-" . $t1;
  $t2 = $t2[$j-1] . $t2;
  $j--;
}
while ($i > 0) {
  $t1 = $t1[$i-1] . $t1;
  $t2 = "-" . $t2;
  $i--;
}
print "$t1\n$t2\n";
}
```

The algorithm for obtaining an alignment between two sequences can also be implemented in R in a straightforward way, as shown in the following R script.

```r
alignment <- function (s1,s2,d) {
  t1 <- strsplit(s1,split="")[[1]]
  t2 <- strsplit(s2,split="")[[1]]
  a1 <- a2 <- c()
  i <- length(t1)+1
  j <- length(t2)+1
  while (i > 1 && j > 1) {
```

```
    if (d[i,j] == d[i,j-1]+1) {
      a1 <- c("-",a1)
      a2 <- c(t2[j-1],a2)
      j <- j-1
    } else if (d[i,j] == d[i-1,j]+1) {
      a1 <- c(t1[i-1],a1)
      a2 <- c("-",a2)
      i <- i-1
    } else {
      a1 <- c(t1[i-1],a1)
      a2 <- c(t2[j-1],a2)
      i <- i-1
      j <- j-1
    }
  }
  while (j > 1) {
    a1 <- c("-",a1)
    a2 <- c(t2[j-1],a2)
    j <- j-1
  }
  while (i > 1) {
    a1 <- c(t1[i-1],a1)
    a2 <- c("-",a2)
    i <- i-1
  }
  print(paste(a1,collapse=""))
  print(paste(a2,collapse=""))
}

> s1 <- "GCTTCCGGCTCGTATAATGTGTGG"
> s2 <- "TGCTTCTGACTATAATAG"
> d <- levenshtein.distance(s1,s2)
> alignment(s1,s2,d)
[1] "-GCTTCC-GG-CTCGTATAAT-GTGTGG"
[1] "TGCTTC-TG-ACT---ATAATAG-----"
> d <- edit.distance(s1,s2)
> alignment(s1,s2,d)
[1] "-GCTTCCGGCTCGTATAATGTGTGG"
[1] "TGCTTCTGACT---ATAATA-G---"
```

The assessment of similarities and differences between two sequences based on the computation of an edit distance or an alignment of the two sequences can also reflect the relative frequencies with which nucleotide substitutions (for DNA and RNA sequences) or amino acid substitutions (for protein sequences) take place. This can be achieved by assigning a weight or score to each edit operation, depending on either the type of edit operation (element insertion, deletion, substitution) or the actual elements (nucleotides, amino acids) involved in the edit operation. These generalized forms of edit distance and alignment can be computed by a straightforward extension to the edit distance recurrences and corresponding algorithms, where the particular score or weight of the edit operation upon the actual elements is substituted for the summand value 1.

Both the Levenshtein distance and the edit distance between two sequences give a *global* alignment of the sequences, that is, an alignment in which the overall number or the total score or weight of the insertions, deletions, and mismatches is as small as possible. In a *local* alignment, on the other hand, only some subsequences of the two sequences are aligned: those subsequences that give the smallest possible edit distance. Two sequences might actually have a large (global) edit distance but still contain subsequences at small (local) edit distance.

In the formulation of local alignment in terms of edit distance, however, a local alignment over short subsequences cannot always be distinguished from a local alignment over longer subsequences. For instance, the edit distance between two sequences that contain identical subsequences might be the same, no matter the length of the common subsequences, while the sequences are more similar to each other the longer the common subsequences.

The shift from distances to similarities, where matches have a positive weight and insertions, deletions, and mismatches have a negative score, overcomes this problem. Insertions and deletions are also called *gaps*, because they introduce a gap (usually represented as a dash) in a sequence alignment.

In general, the local alignment of two sequences defines stretches of high similarity between the sequences, where a certain subsequence of the first sequence is aligned to a subsequence of the second sequence with a high combined weight or score of matches, mismatches, and gaps.

Example 4.14

Prefix GCTTCCGGCTCGTATAAT of DNA sequence GCTTCCGGCTCG TATAATGTGTGG can be aligned to subsequence GCTTCTGACTATAAT of DNA sequence TGCTTCTGACTATAATAG with 13 matches and only 2 mismatches and 3 gaps, as shown in the following local alignment.

```
GCTTCCGGCTCGTATAAT
|||| |*| *| ||||||
GCTT-CTG-AC-TATAAT
```

Given two sequences S_1 and S_2, assume that the largest possible score when aligning a suffix of prefix $S_1[1, \ldots, i-1]$ to a suffix of prefix $S_2[1, \ldots, j]$ is x, the largest possible score when aligning a suffix of prefix $S_1[1, \ldots, i]$ to a suffix of prefix $S_2[1, \ldots, j-1]$ is y, and the largest possible score when aligning a suffix of $S_1[1, \ldots, i-1]$ to a suffix of $S_2[1, \ldots, j-1]$ is z. Then the largest possible score for aligning a suffix of $S_1[1, \ldots, i]$ to a suffix of $S_2[1, \ldots, j]$ is the largest value among z plus either the match weight, if $S_1[i] = S_2[j]$, or the mismatch weight, if $S_1[i] \neq S_2[j]$; y plus the gap score, for deleting element $S_1[i]$ from S_2; x plus the gap score, for inserting element $S_2[j]$ into S_1; and zero, to account for any negative values. In general, the suffix similarity s between two sequences S_1 and S_2 is given by the recurrence

$$s(S_1[1, \ldots, i], S_2[1, \ldots, j]) =$$
$$= \max \begin{cases} s(S_1[1, \ldots, i-1], S_2[1, \ldots, j]) + gap, \\ s(S_1[1, \ldots, i], S_2[1, \ldots, j-1]) + gap, \\ s(S_1[1, \ldots, i-1], S_2[1, \ldots, j-1]) + match, & \text{if } S_1[i] = S_2[j] \\ s(S_1[1, \ldots, i-1], S_2[1, \ldots, j-1]) + mismatch, & \text{if } S_1[i] \neq S_2[j] \\ 0 \end{cases}$$

where *match* is the positive match score, *mismatch* is the negative mismatch score, *gap* is the negative gap score, and $s(S_1[1, \ldots, i], S_2[1, \ldots, j]) = 0$ if $i = 0$ or $j = 0$.

Computation of this recurrence by dynamic programming involves the use of a dynamic programming table to store each $s(S_1[1, \ldots, i], S_2[1, \ldots, j])$, for $1 \leqslant i \leqslant n_1$ and $1 \leqslant j \leqslant n_2$, where n_1 is the length of S_1 and n_2 is the length of S_2. The largest value of suffix similarity is the total weight or score of an optimal local alignment of the two sequences, and the actual local alignment can then be obtained by tracing the dynamic programming table from each such largest suffix similarity value back to the first entry equal to zero.

The actual values chosen as match, mismatch, and gap score determine the local alignment of two sequences. For instance, with a match score of 1, a mismatch score of 0, and a gap score of 0, the local alignment corresponds to the longest common gapped subsequence, while with a match score of 1, a mismatch score of $-\infty$, and a gap score of $-\infty$, the local alignment corresponds to the longest common subsequence.

Example 4.15

The suffix similarities of the DNA sequences GCTTCCGGCTCGTATAAT GTGTGG and TGCTTCTGACTATAATAG given in the following dynamic programming table, for a match score of 3, a mismatch score of -1, and a gap score of -3, contain three local alignments of the largest total score, 28. Each such local alignment can be obtained by following a path of shaded entries (shown here in one case only, for clarity) from the final (largest suffix similarity) back to an initial (zero suffix similarity) entry of the following dynamic programming table, inserting a dash into S_1 when moving to the left and inserting a dash into S_2 when moving up.

		T	G	C	T	T	C	T	G	A	C	T	A	T	A	A	T	A	G
	0	1	2	3	4	5	6	7	8	9	10	11	12	13	14	15	16	17	18
0	0	0	0	0	0	0	0	0	0	0	0	0	0	0	0	0	0	0	0
G 1	0	0	3	0	0	0	0	0	3	0	0	0	0	0	0	0	0	0	3
C 2	0	0	0	6	3	0	3	0	0	2	3	0	0	0	0	0	0	0	0
T 3	0	3	0	3	9	6	3	6	3	0	1	6	3	3	0	0	3	0	0
T 4	0	3	2	0	6	12	9	6	5	2	0	4	5	6	3	0	3	2	0
C 5	0	0	2	5	3	9	15	12	9	6	5	2	3	4	5	2	0	2	1
C 6	0	0	0	5	4	6	12	14	11	8	9	6	3	2	3	4	1	0	1
G 7	0	0	3	2	4	3	9	11	17	14	11	8	5	2	1	2	3	0	3
G 8	0	0	3	2	1	3	6	8	14	16	13	10	7	4	1	0	1	2	3
C 9	0	0	0	6	3	0	6	5	11	13	19	16	13	10	7	4	1	0	1
T 10	0	3	0	3	9	6	3	9	8	10	16	22	19	16	13	10	7	4	1
C 11	0	0	2	3	6	8	9	6	8	7	13	19	21	18	15	12	9	6	3
G 12	0	0	3	1	3	5	7	8	9	7	10	16	18	20	17	14	11	8	9
T 13	0	3	0	2	4	6	4	10	7	8	7	13	15	21	19	16	17	14	11
A 14	0	0	2	0	1	3	5	7	9	10	7	10	16	18	24	22	19	20	17
T 15	0	3	0	1	3	4	2	8	6	8	9	10	13	19	21	23	25	22	19
A 16	0	0	2	0	0	2	3	5	7	9	7	8	13	16	22	24	22	28	25
A 17	0	0	0	1	0	0	1	2	4	10	8	6	11	13	19	25	23	25	27
T 18	0	3	0	0	4	3	0	4	1	7	9	11	8	14	16	22	28	25	24
G 19	0	0	6	3	1	3	2	1	7	4	6	8	10	11	13	19	25	27	28
T 20	0	3	3	5	6	4	2	5	4	6	3	9	7	13	10	16	22	24	26
G 21	0	0	6	3	4	5	3	2	8	5	5	6	8	10	12	13	19	21	27
T 22	0	3	3	5	6	7	4	6	5	7	4	8	5	11	9	11	16	18	24
G 23	0	0	6	3	4	5	6	3	9	6	6	5	7	8	10	8	13	15	21
G 24	0	0	3	5	2	3	4	5	6	8	5	5	4	6	7	9	10	12	18

In this way, the following local alignment of the two sequences is obtained.

```
GCTTCCGGCTCGTATAAT
|||| |*| *| ||||||
GCTT-CTG-AC-TATAAT
```

In the following description, the dynamic programming table S is filled in for each $0 \leqslant i \leqslant n_1$ and $0 \leqslant j \leqslant n_2$, and the local alignments are obtained by tracing the dynamic programming table from each entry $S[i, j]$ with largest suffix similarity back to a zero suffix similarity entry.

---

procedure local_alignment($S_1, S_2, match, mismatch, gap$)
    $n_1 \leftarrow \text{length}(S_1)$
    $n_2 \leftarrow \text{length}(S_2)$
    for $i \leftarrow 0$ **to** n_1 **do**
        $S[i, 0] \leftarrow 0$
    for $j \leftarrow 1$ **to** n_2 **do**
        $S[0, j] \leftarrow 0$

$$\textbf{for } i \leftarrow 1 \textbf{ to } n_1 \textbf{ do}$$
$$\quad \textbf{for } j \leftarrow 1 \textbf{ to } n_2 \textbf{ do}$$
$$\quad\quad \textbf{if } S_1[i] = S_2[j] \textbf{ then}$$
$$\quad\quad\quad S[i,j] \leftarrow S[i-1,j-1] + match$$
$$\quad\quad \textbf{else}$$
$$\quad\quad\quad S[i,j] \leftarrow S[i-1,j-1] + mismatch$$
$$\quad\quad S[i,j] \leftarrow \max(S[i,j], S[i-1,j] + gap, S[i,j-1] + gap, 0)$$
$$\textbf{for } (i,j) \textbf{ in } arg\ max(S) \textbf{ do}$$
$$\quad T_1 \leftarrow T_2 \leftarrow \text{``''}$$
$$\quad \textbf{while } S[i,j] \neq 0 \textbf{ do}$$
$$\quad\quad \textbf{if } S_1[i] = S_2[j] \textbf{ and } S[i,j] = S[i-1,j-1] + match \textbf{ then}$$
$$\quad\quad\quad T_1 \leftarrow S_1[i-1] \cdot T_1$$
$$\quad\quad\quad T_2 \leftarrow S_2[j-1] \cdot T_2$$
$$\quad\quad\quad i \leftarrow i - 1$$
$$\quad\quad\quad j \leftarrow j - 1$$
$$\quad\quad \textbf{else if } S_1[i] \neq S_2[j] \textbf{ and } S[i,j] = S[i-1,j-1] + mismatch \textbf{ then}$$
$$\quad\quad\quad T_1 \leftarrow S_1[i-1] \cdot T_1$$
$$\quad\quad\quad T_2 \leftarrow S_2[j-1] \cdot T_2$$
$$\quad\quad\quad i \leftarrow i - 1$$
$$\quad\quad\quad j \leftarrow j - 1$$
$$\quad\quad \textbf{else if } S[i,j] = S[i-1,j] + gap \textbf{ then}$$
$$\quad\quad\quad T_1 \leftarrow S_1[i-1] \cdot T_1$$
$$\quad\quad\quad T_2 \leftarrow \text{``-''} \cdot T_2$$
$$\quad\quad\quad i \leftarrow i - 1$$
$$\quad\quad \textbf{else if } S[i,j] = S[i,j-1] + gap \textbf{ then}$$
$$\quad\quad\quad T_1 \leftarrow \text{``-''} \cdot T_1$$
$$\quad\quad\quad T_2 \leftarrow S_2[j-1] \cdot T_2$$
$$\quad\quad\quad j \leftarrow j - 1$$
$$\textbf{output } (T_1, T_2)$$

The previous algorithm for obtaining a local alignment between two sequences can be implemented in Perl in a straightforward way, as shown in the following Perl script.

```perl
sub local_alignment {
  my ($s1, $s2, $match, $mismatch, $gap) = @_;
  my ($n1, $n2) = ( length $s1, length $s2);

  my @s;
  for (my $i = 0; $i <= $n1; $i++) { $s[$i][0] = 0; }
  for (my $j = 0; $j <= $n2; $j++) { $s[0][$j] = 0; }

  my @t1 = split //, $s1;
  my @t2 = split //, $s2;
```

```perl
for ( my $i = 1; $i <= $n1; $i++ ) {
  for ( my $j = 1; $j <= $n2; $j++ ) {
    if ($t1[$i-1] eq $t2[$j-1]) {
      $s[$i][$j] = $s[$i-1][$j-1] + $match;
    } else {
      $s[$i][$j] = $s[$i-1][$j-1] + $mismatch;
    }
    $s[$i][$j] = $s[$i-1][$j] + $gap
      if $s[$i-1][$j] + $gap > $s[$i][$j];
    $s[$i][$j] = $s[$i][$j-1] + $gap
      if $s[$i][$j-1] + $gap > $s[$i][$j];
    $s[$i][$j] = 0 if 0 > $s[$i][$j];
  }
}

my $max = 0;
my @opt;
for ( my $i = 1; $i <= $n1; $i++ ) {
  for ( my $j = 1; $j <= $n2; $j++ ) {
    if ($s[$i][$j] > $max) {
      $max = $s[$i][$j];
      @opt = ();
    }
    if ($s[$i][$j] == $max) {
      $max = $s[$i][$j];
      push @opt, $i, $j;
    }
  }
}

while (@opt) {
  my $i = shift @opt;
  my $j = shift @opt;
  my ($t1, $t2) = ("", "");
  while ($s[$i][$j] != 0) {
    if (($t1[$i-1] eq $t2[$j-1] && $s[$i][$j] eq
        $s[$i-1][$j-1] + $match) || ($t1[$i-1] ne
          $t2[$j-1] &&
        $s[$i][$j] eq $s[$i-1][$j-1] + $mismatch))
          {
      $t1 = $t1[$i-1] . $t1;
      $t2 = $t2[$j-1] . $t2;
      $i--;
      $j--;
```

```
        } elsif ($s[$i][$j] eq $s[$i-1][$j] + $gap) {
          $t1 = $t1[$i-1] . $t1;
          $t2 = "-" . $t2;
          $i--;
        } elsif ($s[$i][$j] eq $s[$i][$j-1] + $gap) {
          $t1 = "-" . $t1;
          $t2 = $t2[$j-1] . $t2;
          $j--;
        }
      }
    print "$t1\n$t2\n";
  }
}
```

The algorithm for obtaining a local alignment between two sequences can also be implemented in R in a straightforward way. Since R array indexes start at 1, however, the values $s(S_1[1,\ldots,i], S_2[1,\ldots,j])$ are stored in positions $i+1, j+1$ of array S. This is all illustrated by the following R script.

```
local.alignment <- function
    (s1,s2,match,mismatch,gap) {
  t1 <- strsplit(s1,split="")[[1]]
  t2 <- strsplit(s2,split="")[[1]]
  n1 <- length(t1)
  n2 <- length(t2)
  s <- array(0,dim=c(n1+1,n2+1))
  s[,1] <- 0
  s[1,] <- 0
  for (i in 2:(n1+1)) {
    for (j in 2:(n2+1)) {
      if (t1[i-1] == t2[j-1])
        s[i,j] <- s[i-1,j-1] + match
      else
        s[i,j] <- s[i-1,j-1] + mismatch
      s[i,j] <- max(
        s[i,j],
        s[i-1,j] + gap,
        s[i,j-1] + gap,
        0)
    }
  }
  opt <- which(s==max(s),arr.ind=TRUE)
  for (sol in 1:nrow(opt)) {
    i <- opt[sol,1]
    j <- opt[sol,2]
    m1 <- c()
```

```
    m2 <- c()
    while (s[i,j] != 0) {
      if ((t1[i-1] == t2[j-1] && s[i,j] ==
            s[i-1,j-1] + match) ||
          (t1[i-1] != t2[j-1] && s[i,j] ==
            s[i-1,j-1] + mismatch)) {
        m1 <- c(t1[i-1],m1)
        m2 <- c(t2[j-1],m2)
        i <- i - 1
        j <- j - 1
      } else if (s[i,j] == s[i-1,j] + gap) {
        m1 <- c(t1[i-1],m1)
        m2 <- c("-",m2)
        i <- i - 1
      } else if (s[i,j] == s[i,j-1] + gap) {
        m1 <- c("-",m1)
        m2 <- c(t2[j-1],m2)
        j <- j - 1
      }
    }
    print(paste(m1,collapse=""))
    print(paste(m2,collapse=""))
  }
}
```

The three local alignments of the DNA sequences GCTTCCGGCTCG
TATAATGTGTGG and TGCTTCTGACTATAATAG for a match score of
3, a mismatch score of -1, and a gap score of -3 can then be obtained as
follows.

```
> s1 <- "GCTTCCGGCTCGTATAATGTGTGG"
> s2 <- "TGCTTCTGACTATAATAG"
> local.alignment(s1,s2,3,-1,-3)
[1] "GCTTCCGGCTCGTATAAT"
[1] "GCTT-CTG-AC-TATAAT"
[1] "GCTTCCGGCTCGTATA"
[1] "GCTTCTGACTATAATA"
[1] "GCTTCCGGCTCGTATAAT-G"
[1] "GCTT-CTG-AC-TATAATAG"
```

These sequences share GCTTCTGACTATAATAG, of length 17, as the
longest common gapped subsequence, which can be obtained with a match
score of 1, a mismatch score of 0, and a gap score of 0, as follows.

```
> local.alignment(s1,s2,1,0,0)
[1] "GCTTCCGGCTCGTATAAT-G"
[1] "GCTT-CTG-AC-TATAATAG"
```

```
[1] "GCTTCCGGCTCGTATAAT-GT"
[1] "GCTT-CTG-AC-TATAATAG-"
[1] "GCTTCCGGCTCGTATAATGTG"
[1] "GCTT-CTG-AC-TATAAT-AG"
[1] "GCTTCCGGCTCGTATAATGTGT"
[1] "GCTT-CTG-AC-TATAAT-AG-"
[1] "GCTTCCGGCTCGTATAATGTGTG"
[1] "GCTT-CTG-AC-TATAA--T-AG"
[1] "GCTTCCGGCTCGTATAATGTGTGG"
[1] "GCTT-CTG-AC-TATAA----TAG"
```

Further, the two sequences share TATAAT, of length 6, as the longest common subsequence, which can be obtained with a match score of 1, a mismatch score of $-\infty$, and a gap score of $-\infty$, as follows.

```
> local.alignment(s1,s2,1,-Inf,-Inf)
[1] "TATAAT"
[1] "TATAAT"
```

The gaps introduced by insertions and deletions in the local alignment of two sequences may be scattered through the sequences, but they may also stick together, forming long runs of consecutive gaps. The distribution of gaps in the local alignment of two sequences can be influenced by distinguishing between *gap opening* and *gap extension* scores, where the combined weight of k consecutive gaps is equal to the gap opening score plus k times the gap extension score.

With the Perl package Bio::Tools::dpAlign (pairwise dynamic programming sequence alignment) of BioPerl, which provides a fast implementation of DNA sequence alignment, the local alignment of two DNA sequences can be obtained as follows.

```
use Bio::Tools::dpAlign;
use Bio::AlignIO;

my $s1 = "GCTTCCGGCTCGTATAATGTGTGG";
my $s2 = "TGCTTCTGACTATAATAG";
my $seq1 = Bio::Seq->new(-seq => $s1);
my $seq2 = Bio::Seq->new(-seq => $s2);

my $factory = new Bio::Tools::dpAlign;
my $out = $factory->pairwise_alignment($seq1, $seq2);
my $alnout = new Bio::AlignIO(-format => "pfam");
$alnout->write_aln($out);
```

The local alignment of the DNA sequences GCTTCCGGCTCGTATAAT GTGTGG and TGCTTCTGACTATAATAG, with the default match score of 3, mismatch score of -1, gap opening score of -3, and gap extension score of -1, is then as follows.

```
seq1/1-18    GCTTCCGGCTCGTATAAT
seq2/2-16    GCTTCTGACT---ATAAT
```

Also, with the Perl package `Bio::Tools::pSW` (pairwise Smith Waterman) of BioPerl, which provides an interface to a fast implementation in C language of protein sequence alignment, the local alignment of two protein sequences can be obtained as follows.

```
use Bio::Tools::pSW;
use Bio::AlignIO;

my $s1 = "MVLSPADKTNVKAAWGKVGAHAGEYGAEALERMFLSFPTTKTY
FPHFDLSHGSAQVKGHGKKVADALTNAVAHVDDMPNALSALSDLHAHKLRVDP
VNFKLLSHCLLVTLAAHLPAEFTPAVHASLDKFLASVSTVLTSKYR";
my $s2 = "MVHLTPEEKSAVTALWGKVNVDEVGGEALGRLLVVYPWTQRFF
ESFGDLSTPDAVMGNPKVKAHGKKVLGAFSDGLAHLDNLKGTFATLSELHCDK
LHVDPENFRLLGNVLVCVLAHHFGKEFTPPVQAAYQKVVAGVANALAHKYH";
my $seq1 = new Bio::Seq(-seq => $s1);
my $seq2 = new Bio::Seq(-seq => $s2);

my $factory = new Bio::Tools::pSW;
my $out = $factory->pairwise_alignment($seq1, $seq2);
my $alnout = new Bio::AlignIO(-format => "pfam");
$alnout->write_aln($out);
```

The local alignment of the human hemoglobin α and β protein sequences with the default amino acid substitution matrix BLOSUM 62, gap opening score of -12, and gap extension score of -2 is then as follows.

```
seq1/3-142    LSPADKTNVKAAWGKVGAHAGEYGAEALERMFLSFPTTKT
seq2/4-147    LTPEEKSAVTALWGKV--NVDEVGGEALGRLLVVYPWTQR

YFPHF-DLSH-----GSAQVKGHGKKVADALTNAVAHVDDMPNALSALSDLHA
FFESFGDLSTPDAVMGNPKVKAHGKKVLGAFSDGLAHLDNLKGTFATLSELHC

HKLRVDPVNFKLLSHCLLVTLAAHLPAEFTPAVHASLDKFLASVSTVLTSKYR
DKLHVDPENFRLLGNVLVCVLAHHFGKEFTPPVQAAYQKVVAGVANALAHKYH
```

Bibliographic Notes

Suffix trees are a compact representation of the suffixes of a sequence. They were introduced by Weiner (1973). Improved algorithms for constructing suffix trees were proposed by McCreight (1976), Ukkonen (1992), and Farach

(1997). See also (Giegerich and Kurtz 1997). Excellent textbook presentations of suffix trees include (Gusfield 1997; Smyth 2003).

Suffix arrays were introduced by Manber and Myers (1993) and independently by Gonnet et al. (1992) under the name of *PAT arrays*, in an effort to reduce the space used by suffix trees while keeping their functionality. Fast algorithms for the computation of suffix arrays include (Kärkkäinen et al. 2006; Kim et al. 2005; Ko and Aluru 2005; Schürmann and Stoye 2007). The bottleneck of suffix array computation lies in sorting the suffixes of a sequence, for which fast algorithms were proposed by Ahlswede et al. (2006) and Maniscalco and Puglisi (2008).

The full functionality of suffix trees is maintained when the suffix array is enhanced with a table of longest common prefixes between consecutive suffixes in lexicographical order, known as an *LCP array*, and another additional table (Abouelhoda et al. 2002; 2004). Fast algorithms for computing the LCP array, given the suffix array of the sequence, were proposed by Kasai et al. (2001) and Manzini (2004).

Compressed suffix arrays, introduced by Grossi and Vitter (2005), are still another compact representation of the suffixes of a sequence. Fast algorithms for constructing compressed suffix arrays were proposed by Hon et al. (2007), Mäkinen (2003), and Rao (2002). See also (Firth et al. 2005). The notion of the *generalized suffix array* was introduced by Shi (1996) as the suffix array of the concatenation of two or more sequences.

The *Hamming distance* was introduced in the context of coding theory by Hamming (1950). The *Levenshtein distance* was introduced by Levenshtein (1966), also in the context of coding theory. The edit distance was introduced by Wagner and Fischer (1974), and the space used by the dynamic programming computation of the edit distance was reduced by Hirschberg (1975; 1977). See also (Navarro 2001).

Global sequence alignment was introduced by Needleman and Wunsch (1970). Local sequence alignment was introduced by Smith and Waterman (1981). A faster algorithm for local alignment was proposed by Gotoh (1982). The space used by the dynamic programming computation of the sequence alignment was reduced by Myers and Miller (1988). FASTA (Pearson and Lipman 1988) and BLAST (Altschul et al. 1990) count among the most widely used programs for DNA and protein sequence comparison against sequences stored in databases.

Multiple sequence alignment is reviewed in (Gotoh 1999; Notredame 2007). See also (Apostolico and Guerra 1987; Apostolico and Giancarlo 1998).

Nucleotide substitution matrices depend on a sequence evolution model. The first models of DNA sequence evolution were proposed by Jukes and Cantor (1969), Kimura (1980), and Felsenstein (1981).

The most widely used amino acid substitution matrices are the PAM (Point Accepted Mutation) series of matrices (Dayhoff et al. 1978) for global sequence alignment and the BLOSUM (Block Substitution Matrix) series (Henikoff and

Henikoff 1992) for local sequence alignment. See also (Altschul 1991; Eddy 2004a).

The influence of gap opening and gap extension scores in the local alignment of two sequences was reviewed by Vingron and Waterman (1994). The edit distance with parametric substitution weight was studied by Bunke and Csirik (1995). Sequence alignment with parametric match, mismatch, gap opening, and gap extension scores was studied by Gusfield et al. (1994).

Part II

Tree Pattern Matching

Chapter 5

Trees

Trees and also graphs count among the most useful mathematical abstractions and, at the same time, the most common combinatorial structures in computer science and computational biology. Basic notions underlying combinatorial algorithms on trees, such as counting, generation, and traversal algorithms, as well as appropriate data structures for the representation of trees, are the subject of this introductory chapter.

5.1 Trees in Mathematics

The notion of tree most often found in discrete mathematics is that of an unrooted tree, that is, a connected graph without cycles. Any two nodes are thus connected by exactly one path in a tree.

Some applications of trees in mathematics involve *labeled* trees, where nodes and branches may have additional attributes such as, in the case of computational biology, taxa names and evolutionary distances or bootstrap values.

Example 5.1
The following four trees are identical as unlabeled trees, but they are all different labeled trees.

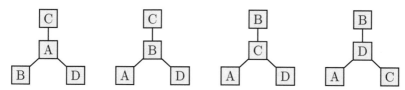

5.1.1 Counting Labeled Trees

Determining the number of possible trees is an important problem in mathematics and computer science, and it becomes even more important in computational biology, where it is essential to the uniform generation of random trees and to the validation of trees produced by various phylogenetic reconstruction methods. Here, *counting* refers to determining the number of possible trees that have certain properties, while *generation* is the process of obtaining the actual trees with these properties such as, for instance, all labeled trees.

The number of possible labeled trees increases very rapidly with the number of terminal nodes, and for 10 terminal nodes there are already 100 million labeled trees.

There is $2^{2-2} = 2^0 = 1$ way to connect two labeled nodes A, B to make a tree, as illustrated by the following single labeled tree:

Three labeled nodes A, B, C can be connected in $3^{3-2} = 3^1 = 3$ ways to make a tree, as illustrated by the following three labeled trees:

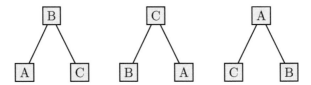

Similarly, four labeled nodes A, B, C, D can be connected in $4^{4-2} = 4^2 = 16$ ways to make a tree, as shown below.

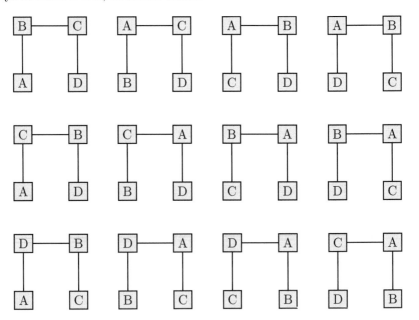

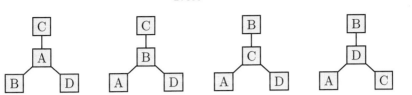

In general, there are n^{n-2} different ways to connect $n \geqslant 2$ labeled nodes to make a tree. The following R script computes the number of trees with $2 \leqslant n \leqslant 11$ labeled nodes.

```
> t(sapply(2:11,function(n)c(n,n^(n-2))))
        [,1]         [,2]
 [1,]    2            1
 [2,]    3            3
 [3,]    4           16
 [4,]    5          125
 [5,]    6         1296
 [6,]    7        16807
 [7,]    8       262144
 [8,]    9      4782969
 [9,]   10    100000000
[10,]   11   2357947691
```

5.2   Trees in Computer Science

While the notion of tree most often found in discrete mathematics is that of an undirected tree, the notion of tree which is most useful in computer science is that of a rooted tree, that is, a directed graph in which there is a distinguished node, called the root of the tree, such that there is a unique path from the root to any node of the tree. The branches of the tree are directed away from the root.

Example 5.2
In the following tree, there is a unique path from the root, A, to every node of the tree: A–B, A–C, A–B–D, A–B–E, A–C–F, A–C–G, and A–C–H.

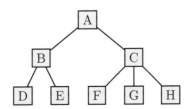

5.2.1 Traversing Rooted Trees

Most algorithms on trees require a systematic method of visiting the nodes of a tree, and combinatorial pattern matching algorithms are no exception. The most common methods for exploring a tree are the *preorder* and the *postorder* traversal.

In a preorder traversal of a rooted tree, parents are visited before children, and siblings are visited in left-to-right order. The root of the tree is visited first, followed by a preorder traversal of the subtree rooted in turn at each of the children of the root: the subtree rooted at the first child is traversed first, followed by the subtree rooted at the next sibling, and so on.

In a postorder traversal of a rooted tree, on the other hand, children are visited before parents, and siblings are also visited in left-to-right order. A postorder traversal of the subtree rooted in turn at each of the children is performed first, and the root of the tree is visited last. Again, the subtree rooted at the first child is traversed first, followed by the subtree rooted at the next sibling, and so on.

Example 5.3

In a preorder traversal of the following tree, the nodes are visited in the order A, B, D, E, C, F, G, H. In a postorder traversal, the order in which the nodes are visited is D, E, B, F, G, H, C, A.

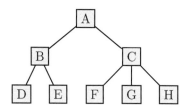

5.3 Trees in Computational Biology

The notion of tree most often found in computational biology is that of a phylogenetic tree, that is, an either unrooted or rooted tree whose terminal nodes are labeled by taxa names. In a classification tree, internal nodes may also be labeled by nested taxa. These trees are often the product of clustering methods that attempt to reconstruct phylogenetic relationships from either distance data (estimates of the divergence between taxonomic units, in terms of number of mutations or some other quantitative measure of evolutionary change), such as the unweighted pair-group method with arithmetic mean (UPGMA) or the neighbor-joining (NJ) method, or DNA sequence data (making some assumption about the underlying model of DNA substitution), such as maximum parsimony, maximum likelihood, and Bayesian methods. However, trees also arise as a mathematical model of RNA secondary structures without pseudo-knots.

The primary structure of ribonucleic acid (RNA) can be represented as a sequence over the alphabet of nucleotides: the purines A (adenine) and G (guanine), and the pyrimidines C (cytosine) and U (uracil). Within the cell, RNA molecules do not retain such a linear form but, instead, fold back on themselves in space to form hydrogen bonds between short stretches of complementary nucleotide sequences: the most frequent ones are A–U and G–C, followed by the G–U bond. The resulting secondary and tertiary structures are essential for RNA molecules to perform their biological roles.

The secondary structure of an RNA sequence can be seen as a set of nucleotide pairs $i - j$ with $i < j$ in which each nucleotide takes part in at most one pair, that is, such that $i = i'$ if and only if $j = j'$ for all nucleotide pairs $i - j$ and $i' - j'$. This representation is also known as an arc-annotated sequence, and it is said to have no pseudo-knots when $i < i'$ if and only if $j > j'$ for all nucleotide pairs $i - j$ and $i' - j'$. In an RNA structure without pseudo-knots, two nucleotide pairs $i - j$ and $i' - j'$ are said to be stacking if they are adjacent in the RNA sequence, that is, if $i' = i + 1$ and $j' = j - 1$.

Example 5.4
The secondary structure of the following RNA sequence, in 5′ to 3′ order,

AUAUUACCGUUUCGAAAGCAUCCUGUUGAUGGCUUGGCGGCCAA

corresponds to the arc-annotated sequence shown below.

There are no arc crossings in this arc-annotated sequence, because the secondary structure of this RNA sequence has no pseudo-knots. Among others, the nucleotide pairs C–G at sequence positions 7–40 and 8–39 are stacking.

Several types of RNA secondary structure elements can be distinguished:

Hairpin loops are unpaired stretches of nucleotides located within the two strands of the RNA molecule that end in an unpaired loop. They are defined by stacking nucleotide pairs $i - j$, ..., $i' - j'$ such that the inner nucleotides, that is, those between i' and j', are unpaired.

Internal loops are unpaired stretches of nucleotides located within the two strands of the RNA molecule. They are defined by two non-stacking nucleotide pairs $i - j$ and $i' - j'$ such that all the nucleotides between i and i' and also between j and j' are unpaired.

Bulge loops are unpaired stretches of nucleotides located within one strand of the RNA molecule. Bulges are thus a special class of internal loops which have no nucleotides either between i and i' or between j and j'. In a bulge, either $i' = i + 1$ or $j' = j - 1$.

Multiple bifurcation loops are also unpaired stretches of nucleotides located within the two strands of the RNA molecule, but they are defined by three or more non-stacking nucleotide pairs $i - j$, $i' - j'$, and $i'' - j''$ such that all the nucleotides between i and i' and also between j and j' and between i'' and j'' are unpaired.

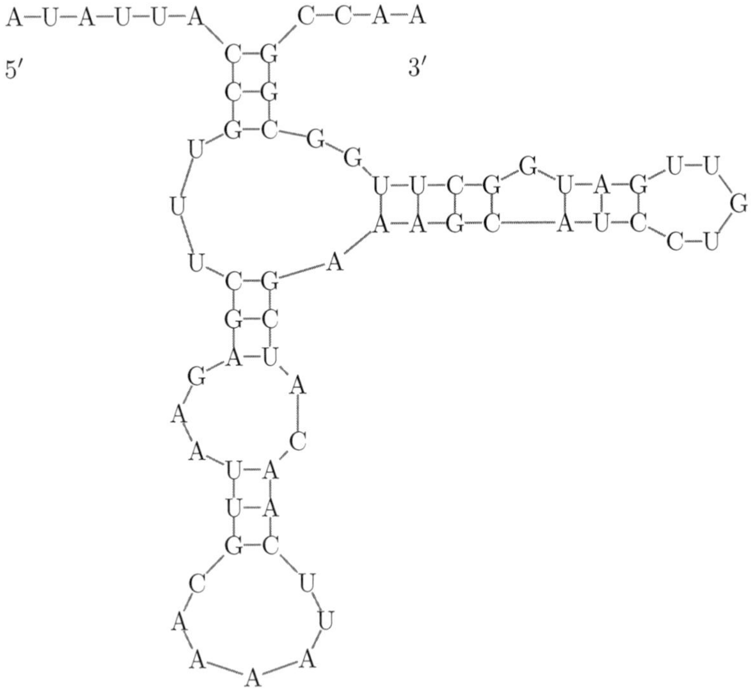

FIGURE 5.1: Clover-leaf representation of an RNA secondary structure.

These secondary structure elements are joined by helical stems of paired nucleotides within the RNA molecule. Several methods are known for computing an RNA secondary structure from sequence data, and the predicted structures depend on either thermodynamic data or phylogenetic comparison, that is, the determination of those structural features that are conserved during evolution.

Example 5.5
The following RNA sequence, in 5' to 3' order,

AUAUUACCGUUUCGAGAAUUGCAAAAUUCAACAUCGAAAGCAUCCUGUUGAUG
GCUUGGCGGCCAA

may fold into the predicted RNA secondary structure shown in Figure 5.1. There is a single-stranded region of 6 and 4 nucleotides followed by a helical stem of 3 paired nucleotides, a bifurcating loop of 6 nucleotides, and, along the vertical branch, a helical stem of 3 paired nucleotides, an internal loop of 5 nucleotides, another helical stem of 3 paired nucleotides, and a hairpin loop of 7 nucleotides, while along the horizontal branch there is a helical stem of 4 paired nucleotides, a small bulge loop of only 1 nucleotide, another helical stem of 3 paired nucleotides, and a hairpin loop of 5 nucleotides.

In an abstract model, RNA secondary structures without pseudo-knots can be represented by means of trees in which the ordering among sibling nodes corresponds to the 5' to 3' order of the RNA sequence. There is a node in such a tree for each secondary structure element of the RNA molecule, labeled by H (hairpin loop), B (bulge), I (internal loop), M (multiple bifurcation loop), or S (single-stranded region). Branches in the tree correspond to the helical stems that join secondary structure elements in the RNA molecule.

Example 5.6
The following ordered labeled tree is an abstract representation of the RNA secondary structure from Example 5.5.

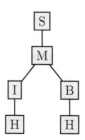

In a more detailed model, RNA secondary structures without pseudo-knots are also represented by means of trees in which the ordering among sibling nodes corresponds to the 5' to 3' order of the RNA sequence, but where there is a node for each unpaired nucleotide (the leaves) or paired nucleotide (the internal nodes of the tree). Branches in the tree correspond to consecutive helical stems, and they also join the nucleotides of a secondary structure element with the previous secondary structure element along the RNA sequence. The whole tree is rooted at a special, additional node.

Example 5.7
The following ordered labeled tree is a mode detailed representation of the RNA secondary structure from Examples 5.5 and 5.6.

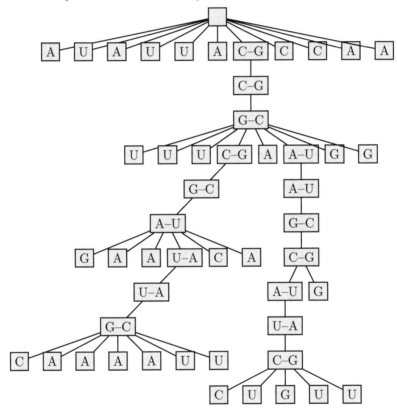

Enough has been said about the tree representation of RNA secondary structures. The evolutionary relationships among a group of organisms are often illustrated by means of a phylogenetic tree, also called a *cladogram* or a *dendrogram*. The nodes of a phylogenetic tree represent taxonomic units, which can be species or taxa, higher or nested taxa, populations, individuals, or genes. The branches of a phylogenetic tree define the evolutionary relationships among the taxonomic units, in such a way that children nodes descend from their parent. Such a pattern of ancestry and descent relationships is called the *topology* of the phylogenetic tree.

The root of a phylogenetic tree represents the most recent common ancestor of the taxonomic units at the leaves of the tree, and it is called *fully resolved* if every internal node in the tree has exactly two children. In the case of ancient ancestry, though, such information is not always available. Moreover, most phylogenetic tree reconstruction methods yield unrooted trees. In such a case, the phylogenetic tree can be rooted by choosing an *outgroup* and then placing the root between the outgroup and the node connecting it to the *ingroup*. An outgroup is a taxonomic unit for which there is additional knowledge (such as taxonomic or paleontological information) about its divergence from the common ancestor prior to all the other taxonomic units, which then become the ingroup of the phylogenetic tree.

Example 5.8

The phylogenetic relationships among pandas, bears, and raccoons, assessed using mitochondrial DNA sequence evolution, indicate an early divergence (about 30 million years ago) of the red panda from the raccoon, whereas these species diverged from the outgroup of other carnivore families about 35 million years ago.

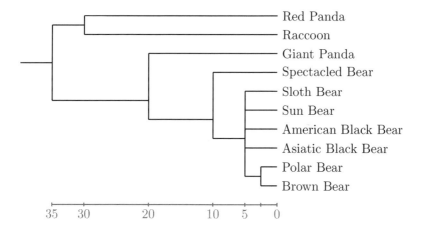

5.3.1   The Newick Linear Representation

The Newick format is the *de facto* standard for representing phylogenetic trees, and it is quite convenient since it makes it possible to describe a whole tree in linear form in a unique way once the tree is drawn or the ordering among children nodes is fixed. The Newick description of a tree is a string of nested parentheses annotated with taxa names and possibly also with branch lengths or bootstrap values (which measure how consistently the phylogenetic tree topology is supported by the underlying data), as illustrated by the following simple example:

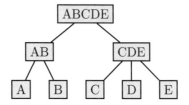

`((A,B)AB,(C,D,E)CDE)ABCDE;`

The Newick description of a given tree can be obtained by traversing the tree in postorder and writing down the name or label of the node when visiting a terminal (taxon) node, a left parenthesis (preceded by a comma unless the node is the first child of its parent) when visiting a non-terminal node for the first time, and a right parenthesis followed by the name or label of the node (if any) when visiting a non-terminal node for the second time, that is, after having visited all its descendants. The name of a node is preceded by a comma unless it is the first child of its parent, and it is followed by a colon and the length (if any) of the branch from its parent. The description of the tree is terminated with a semicolon. The following table summarizes the process of obtaining the Newick description of the previous tree:

first visit non-terminal node ABCDE	`(`
first visit non-terminal node AB	`((`
visit terminal node A	`((A`
visit terminal node B	`((A,B`
second visit non-terminal node AB	`((A,B)AB`
first visit non-terminal node CDE	`((A,B)AB,(`
visit terminal node C	`((A,B)AB,(C`
visit terminal node D	`((A,B)AB,(C,D`
visit terminal node E	`((A,B)AB,(C,D,E`
second visit non-terminal node CDE	`((A,B)AB,(C,D,E)CDE`
second visit non-terminal node ABCDE	`((A,B)AB,(C,D,E)CDE)ABCDE;`

In the Newick representation of an unrooted phylogenetic tree, there are at least three siblings connected to some internal node, as illustrated by the following alternative representations of the same simple example:

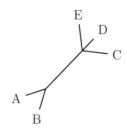

```
((A,B),C,D,E);
(A,B,(C,D,E));
```

5.3.2 Counting Phylogenetic Trees

The number of possible phylogenetic trees also increases very rapidly with the number of terminal nodes, and for 10 terminal nodes there are already more than 2 million fully resolved unrooted phylogenetic trees and more than 34 million fully resolved rooted phylogenetic trees.

There is $(2 \cdot 3 - 5)!/(2^{3-3}(3-3)!) = (6-5)!/(2^0 0!) = 1$ way to connect three labeled nodes A, B, C to make an unrooted phylogenetic tree, as illustrated by the following single fully resolved phylogenetic tree:

Four labeled nodes A, B, C, D can be connected in $(2 \cdot 4 - 5)!/(2^{4-3}(4-3)!) = 3!/(2^1 1!) = 3$ different ways to make a fully resolved unrooted phylogenetic tree, as illustrated by the following three fully resolved phylogenetic trees:

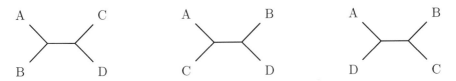

In general, a fully resolved unrooted phylogenetic tree with $n \geqslant 3$ terminal nodes has $n - 2$ internal nodes and $2n - 3$ internal branches, each of which can be split to accomodate a branch to a new terminal node. Thus, the number $U(n)$ of fully resolved unrooted phylogenetic trees with $n \geqslant 3$ terminal nodes is $U(n) = U(n-1)(2(n-1) - 3) = U(n-1)(2n-5) = \Pi_{i=3}^{n}(2i - 5) = (2n-5)!/(2^{n-3}(n-3)!)$. The following R script computes the number $U(n)$ of fully resolved unrooted phylogenetic trees with $2 \leqslant n \leqslant 12$ terminal nodes.

```
> t(sapply(3:12,function(n)
  c(n,factorial(2*n-5)/(2^{n-3}*factorial(n-3))))))
       [,1]      [,2]
[1,]     3         1
[2,]     4         3
[3,]     5        15
[4,]     6       105
[5,]     7       945
[6,]     8     10395
```

[7,]	9	135135
[8,]	10	2027025
[9,]	11	34459425
[10,]	12	654729075

On the other hand, each of the internal branches of a fully resolved unrooted phylogenetic tree can be split to accommodate a branch to the root and, thus, the number $T(n - 1)$ of rooted phylogenetic trees with $n - 1 \geqslant 3$ terminal nodes is the same as the number $U(n)$ of unrooted phylogenetic trees with $n \geqslant 3$ terminal nodes. Thus, the number $T(n)$ of rooted phylogenetic trees with $n \geqslant 2$ terminal nodes is $T(n) = (2n - 3)!/(2^{n-2}(n - 2)!)$. The following R script computes the number of fully resolved rooted phylogenetic trees with $2 \leqslant n \leqslant 11$ terminal nodes.

```
> t(sapply(2:11,function(n)
    c(n,factorial(2*n-3)/(2^{n-2}*factorial(n-2))))))
           [,1]        [,2]
   [1,]      2           1
   [2,]      3           3
   [3,]      4          15
   [4,]      5         105
   [5,]      6         945
   [6,]      7       10395
   [7,]      8      135135
   [8,]      9     2027025
   [9,]     10    34459425
  [10,]     11   654729075
```

5.3.3 Generating Phylogenetic Trees

All the fully resolved rooted phylogenetic trees on $n \geqslant 2$ terminal nodes can be generated by taking each of the fully resolved rooted phylogenetic trees on $n - 1$ terminal nodes in turn and then either splitting one of the branches to the n-th terminal node or joining it together with the n-th terminal node at a new root.

Example 5.9
The three fully resolved rooted phylogenetic trees on three terminal nodes can be generated by taking the only fully resolved rooted phylogenetic tree topology on two terminal nodes,

and either splitting one of the two branches to the new terminal node C or joining (A, B) with C at a new root:

In the same way, the 15 fully resolved rooted phylogenetic trees on 4 terminal nodes can be generated by taking each of these three fully resolved rooted phylogenetic tree topologies on three terminal nodes and either splitting one of the four branches to the new terminal node D or joining the tree with D at a new root:

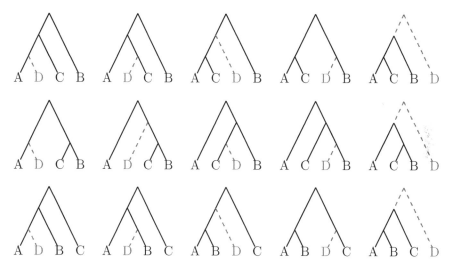

Such an algorithm for generating fully resolved rooted phylogenetic trees is implemented in the BioPerl modules for phylogenetic networks. The following Perl script uses them to generate the 15 fully resolved rooted phylogenetic trees on 4 terminal nodes.

```
use Bio::PhyloNetwork;
use Bio::PhyloNetwork::TreeFactory;

my $factory=Bio::PhyloNetwork::TreeFactory->new(
  -leaves=>[qw(A B C D)]
);

while (my $net=$factory->next_network()) {
  print $net->eNewick()."\n";
}
```

Running the previous Perl script produces the following output, which consists of the 15 fully resolved rooted phylogenetic trees on four terminal nodes labeled A, B, C, D, in Newick format:

```
(((D,A),C),B);
((C,A),(D,B));
(((D,C),A),B);
(((A,C),B),D);
((D,(A,C)),B);
((D,A),(C,B));
(A,((D,B),C));
(A,((D,C),B));
((A,(C,B)),D);
(A,(D,(B,C)));
(C,(B,(D,A)));
(C,(A,(D,B)));
((D,C),(A,B));
((D,(A,B)),C);
((C,(B,A)),D);
```

5.3.4 Representing Trees in Perl

There are many ways in which trees can be represented in Perl and, as a matter of fact, many different Perl modules implementing various types of trees are available for download from CPAN, the Comprehensive Perl Archive Network, at http://www.cpan.org/. Among them, let us focus on the Bio-Perl tree representation, which is essentially an object-oriented representation of phylogenetic trees.

A phylogenetic tree is represented in BioPerl as a Bio::Tree::Tree object, whose nodes are in turn represented as Bio::Tree::Node objects. The branches of the tree are represented as pointers to the parent and to each of the children, which are thus associated with the nodes of the tree. These pointers can be accessed by means of such methods as ancestor, each_Descendent, and get_all_Descendents.

For instance, the phylogenetic tree with Newick string ((A,B)C); can be obtained by first creating one Bio::Tree::Node object for each of the labeled terminal nodes, then creating an internal node for the (A, B) clade and adding branches to its children, creating another node for the root, again adding branches to its children, and, finally, creating a Bio::Tree::Tree object with the latter as root. This is all shown in the following Perl script.

```perl
use Bio::Tree::Tree;
use Bio::Tree::Node;

my $nodeA = Bio::Tree::Node->new(-id => "A");
my $nodeB = Bio::Tree::Node->new(-id => "B");
my $nodeC = Bio::Tree::Node->new(-id => "C");

my $nodeAB = Bio::Tree::Node->new();
```

```
$nodeAB->add_Descendent($nodeA);
$nodeAB->add_Descendent($nodeB);

my $nodeABC = Bio::Tree::Node->new();
$nodeABC->add_Descendent($nodeAB);
$nodeABC->add_Descendent($nodeC);

my $tree = Bio::Tree::Tree->new(-root => $nodeABC);
```

The same phylogenetic tree can be obtained by adding the branches to the children when creating each of the internal nodes, as illustrated by the following Perl script.

```
use Bio::Tree::Tree;
use Bio::Tree::Node;

my $nodeA = Bio::Tree::Node->new(-id => "A");
my $nodeB = Bio::Tree::Node->new(-id => "B");
my $nodeC = Bio::Tree::Node->new(-id => "C");

my $nodeAB = Bio::Tree::Node->new(
  -descendents => [$nodeA,$nodeB]
);

my $nodeABC = Bio::Tree::Node->new(
  -descendents => [$nodeAB,$nodeC]
);

my $tree = Bio::Tree::Tree->new(-root => $nodeABC);
```

However, a Bio::Tree::Tree object can be also obtained from a Newick string with the help of the Bio::TreeIO module in BioPerl, as shown in the following Perl script, where the Newick string is stored in a file.

```
use Bio::TreeIO;

my $input = new Bio::TreeIO(
  -file => "tree.tre",
  -format => "newick"
);

my $tree = $input->next_tree;
```

Furthermore, a Bio::Tree::Tree object can be reconstructed from distance data, using, for instance, the neighbor-joining method,

```
use Bio::Matrix::IO;
use Bio::Tree::DistanceFactory;
```

```
my $parser = new Bio::Matrix::IO(
    -format => 'phylip',
    -file   => "distances.mat");
my $mat = $parser->next_matrix;

my $dfactory = Bio::Tree::DistanceFactory->new(
    -method => "NJ");
my $tree = $dfactory->make_tree($mat);
```

and it can also be reconstructed from aligned DNA sequence data, using, for instance, the Kimura model of DNA substitution together with neighbor-joining,

```
use Bio::AlignIO;
use Bio::Align::DNAStatistics;
use Bio::Tree::DistanceFactory;

my $io = Bio::AlignIO->new(-file => "file.aln",
    -format => "clustalw");
my $fact = Bio::Tree::DistanceFactory->new(
    -method => "NJ");
my $stat = Bio::Align::DNAStatistics->new;

my $aln = $io->next_aln;
my $mat = $stat->distance(-method => "Kimura",
    -align => $aln);
my $tree = $fact->make_tree($mat);
```

The representation of phylogenetic trees in BioPerl includes additional methods for performing various operations on trees and their nodes; for instance, to access the root of a phylogenetic tree,

```
my $root = $tree->get_root_node;
```

to access the terminal nodes of a phylogenetic tree,

```
my @taxa = $tree->get_leaf_nodes;
```

to access the parent of a node other than the root of a phylogenetic tree,

```
my @parent = $node->ancestor;
```

to access the children of a node in a phylogenetic tree,

```
my @children = $node->each_Descendent;
```

to obtain the full lineage of a node, that is, all the ancestors of the node starting from the root of the phylogenetic tree,

```
my @nodes = $tree->get_lineage_nodes($node);
```

to access all the descendants of a node in a preorder traversal of a phylogenetic tree,

```
my @descendents = $node->get_all_Descendents;
```

and to access all the descendants of a node in a postorder traversal of a phylogenetic tree,

```
use Bio::Tree::Compatible;

my @descendents = @{ $tree->postorder_traversal }
```

Phylogenetic trees can be displayed using BioPerl in a variety of ways, such as in Newick format,

```
use Bio::TreeIO;

my $output = new Bio::TreeIO(-format => "newick");

$output->write_tree($tree);
```

drawn in Scalable Vector Graphics (SVG) format,

```
use Bio::TreeIO;

my $output = new Bio::TreeIO(
  -file => ">output.svg",
  -format => "svggraph"
);

$output->write_tree($tree);
```

and drawn as a rectangular cladogram, with horizontal orientation and ancestral nodes centered over their descendants, in Encapsulated PostScript (EPS) format,

```
use Bio::Tree::Draw::Cladogram;

my $obj = Bio::Tree::Draw::Cladogram->new(
  -tree => $tree
);

$obj->print(-file => "cladogram.eps");
```

5.3.5 Representing Trees in R

There are also many ways in which trees can be represented in R and, as a matter of fact, many different R contributed packages implementing various types of trees are available for download from CRAN, the Comprehensive R

Archive Network, at `http://cran.r-project.org/`. Among them, let us focus on the APE (Analysis of Phylogenetics and Evolution) tree representation, which is essentially a matrix-based representation of phylogenetic trees.

A phylogenetic tree in represented in the R package APE as a list of class `phylo` consisting of three elements: a numeric matrix `edge` with two columns and one row for each branch in the tree, a character vector `tip.label` with the labels of the internal nodes, and the number `Nnode` of internal nodes in the tree. In the edge matrix, all internal nodes appear in the first column at least twice, and values corresponding to terminal nodes appear only in the second column.

This representation is shared by rooted and unrooted phylogenetic trees. For the former, in a tree with n terminal nodes the root is numbered $n + 1$, while for the latter, the numbering of the root is arbitrary. In the edge matrix, all nodes (except the root) appear exactly once in the second column, corresponding to the only branch from their parent.

For instance, the phylogenetic tree with Newick string `((A,B)C);` has three terminal nodes numbered 1, 2, 3 and labeled A, B, C; two internal nodes numbered 4, 5; and branches 4–5, 5–1, 5–2, 4–3. The following R script computes the representation of such a phylogenetic tree, where the edge matrix is given by default in column order.

```
> library(ape)
Loading required package: gee
Loading required package: nlme
Loading required package: lattice
> tree <- list(
    edge = matrix(c(4,5,5,4,5,1,2,3),4,2),
    tip.label = c("A","B","C"),
    Nnode = 2)
> class(tree) <- "phylo"
> tree

Phylogenetic tree with 3 tips and 2 internal nodes.

Tip labels:
[1] "A" "B" "C"

Rooted; no branch lengths.
```

The same tree representation can be obtained by reading in a Newick string, as shown in the following R script.

```
> tree <- read.tree(text="((A,B),C);")
> tree

Phylogenetic tree with 3 tips and 2 internal nodes.
```

```
Tip labels:
[1] "A" "B" "C"

Rooted; no branch lengths.
```

A phylogenetic tree can also be reconstructed with the R package APE from distance data, using, for instance, the neighbor-joining method,

```
> mat <- matrix(scan("distances.mat"),n,n,byrow=T)
> tree <- nj(mat)
```

and it can be reconstructed from aligned DNA sequence data, using, for instance, the Kimura model of DNA substitution together with neighbor-joining,

```
> aln <- read.dna("sequences.aln")
> mat <- dist.dna(aln, model="K80")
> tree <- nj(mat)
```

The branches of a phylogenetic tree can be obtained by accessing the `edge` matrix,

```
> tree$edge
     [,1] [,2]
[1,]    4    5
[2,]    5    1
[3,]    5    2
[4,]    4    3
```

The labels of the terminal nodes can also be obtained by accessing the `tip.label` character vector,

```
> tree$tip.label
[1] "A" "B" "C"
```

In the same way, the number of internal nodes can be obtained by accessing the `Nnode` variable,

```
> tree$Nnode
[1] 2
```

Based on this matrix representation, it is rather easy to code various operations on phylogenetic trees and their nodes and branches; for instance, to obtain the number of terminal nodes,

```
> length(tree$tip.label)
[1] 3
```

to determine the number of branches,

```
> dim(tree$edge)[1]
[1] 4
```

to find (the position, by column order, in the edge matrix of) a node in the tree,

```
> which(tree$edge == 1)
[1] 6
> which(tree$edge == 2)
[1] 7
> which(tree$edge == 3)
[1] 8
> which(tree$edge == 4)
[1] 1 4
> which(tree$edge == 5)
[1] 2 3 5
```

to access the (branches from the) root of the tree,

```
> which(tree$edge[,1]==length(tree$tip.label)+1)
[1] 1 4
> tree$edge[
    which(tree$edge[,1]==length(tree$tip.label)+1),]
    [,1] [,2]
[1,]    4    5
[2,]    4    3
```

and to obtain the parent of a node in the tree,

```
> tree$edge[which(tree$edge[,2] == 1),1]
[1] 5
> tree$edge[which(tree$edge[,2] == 2),1]
[1] 5
> tree$edge[which(tree$edge[,2] == 3),1]
[1] 4
> tree$edge[which(tree$edge[,2] == 5),1]
[1] 4
```

These operations can certainly be wrapped in a function, as illustrated by the following R script.

```
> parent <- function (tree,x)
    tree$edge[which(tree$edge[,2] == x),1]
> parent(tree,1)
[1] 5
> parent(tree,2)
[1] 5
> parent(tree,3)
[1] 4
> parent(tree,5)
[1] 4
```

Phylogenetic trees can also be displayed using R in a variety of ways, such as in Newick format,

```
> write.tree(tree)
[1] "((A,B),C);"
```

with the Newick string stored in a file,

```
> write.tree(tree, file = "tree.tre")
```

and drawn as a rectangular cladogram, with horizontal orientation and ancestral nodes centered over their descendants, in Encapsulated PostScript (EPS) format,

```
> postscript(file="cladogram.eps")
> plot(tree, type = "p")
> dev.off()
```

among several other display options for unrooted and rooted phylogenetic trees as well.

Bibliographic Notes

The correspondence between trees and nested parentheses was first noticed by Cayley (1857; 1881). A modern proof of the result on counting labeled trees can be found in (Shor 1995).

The very idea of representing evolution by means of trees dates back to Charles Darwin (Burkhardt and Smith 1987). A proof of the counting formulas can be found in (Cavalli-Sforza and Edwards 1967). See also (Felsenstein 2004, ch. 3) and (Waterman 1995, p. 346).

The unweighted pair-group method with arithmetic mean (UPGMA) is discussed in detail in (Sneath and Sokal 1973), where it is proved to be correct for ultrametric distance matrices. The neighbor-joining (NJ) method was introduced by Saitou and Nei (1987) and further improved in (Studier and Keppler 1988; Mailund et al. 2006). A proof that NJ is correct for additive distance matrices can be found in (Durbin et al. 1998, Appendix 7.8).

The model of RNA secondary structures without pseudo-knots by means of trees was introduced by Shapiro (1988) and further developed in (Le et al. 1989; Shapiro and Zhang 1990). Arc-annotated sequences were introduced by Evans (1999) as a more general model of RNA secondary structures.

The object-oriented representation of phylogenetic trees in BioPerl (Stajich et al. 2002) is described in detail in (Birney et al. 2009). The matrix-based representation of phylogenetic trees in R is described in more detail in (Paradis 2006).

Chapter 6

Simple Pattern Matching in Trees

Combinatorial pattern matching is the search for exact or approximate occurrences of a given pattern within a given text. When it comes to trees in computational biology, both the pattern and the text are trees and the pattern matching problem becomes one of finding the occurrences of a tree within another tree. For instance, scanning an RNA secondary structure for the presence of a known pattern can help in finding conserved RNA motifs, and finding a phylogenetic tree within another phylogenetic tree can help in assessing their similarities and differences. This will be the subject of the next chapter.

A related pattern matching problem that arises in the analysis of trees consists in finding simpler patterns, that is, paths within a given tree. For instance, finding the path between two nodes of a tree is useful for computing distances in a tree and also for computing distances between two trees. This is the subject of this chapter.

6.1   Finding Paths in Unrooted Trees

Any two nodes are connected by exactly one path in a tree, as long as no branch is to be traversed more than once in the path between the two nodes of the tree.

Example 6.1
In the following fully resolved unrooted phylogenetic tree, the path between terminal nodes A and D involves three branches.

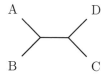

The path A–D–B–C is not valid because the internal branch is traversed more than once: it is traversed three times along this path.

The path between any two given terminal nodes of a fully resolved rooted phylogenetic tree traverses the most recent common ancestor of the nodes in the tree. In an unrooted phylogenetic tree there is, in principle, no explicit information about common ancestors, but it can be rooted by placing the root between some chosen outgroup and the node connecting it to the ingroup. An unrooted phylogenetic tree can also be rooted by placing the root at an arbitrary internal node, or by adding a new root node.

Example 6.2
Rooting the fully resolved unrooted phylogenetic tree with Newick string ((A,B),C,D); (shown to the left) at an internal node gives the rooted phylogenetic trees with Newick string ((A,B),C,D); and (A,B,(C,D)); (shown at the middle), while rooting it at a new node along the branch between the two internal nodes gives the fully resolved rooted phylogenetic tree with Newick string ((A,B),(C,D)); (shown to the right).

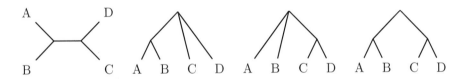

6.1.1   Distances in Unrooted Trees

The distance between any two terminal nodes in an unrooted phylogenetic tree is the length of the path between the two terminal nodes in the tree. In the case of phylogenetic trees with branch lengths, the distance between any two terminal nodes can be calculated as the sum of the length of the branches in the path between the two terminal nodes in the tree.

Example 6.3
The distance between each pair of terminal nodes in the following fully resolved unrooted phylogenetic tree is indicated in the table to the right.

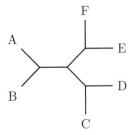

	A	B	C	D	E	F
A	0	2	4	4	4	4
B	2	0	4	4	4	4
C	4	4	0	2	4	4
D	4	4	2	0	4	4
E	4	4	4	4	0	2
F	4	4	4	4	2	0

The representation of phylogenetic trees in BioPerl also includes a method `distance` to compute the distance between any two terminal nodes of a phylogenetic tree with branch lengths, as illustrated by the following Perl script.

```
use Bio::TreeIO;

my $input = new Bio::TreeIO(
  -fh => \*DATA,
  -format => "newick"
);

my $tree = $input->next_tree;

my @nodes = $tree->find_node(-id => "A");
push @nodes, $tree->find_node(-id => "F");

my $dist = $tree->distance(-nodes => \@nodes);

__DATA__
((A:1,B:1):1,(C:1,D:1):1,(E:1,F:1):1):1;
```

The representation of phylogenetic trees in R, on the other hand, also includes a function `cophenetic.phylo` to compute the distance between each pair of terminal nodes and a function `dist.nodes` to compute the distance between each pair of nodes of a phylogenetic tree with branch lengths, as illustrated by the following R script.

```
> library(ape)
> t <- read.tree(text="((A:1,B:1):1,(C:1,D:1):1,(E:1,
  F:1):1):1;")

> cophenetic.phylo(t)
  A B C D E F
A 0 2 4 4 4 4
B 2 0 4 4 4 4
C 4 4 0 2 4 4
D 4 4 2 0 4 4
E 4 4 4 4 0 2
F 4 4 4 4 2 0

> dist.nodes(t)
   1 2 3 4 5 6 7 8 9 10 11
1  0 2 4 4 4 4 2 1 3  3  2
2  2 0 4 4 4 4 2 1 3  3  2
3  4 4 0 2 4 4 2 3 1  3  2
4  4 4 2 0 4 4 2 3 1  3  2
```

5	4	4	4	4	0	2	2	3	3	1	2
6	4	4	4	4	2	0	2	3	3	1	2
7	2	2	2	2	2	2	0	1	1	1	0
8	1	1	3	3	3	3	1	0	2	2	1
9	3	3	1	1	3	3	1	2	0	2	1
10	3	3	3	3	1	1	1	2	2	0	1
11	2	2	2	2	2	2	0	1	1	1	0

6.1.2 The Partition Distance between Unrooted Trees

The similarities and differences between two unrooted phylogenetic trees can be assessed by computing a distance measure between the two trees. The *partition distance* is based on the partition of the taxa induced by each internal branch in the two trees under comparison. While cutting a tree along each of the the external branches partitions the set of taxa in a trivial way (with each partition consisting of a single taxon on one side and the remaining taxa on the other side), cutting along each of the internal branches reveals similarities and differences between the two trees.

Example 6.4
Each of the following fully resolved unrooted phylogenetic trees has six taxa and three internal branches and, thus, can be partitioned in three different ways by cutting one of the internal branches.

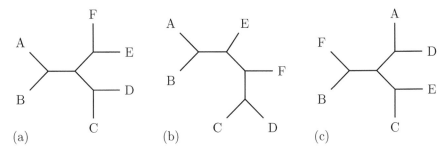

The three partitions of the six taxa in tree (a) are:

- (A, B) and (C, D, E, F)

- (A, B, C, D) and (E, F)

- (A, B, E, F) and (C, D)

while in tree (b), the three partitions are:

- (A, B) and (C, D, E, F)

- (A, B, E) and (C, D, F)

- (A, B, E, F) and (C, D)

and the three partitions of the six taxa in tree (c) are:

- (A, B, D, F) and (C, E)

- (A, C, D, E) and (B, F)

- (A, D) and (B, C, E, F)

The partition distance between two unrooted phylogenetic trees is defined as the size of the symmetric difference of the partitions of the set of taxa obtained when cutting the trees along each of their internal branches, that is, the number of internal branches in the two trees resulting in different partitions of the taxa.

Example 6.5
In the previous example, there are two identical partitions between trees (a) and (b), while neither tree (a) nor tree (b) share any partition with tree (c). There are two different partitions between trees (a) and (b), the partition (A, B, C, D) and (E, F) in tree (a) and the partition (A, B, E) and (C, D, F) in tree (b) and, thus, the partition distance between trees (a) and (b) is 2, while the partition distance between trees (a) and (c) and between trees (b) and (c) is 6, and, therefore, tree (a) is more similar to tree (b) than to tree (c).

The partition distance between two unrooted phylogenetic trees can be computed by first obtaining, for each of the two phylogenetic trees, the partition of the taxa induced by each of their internal branches. In the rooted representation of an unrooted tree, the partition of the taxa induced by an internal branch (v, w) consists of the labels of all terminal nodes which are descendants of node w and the labels of all other terminal nodes.

```
function partition(T)
    P ← ∅
    for each internal node v of T do
        A ← taxa of all descendants of v in T
        B ← taxa of all other leaves of T
        P ← P ∪ {(A, B)}
    return P
```

Once the partition of the taxa in each of the two phylogenetic trees induced by each of their internal branches is known, the partition distance can be computed by counting the number of partitions of the taxa in each of the trees that do not belong to the partitions of the taxa in the other tree.

```
function partition distance(T₁, T₂)
    P₁ ← partition(T₁)
    P₂ ← partition(T₂)
    d ← 0
    for (A, B) ∈ P₁ do
        if (A, B) ∉ P₂ then
            d ← d + 1
    for (A, B) ∈ P₂ do
        if (A, B) ∉ P₁ then
            d ← d + 1
    return d
```

The representation of phylogenetic trees in BioPerl does not include any method to compute the partition distance between two unrooted phylogenetic trees. However, the partition of the taxa in a phylogenetic tree induced by each internal branch (which, in the rooted representation of the unrooted tree, are actually induced by each internal node) can be obtained by finding the labels of all terminal nodes which are descendants of the internal node, finding the labels of all other terminal nodes, and then joining them into a string, using some appropriate delimiter symbol.

Notice that in the rooted representation of the unrooted tree, the two nodes connected by an internal branch may induce the same partition of the taxa. These repeated partitions are removed with the help of a hash, as illustrated by the following Perl script.

```perl
sub partition {
  my $tree = shift;
  my %partition = ();
  for my $node ($tree->get_nodes) {
    next if $node->is_Leaf; # discard terminal nodes
    next unless $node->ancestor; # discard root
    my @a = sort { $a cmp $b } map { $_->id }
      grep {$_->is_Leaf } $node->get_all_Descendents;
    my $aa = join ",", @a;
    my @b = ();
    my %count = ();
    foreach my $label (@a, map { $_->id } $tree->
      get_leaf_nodes) {
      $count{$label}++
    }
    foreach my $label (keys %count) {
      push @b, $label if $count{$label} == 1;
    }
    my $bb = join ",", sort { $a cmp $b } @b;
    my $p = $aa le $bb ? "$aa:$bb" : "$bb:$aa";
```

```
    $partition{$p} = 1;
  }
  my @partition = sort keys %partition;
  return \@partition;
}
```

Given the partition of the taxa in each of the two phylogenetic trees induced by each of their internal branches, the partition distance can be computed by counting the number of partitions of the taxa in each of the trees that do not belong to the partitions of the taxa in the other tree. Such a symmetric difference can be easily obtained by counting the number of occurrences of each partition and then selecting only those partitions that occur at most once, as illustrated by the following Perl script.

```
sub partition_distance {
  my $tree1 = shift;
  my $tree2 = shift;
  my @partition1 = @{ partition($tree1) };
  my @partition2 = @{ partition($tree2) };
  my $dist = 0;
  my %count = ();
  foreach my $p (@partition1, @partition2) {
    $count{$p}++
  }
  foreach my $p (keys %count) {
    $dist++ if $count{$p} == 1;
  }
  return $dist;
}
```

The representation of phylogenetic trees in R, on the other hand, also includes a function `dist.topo` to compute the partition distance between two unrooted phylogenetic trees, as illustrated by the following R script.

```
> library(ape)

> t1 <- read.tree(text="((A,B),(C,D),(E,F));")
> t2 <- read.tree(text="(((A,B),E),(C,D),F);")
> t3 <- read.tree(text="((F,B),(C,E),(D,A));")

> dist.topo(t1,t2)
[1] 2
> dist.topo(t1,t3)
[1] 6
> dist.topo(t2,t3)
[1] 6
```

6.1.3 The Nodal Distance between Unrooted Trees

The *nodal distance*, also called *path difference metric*, is based on the distances between each two terminal nodes in the two trees under comparison. Let $D(T)$ be the length $n(n-1)/2$ vector of nodal distances between each pair of terminal nodes of an unrooted phylogenetic tree T, that is,

$$D(T) = (d_T(1,2), d_T(1,3), \ldots, d_T(1,n), d_T(2,3), \ldots, d_T(n-1,n)),$$

where the n terminal nodes of T are numbered $1, \ldots, n$. The nodal distance $d_N(T_1, T_2)$ between two unrooted phylogenetic trees T_1 and T_2 is the sum of the absolute differences between their vectors of nodal distances, that is,

$$d_N(T_1, T_2) = \sum_{\substack{1 \leqslant i < n \\ i < j \leqslant n}} |d_{T_1}(i,j) - d_{T_2}(i,j)|$$

Example 6.6

The distances between each two terminal nodes in the following fully resolved unrooted phylogenetic trees are given in the tables next to each of the trees.

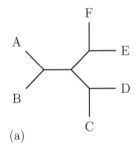

	A	B	C	D	E	F
A	0	2	4	4	4	4
B	2	0	4	4	4	4
C	4	4	0	2	4	4
D	4	4	2	0	4	4
E	4	4	4	4	0	2
F	4	4	4	4	2	0

(a)

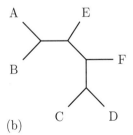

	A	B	C	D	E	F
A	0	2	5	5	3	4
B	2	0	5	5	3	4
C	5	5	0	2	4	3
D	5	5	2	0	4	3
E	3	3	4	4	0	3
F	4	4	3	3	3	0

(b)

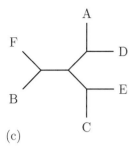

(c)

	A	B	C	D	E	F
A	0	4	4	2	4	4
B	4	0	4	4	4	2
C	4	4	0	4	2	4
D	2	4	4	0	4	4
E	4	4	2	4	0	4
F	4	2	4	4	4	0

Their vectors of nodal distances are given in the following table, together with the absolute differences between each pair of vectors. The nodal distance between phylogenetic trees (a) and (b) is 9, between (a) and (c) is 12, and between (b) and (c) is 19. Trees (a) and (b) are thus more similar to each other than they are to tree (c).

	(a)	(b)	(c)	$|(a) - (b)|$	$|(a) - (c)|$	$|(b) - (c)|$
AB	2	2	4	0	2	2
AC	4	5	4	1	0	1
AD	4	5	2	1	2	3
AE	4	3	4	1	0	1
AF	4	4	4	0	0	0
BC	4	5	4	1	0	1
BD	4	5	4	1	0	1
BE	4	3	4	1	0	1
BF	4	4	2	0	2	2
CD	2	2	4	0	2	2
CE	4	4	2	0	2	2
CF	4	3	4	1	0	1
DE	4	4	4	0	0	0
DF	4	3	4	1	0	1
EF	2	3	4	1	2	1
				9	12	19

The nodal distance between two unrooted phylogenetic trees can be obtained by computing the distance between each pair of terminal nodes in each of the trees and then computing the absolute difference between the two vectors of nodal distances.

```
function nodal distance(T₁, T₂)
    L ← terminal node labels in T₁ and T₂
    n ← length(L)
    d ← 0
    for i ← 1, . . . , n − 1 do
        i₁ ← terminal node of T₁ labeled L[i]
        i₂ ← terminal node of T₂ labeled L[i]
```

```
for j ← i + 1, . . . , n do
    j₁ ← terminal node of T₁ labeled L[j]
    j₂ ← terminal node of T₂ labeled L[j]
    d₁ ← distance(T₁, i₁, j₁)
    d₂ ← distance(T₂, i₂, j₂)
    d ← d + |d₁ − d₂|
return d
```

The representation of phylogenetic trees in BioPerl does not include any method to compute the nodal distance between phylogenetic trees. However, the `distance` method can be used to compute the vector of distances between each pair of terminal nodes in an unrooted phylogenetic tree, as illustrated by the following Perl script.

```
sub distances {
  my $tree = shift;
  my @leaves = sort {$a->id cmp $b->id} $tree->
    get_leaf_nodes;
  my @labels = map { $_->id } @leaves;
  my $n = scalar @labels;
  my @dist;
  for my $i (1..$n-1) {
    my @nodes = $tree->find_node(-id => $labels[$i
      -1]);
    for my $j ($i+1..$n) {
      push @nodes, $tree->find_node(-id => $labels[$j
        -1]);
      push @dist, $tree->distance(-nodes => \@nodes);
      pop @nodes;
    }
  }
  return \@dist;
}
```

The nodal distance between two unrooted phylogenetic trees can then be computed by performing a simultaneous traversal of their vectors of distances, as implemented by the `nodal_distance` method in the following Perl script.

```
sub nodal_distance {
  my $tree1 = shift;
  my $tree2 = shift;
  my @dist1 = @{ distances($tree1) };
  my @dist2 = @{ distances($tree2) };
  my ($dist, $d1, $d2);
  while (@dist1 and @dist2) {
    $d1 = pop @dist1;
```

```
    $d2 = pop @dist2;
    $dist += abs($d1-$d2);
  }
  return $dist;
}
```

The representation of phylogenetic trees in R does not include any method to compute the nodal distance between phylogenetic trees, either. However, a `nodal.distance` function can easily be defined using the `cophenetic.phylo` function to compute the distance between each two terminal nodes, as illustrated by the following Perl script. The matrices of distances between each pair of terminal nodes in each of the phylogenetic trees are first rearranged by sorting the rows and columns by node label, in order to ensure the two trees coincide in the numbering of terminal nodes.

```
> library(ape)
> nodal.distance <- function (t1,t2) {
    m1<-cophenetic.phylo(t1)
    m1<-m1[order(rownames(m1)),order(colnames(m1))]
    n1<-m1[upper.tri(m1)]
    m2<-cophenetic.phylo(t2)
    m2<-m2[order(rownames(m2)),order(colnames(m2))]
    n2<-m2[upper.tri(m2)]
    sum(abs(n1-n2))
}

> t1 <- compute.brlen(
  read.tree(text="((A,B),(C,D),(E,F));"),
  1)
> t2 <- compute.brlen(
  read.tree(text="(((A,B),E),(C,D),F);"),
  1)
> t3 <- compute.brlen(
  read.tree(text="((F,B),(C,E),(D,A));"),
  1)

> nodal.distance(t1,t2)
[1] 9
> nodal.distance(t1,t3)
[1] 12
> nodal.distance(t2,t3)
[1] 19
```

6.2 Finding Paths in Rooted Trees

Any two nodes are connected by exactly one path in a rooted tree, as long as no branch is to be traversed more than once in the path between the two nodes, and such a unique path traverses the most recent common ancestor of the two nodes in the tree.

The most recent common ancestor of two terminal nodes in a rooted phylogenetic tree can be found by obtaining first the lineages (paths to the root) of the two terminal nodes in the tree and then finding the first node in one of the lineages than also belongs to the other lineage. In the following description, the lineage of node i in a rooted phylogenetic tree T is stored in a list L and then the nodes in the lineage of node j are tested one after the other until a node is found that also belongs to the lineage of node i.

```
function mrca(T, i, j)
    L ← {i}
    k ← i
    while T[k] ≠ root(T) do
        k ← parent(T, k)
        L ← L ∪ {k}
    k ← j
    while k ≠ root(T) do
        if k ∈ L then
            return k
        else
            k ← parent(T, k)
```

Example 6.7

In the following fully resolved rooted phylogenetic tree, the most recent common ancestor of each pair of terminal nodes is indicated in the table to the right. The four terminal nodes labeled A through D are numbered 1 through 4, respectively, and the three internal nodes are numbered 5 through 7, for reference.

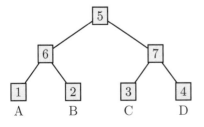

	A	B	C	D
A	1	6	5	5
B	6	2	5	5
C	5	5	3	7
D	5	5	7	4

The representation of phylogenetic trees in BioPerl has a method `get_lca` to find the most recent common ancestor of any subset of the nodes of a phylogenetic tree, as illustrated by the following Perl script.

```
use Bio::TreeIO;

my $input = new Bio::TreeIO(
  -fh => \*DATA,
  -format => "newick"
);
my $tree = $input->next_tree;

my @nodes = $tree->find_node(-id => "A");
push @nodes, $tree->find_node(-id => "C");

my $mrca = $tree->get_lca(-nodes => \@nodes);

__DATA__
((A,B),(C,D));
```

The representation of phylogenetic trees in R, on the other hand, includes a function `mrca` to find the most recent common ancestor of each pair of terminal nodes or the most recent common ancestor of each pair of nodes of a phylogenetic tree, as illustrated by the following R script.

```
> library(ape)
> t <- read.tree(text="((A,B),(C,D));")

> mrca(t)
  A B C D
A 1 6 5 5
B 6 2 5 5
C 5 5 3 7
D 5 5 7 4

> mrca(t,full=TRUE)
  1 2 3 4 5 6 7
1 1 6 5 5 5 6 5
2 6 2 5 5 5 6 5
3 5 5 3 7 5 5 7
4 5 5 7 4 5 5 7
5 5 5 5 5 5 5 5
6 6 6 5 5 5 6 5
7 5 5 7 7 5 5 7
```

6.2.1 Distances in Rooted Trees

The distance between any two terminal nodes in a rooted phylogenetic tree is the sum of the length of the paths between the two nodes and their most recent common ancestor in the tree. In the case of phylogenetic trees with branch lengths, the distance between any two terminal nodes can be calculated as the sum of the lengths of the branches along these paths.

The `distance` method provided by the representation of phylogenetic trees in BioPerl also allows one to compute the distance between any two terminal nodes of a rooted phylogenetic tree with branch lengths, as illustrated by the following Perl script.

```
use Bio::TreeIO;

my $input = new Bio::TreeIO(
  -fh => \*DATA,
  -format => "newick"
);

my $tree = $input->next_tree;

my @nodes = grep { $_->id =~ /A|C/ } $tree->get_nodes
  ;

my $dist = $tree->distance(-nodes => \@nodes);

__DATA__
((A:1,B:1):1,(C:1,D:1):1):1;
```

The `cophenetic.phylo` function provided by the representation of phylogenetic trees in R, on the other hand, also allows one to compute the distance between each pair of terminal nodes of a rooted phylogenetic tree with branch lengths, while the `dist.nodes` function also allows one to compute the distance between each pair of nodes of a rooted phylogenetic tree with branch lengths, as illustrated by the following R script.

```
> library(ape)
> t <- compute.brlen(read.tree(text="((A,B),(C,D));")
  ,1)

> cophenetic.phylo(t)
  A B C D
A 0 2 4 4
B 2 0 4 4
C 4 4 0 2
D 4 4 2 0
```

```
> dist.nodes(t)
  1 2 3 4 5 6 7
1 0 2 4 4 2 1 3
2 2 0 4 4 2 1 3
3 4 4 0 2 2 3 1
4 4 4 2 0 2 3 1
5 2 2 2 2 0 1 1
6 1 1 3 3 1 0 2
7 3 3 1 1 1 2 0
```

6.2.2 The Partition Distance between Rooted Trees

The partition distance between two rooted phylogenetic trees can also be computed by first obtaining, for each of the two phylogenetic trees, the partition of the taxa induced by each of their internal branches (where the partition of the taxa induced by an internal branch (v, w) consists of the labels of all terminal nodes which are descendants of node w and the labels of all other terminal nodes) and then counting the number of partitions of the taxa in each of the trees that do not belong to the partitions of the taxa in the other tree.

The `partition_distance` Perl method presented above, as well as the `dist.topo` R function, can also be used to compute the partition distance between two rooted phylogenetic trees.

6.2.3 The Nodal Distance between Rooted Trees

The nodal distance between two rooted phylogenetic trees can also be obtained by computing the distance between each pair of terminal nodes in each of the trees and then computing the absolute difference between the two vectors of nodal distances. Both the `nodal_distance` Perl method and the `nodal.distance` R function presented above can also be used to compute the nodal distance between two fully resolved rooted phylogenetic trees.

When the rooted phylogenetic trees are not fully resolved, however, the nodal distance fails to be a metric on the space of rooted phylogenetic trees. For instance, the following two rooted phylogenetic trees have nodal distance zero, but they are non-isomorphic. Notice that in the representation of rooted phylogenetic trees in R, the length of the branch to the root has to be set to 0 if the root is not fully resolved; otherwise, the tree is interpreted as unrooted.

Example 6.8

The rooted phylogenetic trees with Newick string `((A,B),C,D);` (left) and `(A,B,(C,D));` (right) have the same vectors of nodal distances and, thus, their nodal distance is 0.

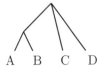

A   B   C   D   A   B   C   D

```
> t1 <- compute.brlen(read.tree(text="((A,B),C,D);")
  ,1)
> t1$root.edge <- 0
> t2 <- read.tree(text="(A:1,B:1,(C:1,D:1):1):0;")
> cophenetic.phylo(t1)
  A B C D
A 0 2 3 3
B 2 0 3 3
C 3 3 0 2
D 3 3 2 0
> cophenetic.phylo(t2)
  A B C D
A 0 2 3 3
B 2 0 3 3
C 3 3 0 2
D 3 3 2 0
> nodal.distance(t1,t2)
[1] 0
```

Bibliographic Notes

The partition distance between unrooted phylogenetic trees was introduced by Robinson and Foulds (1981) and further studied in (Penny and Hendy 1985; Rzhetsky and Nei 1992). Improved algorithms for computing the partition distance can be found in Day (1985); Pattengale et al. (2007).

The nodal distance between phylogenetic trees was first studied by Williams and Clifford (1971) and later rediscovered by Bluis and Shin (2003), and it was proved to be a metric on the space of unrooted phylogenetic trees in (Zaretskii 1965). Further properties of the nodal distance (also called *path difference metric*) were studied by Steel and Penny (1993).

Efficient algorithms for finding most recent common ancestors in rooted trees can be found in (Bender and Farach-Colton 2000; Bender et al. 2005).

The partition distance and the nodal distance are two widely used measures to assess the similarities and differences between two phylogenetic trees. Other distances between unrooted phylogenetic trees include the nearest neighbor interchange distance, the subtree transfer distance, and the quartet distance.

Further distances between rooted phylogenetic trees include the transposition distance.

A nearest neighbor interchange operation is the swap of two subtrees that are separated by an internal edge in a fully resolved unrooted phylogenetic tree, and the nearest neighbor interchange distance between two fully resolved unrooted phylogenetic trees is the smallest number of nearest neighbor interchange operations needed to transform one tree into the other. The nearest neighbor interchange distance was introduced independently in (Robinson and Foulds 1971) and (Moore et al. 1973). See also (Smith and Waterman 1980). Computing the nearest neighbor interchange distance is NP-hard (DasGupta et al. 1997; Křivánek 1986).

A more general operation is the transfer of a subtree from one place to another, and the subtree transfer distance between two fully resolved unrooted phylogenetic trees is the smallest number of subtree transfer operations needed to transform one tree into the other. The subtree transfer distance was introduced by Hein (1990) and further studied in (Hein 1993). Computing the subtree transfer distance is also NP-hard (Hein et al. 1996). See also (Allen and Steel 2001).

The quartet distance is the number of quartets (subtrees induced by four terminal nodes) that differ between two fully resolved unrooted phylogenetic trees. The quartet distance was introduced in (Estabrook et al. 1985). Unlike nearest neighbor interchange and subtree transfer, the quartet distance can be computed in polynomial time (Brodal et al. 2003; Bryant et al. 2000; Christiansen et al. 2006).

The transposition distance between two fully resolved rooted phylogenetic trees is the smallest number of transpositions needed to transform the matching representation of one tree into the matching representation of the other one. The transposition distance, which was introduced in (Valiente 2005), can be computed in polynomial time.

Chapter 7

General Pattern Matching in Trees

Combinatorial pattern matching is the search for exact or approximate occurrences of a given pattern within a given text. When it comes to trees in computational biology, both the pattern and the text are trees and the pattern matching problem becomes one of finding the occurrences of a tree within another tree. For instance, scanning an RNA secondary structure for the presence of a known pattern can help in finding conserved RNA motifs, and finding a phylogenetic tree within another phylogenetic tree can help in assessing their similarities and differences. This is the subject of this chapter.

7.1   Finding Subtrees

There are several ways in which a tree can be contained in another tree. In the most general sense, a subtree of a given (unrooted or rooted) tree is a connected subgraph of the tree, while in the case of rooted trees, a distinction can be made between *top-down* and *bottom-up* subtrees.

A bottom-up subtree of a given rooted tree is the whole subtree rooted at some node of the tree, and a connected subgraph of a rooted tree is called a top-down subtree if the parent of all nodes in the subtree (up to, and including, the most recent common ancestor of all nodes in the subtree) also belongs to the subtree. Further, the subtree of an (unrooted or rooted) tree induced by a set of terminal nodes is the unique connected subgraph that contains the set of terminal nodes but does not include any other connected subgraph of the given tree with these terminal nodes, where elementary paths (paths of two or more edges without internal branching) are contracted to single edges in a subgraph.

In the following example, the elementary path of three edges between the terminal node labeled C and the most recent common ancestor of the terminal nodes labeled C and E is contracted to a single edge in the subtree induced by the terminal nodes labeled A, C, and E.

Example 7.1

In the following fully resolved rooted phylogenetic tree, a top-down subtree is shown highlighted (left), and the bottom-up subtree rooted at the most recent common ancestor of the terminal nodes labeled B, C, D, and E is also shown (middle), together with the subtree induced by the terminal nodes labeled A, C, and E (right).

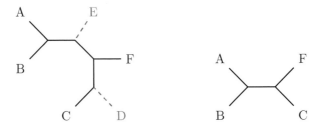

Example 7.2
In the following fully resolved unrooted phylogenetic tree (left), the subtree induced by the terminal nodes labeled A, B, C, and F is shown highlighted before contraction of elementary paths. The subtree resulting from contraction of elementary paths is shown to the right.

7.1.1   Finding Subtrees Induced by Triplets

Finding subtrees induced by triplets of terminal nodes in rooted trees is an interesting problem, because a rooted tree can be reconstructed in a unique way from the set of all its triplet topologies. Given a triplet of terminal nodes, there are only three possible induced subtrees if the phylogenetic tree is fully resolved.

Example 7.3
There are three fully resolved rooted phylogenetic trees on three terminal nodes.

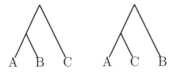

The subtree induced by a triplet of terminal nodes of a fully resolved rooted phylogenetic tree can be obtained by removing all other terminal nodes and then contracting any elementary paths. However, a more efficient algorithm consists in first finding the most recent common ancestor of each pair of terminal nodes from the given triplet and then building the induced subtree by distinguishing among the three possible topologies on the basis of the relationship among the most recent common ancestors. In the following description, the three possible triplet topologies are distinguished by $mrca(i,k) = mrca(j,k)$, $mrca(i,j) = mrca(j,k)$, and $mrca(i,j) = mrca(i,k)$, and their Newick string is output.

```
function triplet(T, i, j, k)
    ij ← mrca(T, i, j)
    ik ← mrca(T, i, k)
    jk ← mrca(T, j, k)
    if ik = jk then
        return ((i, j), k);
    else
        if ij = jk then
            return ((i, k), j);
        else
            return ((j, k), i);
```

The representation of phylogenetic trees in BioPerl does not include any method to compute subtrees induced by triplets of terminal nodes. Nevertheless, the subtree of a fully resolved rooted phylogenetic tree induced by a triplet of terminal nodes can be easily obtained by using the `get_lca` method to find the most recent common ancestor of the terminal nodes in the given triplet, as illustrated by the following Perl script.

```
use Bio::TreeIO;

my $input = new Bio::TreeIO(
    -fh => \*DATA,
    -format => "newick"
);

my $tree = $input->next_tree;
```

```perl
sub triplet {
  my $tree = shift;
  my $i = shift;
  my $j = shift;
  my $k = shift;

  my @ij = grep { $_->id =~ /$i|$j/ } $tree->
     get_leaf_nodes;
  my @ik = grep { $_->id =~ /$i|$k/ } $tree->
     get_leaf_nodes;
  my @jk = grep { $_->id =~ /$j|$k/ } $tree->
     get_leaf_nodes;

  my $ij = $tree->get_lca(-nodes => \@ij);
  my $ik = $tree->get_lca(-nodes => \@ik);
  my $jk = $tree->get_lca(-nodes => \@jk);

  my $str;
  if ($ik == $jk) {
    $str = "(($i,$j),$k);";
  } else {
    if ($ij == $jk) {
      $str = "(($i,$k),$j);";
    } else {
      $str = "(($j,$k),$i);";
    }
  }

  return $str;
}

my $str = triplet($tree,"A","C","E");

__DATA__
(A,(((B,C),D),E));
```

The representation of phylogenetic trees in R does not include any method to compute subtrees induced by triplets of terminal nodes either. However, a `triplet` function can easily be defined using the `mrca` function, as illustrated by the following R script.

```r
> library(ape)

> triplet <- function (t,i,j,k) {
    ij <- mrca(t)[i,j]
    ik <- mrca(t)[i,k]
```

```
    jk <- mrca(t)[j,k]
    if (ik == jk)
       paste("((",i,",",j,"),",k,");",sep="")
    else
       if (ij == jk)
          paste("((",i,",",k,"),",j,");",sep="")
       else
          paste("((",j,",",k,"),",i,");",sep="")
}

> t <- read.tree(text="(A,(((B,C),D),E));")

> triplet(t,"A","C","E")
[1] "((C,E),A);"
```

7.1.2 Finding Subtrees Induced by Quartets

In unrooted trees, finding subtrees induced by quartets of terminal nodes is
also an interesting problem, because an unrooted tree can be reconstructed in
a unique way from the set of all its quartet topologies. Given a quartet of terminal nodes, there are only three possible induced subtrees if the phylogenetic
tree is fully resolved.

Example 7.4
There are three fully resolved unrooted phylogenetic trees on four terminal
nodes.

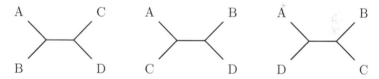

The subtree induced by a quartet of terminal nodes of a fully resolved unrooted phylogenetic tree can be obtained by removing all other terminal nodes
and then contracting any elementary paths. In the following description, the
removal of a terminal node causes the removal of the corresponding external
branch, while the removal of an internal node originates the contraction of
the corresponding internal branches, and the Newick string of the resulting
induced subtree is output.

function quartet(T, i, j, k, ℓ)
   for each node v of T **do**
      if v is a terminal node labeled i, j, k or ℓ **then**

> $w \leftarrow$ node of T adjacent to node v
> remove nodes v and w from T

The representation of phylogenetic trees in BioPerl does not include any method to compute subtrees induced by quartets of terminal nodes. Nevertheless, the subtree of a fully resolved unrooted phylogenetic tree induced by a quartet of terminal nodes can be easily obtained by using the `remove_Node` method to remove terminal nodes and contract the corresponding elementary paths from a copy of an unrooted phylogenetic tree, obtained using the `clone` method from the `Clone` module, as illustrated by the following Perl script.

```perl
use Clone qw(clone);

sub quartet {
  my ($tree,$i,$j,$k,$l) = @_;
  my $copy = clone $tree;

  map { $copy->remove_Node($_) }
    grep { !($_->id =~ /$i|$j|$k|$l/) }
    $copy->get_leaf_nodes;

  return $copy;
}
```

The representation of phylogenetic trees in R does not include any method to compute subtrees induced by quartets of terminal nodes, either. However, a `quartet` function can easily be defined using the `drop.tip` function to remove terminal branches and the corresponding internal branches from an unrooted phylogenetic tree, as illustrated by the following R script.

```r
> library(ape)

> quartet <- function (t,i,j,k,l) {
  unroot(drop.tip(t,setdiff(t$tip.label,c(i,j,k,l))))
}

> t <- read.tree(text="((A,B),(C,D),(E,F));")
> q <- quartet(t,"A","B","C","F")
> write.tree(q)
[1] "((A,B),C,F);"
```

7.2 Finding Common Subtrees

Subtrees shared by two trees reveal information common to the two trees. As there are several ways in which a tree can be contained in another tree, common subtrees can be bottom-up, top-down, or induced by a set of terminal nodes. Further, in order to reveal the most of their shared information, it is interesting to find common subtrees of largest size between two given trees.

7.2.1 Maximum Agreement of Rooted Trees

Two (unrooted or rooted) phylogenetic trees are said to agree on a set of terminal nodes if their subtrees induced by that set of terminal nodes are isomorphic, and a maximum agreement subtree of two phylogenetic trees is a common subtree induced by a set of terminal nodes of largest possible size.

Example 7.5
The following fully resolved rooted phylogenetic trees agree on the set of terminal nodes labeled A, B, D, and E, and this is the largest set of terminal nodes on which they agree. Thus, their maximum agreement subtree has four terminal nodes.

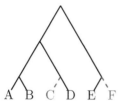 

Example 7.6
The following fully resolved unrooted phylogenetic trees also agree on the set of terminal nodes labeled A, B, D, and E, and this is the largest set of terminal nodes on which they agree. Their maximum agreement subtree also has four terminal nodes.

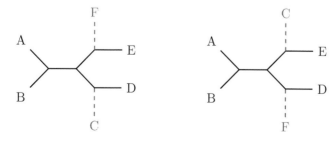

A maximum agreement subtree of two fully resolved rooted phylogenetic trees can be obtained by matching one of the trees to the other one in such a way that the number of common terminal nodes is maximized. If one of the trees is just a terminal node, the maximum agreement subtree will consist of a single node (a tree of size one) if there is a terminal node labeled the same in the other tree, and it will be empty (of size zero) otherwise. In general, when the two trees have three or more nodes each, the maximum agreement subtree will result from either matching the subtree rooted at a child of the root in one of the trees to the other tree or matching the two subtrees in one tree to the two subtrees in the other tree.

Thus, in general, the number of terminal nodes $M(T_1, T_2)$ of a maximum agreement subtree of two fully resolved rooted phylogenetic trees T_1 and T_2 is given by the recurrence

$$M(T_1[v_1], T_2[v_2]) = \max \begin{cases} M(T_1[\ell_1], T_2[\ell_2]) + M(T_1[r_1], T_2[r_2]) \\ M(T_1[\ell_1], T_2[r_2]) + M(T_1[r_1], T_2[\ell_2]) \\ M(T_1[\ell_1], T_2[v_2]) \\ M(T_1[r_1], T_2[v_2]) \\ M(T_1[v_1], T_2[\ell_2]) \\ M(T_1[v_1], T_2[r_2]) \end{cases}$$

where $T_1[v_1]$ is the subtree of T_1 rooted at node v_1, with left child ℓ_1 and right child r_1, and $T_2[v_2]$ is the subtree of T_2 rooted at node v_2, with left child ℓ_2 and right child r_2. Notice that in phylogenetic trees, the distinction between left and right children is arbitrary, and it is just a way to refer to each of the two children of an internal node in a fully resolved rooted phylogenetic tree. The six cases are as follows.

- Match $T_1[\ell_1]$ to $T_2[\ell_2]$ and $T_1[r_1]$ to $T_2[r_2]$

- Match $T_1[\ell_1]$ to $T_2[r_2]$ and $T_1[r_1]$ to $T_2[\ell_2]$

- Match $T_1[\ell_1]$ (which contains all the common terminal nodes) to $T_2[v_2]$

- Match $T_1[r_1]$ (which contains all the common terminal nodes) to $T_2[v_2]$

- Match $T_1[v_1]$ to $T_2[\ell_2]$ (which contains all the common terminal nodes)

- Match $T_1[v_1]$ to $T_2[r_2]$ (which contains all the common terminal nodes)

The number of terminal nodes of a maximum agreement subtree of two fully resolved rooted phylogenetic trees can be obtained by computing the number of terminal nodes of a maximum agreement subtree of the trees rooted in turn at each node of the given trees, using dynamic programming. It suffices to perform the computation upon the two trees in postorder.

function mast(T_1, T_2)
    for each node v_1 of T_1 in postorder **do**
        for each node v_2 of T_2 in postorder **do**
            $M[v_1, v_2] \leftarrow 0$
            if v_1 is a terminal node **then**
                if v_2 is a terminal node **then**
                    if v_1 and v_2 are labeled the same **then**
                        $M[v_1, v_2] \leftarrow 1$
                else
                    if $T_2[v_2]$ has a terminal node labeled as v_1 **then**
                        $M[v_1, v_2] \leftarrow 1$
            else
                if v_2 is a terminal node **then**
                    if $T_1[v_1]$ has a terminal node labeled as v_2 **then**
                        $M[v_1, v_2] \leftarrow 1$
            else
                $\ell_1, r_1 \leftarrow$ *children of* v_1
                $\ell_2, r_2 \leftarrow$ *children of* v_2
                $m_1 \leftarrow M[\ell_1, \ell_2] + M[r_1, r_2]$
                $m_2 \leftarrow M[\ell_1, r_2] + M[r_1, \ell_2]$
                $m_3 \leftarrow M[v_1, \ell_2]$
                $m_4 \leftarrow M[v_1, r_2]$
                $m_5 \leftarrow M[\ell_1, v_2]$
                $m_6 \leftarrow M[r_1, v_2]$
                $M[v_1, v_2] \leftarrow \max\{m_1, m_2, m_3, m_4, m_5, m_6\}$
    return $M[root(T_1), root(T_2)]$

Example 7.7

A maximum agreement subtree of the fully resolved rooted phylogenetic trees
with Newick string $(((A,B),(C,D)),(E,F));$ and $(((A,B),(F,D)),(E,C));$
has 4 terminal nodes, as shown in the last entry of the following dynamic
programming table.

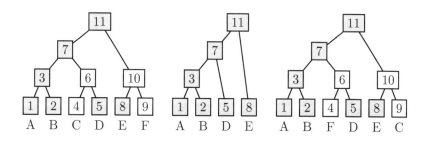

	1	2	3	4	5	6	7	8	9	10	11
1	1	0	1	0	0	0	1	0	0	0	1
2	0	1	1	0	0	0	1	0	0	0	1
3	1	1	2	0	0	0	2	0	0	0	2
4	0	0	0	0	0	0	0	0	1	1	1
5	0	0	0	0	1	1	1	0	0	0	1
6	0	0	0	0	1	1	1	0	1	1	2
7	1	1	2	0	1	1	3	0	1	1	3
8	0	0	0	0	0	0	0	1	0	1	1
9	0	0	0	1	0	1	1	0	0	0	1
10	0	0	0	1	0	1	1	1	0	1	2
11	1	1	2	1	1	2	3	1	1	2	4

The representation of phylogenetic trees in BioPerl does not include any method to compute the number of terminal nodes of a maximum agreement subtree of two fully resolved rooted phylogenetic trees. However, the previous algorithm can easily be implemented using the `postorder_traversal` method from the `Bio::Tree::Compatible` module to perform the computation upon the two trees in postorder, as shown in the following Perl script.

```perl
sub mast_size {
  my $tree1 = shift;
  my $tree2 = shift;
  my (@m, $m);
  for my $node1 (@{ $tree1->Bio::Tree::Compatible::
    postorder_traversal }) {
    my $n1 = $node1->internal_id;
    for my $node2 (@{ $tree2->Bio::Tree::Compatible::
      postorder_traversal }) {
      my $n2 = $node2->internal_id;
      if ($node1->is_Leaf) {
        if ($node2->is_Leaf) {
          $m = ($node1->id eq $node2->id ? 1 : 0);
        } else {
          my $label1 = $node1->id;
          $m = (grep { /$label1/ } map { $_->id }
            grep { $_->is_Leaf } $node2->
            get_all_Descendents) ? 1 : 0;
        }
      } else {
        if ($node2->is_Leaf) {
          my $label2 = $node2->id;
          $m = (grep { /$label2/ } map { $_->id }
            grep { $_->is_Leaf } $node1->
            get_all_Descendents) ? 1 : 0;
        } else {
```

```
      my ($l1,$r1) = map { $_->internal_id }
          $node1->each_Descendent;
      my ($l2,$r2) = map { $_->internal_id }
          $node2->each_Descendent;
      $m = $m[$l1][$l2] + $m[$r1][$r2];
      if ($m[$l1][$r2] + $m[$r1][$l2] > $m) {
        $m = $m[$l1][$r2] + $m[$r1][$l2];
      }
      if ($m[$n1][$l2] > $m) {
        $m = $m[$n1][$l2];
      }
      if ($m[$n1][$r2] > $m) {
        $m = $m[$n1][$r2];
      }
      if ($m[$l1][$n2] > $m) {
        $m = $m[$l1][$n2];
      }
      if ($m[$r1][$n2] > $m) {
        $m = $m[$r1][$n2];
      }
    }
  }
  $m[$n1][$n2] = $m;
  }
 }
 return $m;
}
```

The representation of phylogenetic trees in R does not include any method
to compute the number of terminal nodes of a maximum agreement subtree
of two fully resolved rooted phylogenetic trees, either. However, the previous
algorithm can easily be implemented with the help of a function to obtain the
parent of a node in a tree,

```
> parent <- function (tree,node)
  tree$edge[which(tree$edge[,2]==node),1]
```

to access the children of a node in a tree,

```
> children <- function (tree,node)
  tree$edge[which(tree$edge[,1]==node),2]
```

to determine if a node is the root of a tree,

```
> is.root <- function (tree,node)
  length(tree$tip.label)+1==node
```

to determine if a node is an ancestor of another node in a tree,

```
> is.ancestor <- function(tree,anc,des) {
  if (anc==des) return(TRUE)
  if (is.root(tree,des)) return(FALSE)
  else return(is.ancestor(tree,anc,parent(tree,des)))
}
```

and to perform a postorder traversal of a tree,

```
> postorder <- function (tree) {
  r <- length(tree$tip.label)+1
  postorder.traversal(tree,r,c())
}
> postorder.traversal <- function (tree,v,res) {
  for (w in t1$edge[which(tree$edge[,1]==v),2])
    res <- postorder.traversal(tree,w,res)
  c(res,v)
}
```

Now, the previous algorithm for computing the number of terminal nodes of a maximum agreement subtree of two fully resolved rooted phylogenetic trees can be implemented as shown in the following R script.

```
> mast.size <- function (t1,t2) {
  po1 <- postorder(t1)
  po2 <- postorder(t2)
  m <- matrix(0,nrow=length(po1),ncol=length(po2),
    dimnames=list(po1,po2))
  for (i in po1) {
    for (j in po2) {
      if (length(t1$edge[which(t1$edge[,1]==i),1])
        ==0) {
        if (length(t2$edge[which(t2$edge[,1]==j),1])
          ==0) {
          if (t1$tip.label[i]==t2$tip.label[j]) m[i,j
            ] <- 1
        } else {
          jj <- which(t2$tip.label==t1$tip.label[i])
          if (is.ancestor(t2,j,jj)) m[i,j] <- 1
        }
      } else {
        if (length(t2$edge[which(t2$edge[,1]==j),1])
          ==0) {
          ii <- which(t1$tip.label==t2$tip.label[j])
          if (is.ancestor(t1,i,ii)) m[i,j] <- 1
        } else {
          l1 <- children(t1,i)[1]
          r1 <- children(t1,i)[2]
```

```
          l2 <- children(t2,j)[1]
          r2 <- children(t2,j)[2]
          m[i,j] <- max(m[l1,l2]+m[r1,r2],m[l1,r2]+m[
             r1,l2],m[i,l2],m[i,r2],m[l1,j],m[r1,j])
        }
      }
    }
  }
  m <- m[po1,po2]
  dimnames(m) <- list(po1,po2)
  m
}
```

Recall that with the representation of a phylogenetic tree in the R package APE, the terminal nodes are numbered $1, \ldots, n$ and the root is numbered $n+1$. Therefore, in a postorder traversal of the fully resolved rooted phylogenetic trees from the previous example, the numbering of the nodes is 1, 2, 9, 3, 4, 10, 8, 5, 6, 11, 7.

```
> t1 <- read.tree(text="(((A,B),(C,D)),(E,F));")
> t2 <- read.tree(text="(((A,B),(F,D)),(E,C));")
> mast.size(t1,t2)
```

	1	2	9	3	4	10	8	5	6	11	7
1	1	0	1	0	0	0	1	0	0	0	1
2	0	1	1	0	0	0	1	0	0	0	1
9	1	1	2	0	0	0	2	0	0	0	2
3	0	0	0	0	0	0	0	0	1	1	1
4	0	0	0	0	1	1	1	0	0	0	1
10	0	0	0	0	1	1	1	0	1	1	2
8	1	1	2	0	1	1	3	0	1	1	3
5	0	0	0	0	0	0	0	1	0	1	1
6	0	0	0	1	0	1	1	0	0	0	1
11	0	0	0	1	0	1	1	1	0	1	2
7	1	1	2	1	1	2	3	1	1	2	4

An actual maximum agreement subtree of two fully resolved rooted phylogenetic trees can be obtained as the subtree (of any of the two trees) induced by a set of common terminal nodes of largest size on which the two trees agree. Now, the set of terminal nodes in a maximum agreement subtree of two fully resolved rooted phylogenetic trees can be obtained by computing the set of terminal nodes in a maximum agreement subtree of the trees rooted in turn at each node of the given trees, using dynamic programming.

function $\text{mast}(T_1, T_2)$
    for each node v_1 of T_1 in postorder **do**
        for each node v_2 of T_2 in postorder **do**

$M[v_1, v_2] \leftarrow \emptyset$
if v_1 is a terminal node **then**
    if v_2 is a terminal node **then**
        if v_1 and v_2 are labeled the same **then**
            $M[v_1, v_2] \leftarrow \{label\ of\ v_1\}$
    else
        if $T_2[v_2]$ has a terminal node labeled as v_1 **then**
            $M[v_1, v_2] \leftarrow \{label\ of\ v_1\}$
else
    if v_2 is a terminal node **then**
        if $T_1[v_1]$ has a terminal node labeled as v_2 **then**
            $M[v_1, v_2] \leftarrow \{label\ of\ v_2\}$
    else
        $\ell_1, r_1 \leftarrow children\ of\ v_1$
        $\ell_2, r_2 \leftarrow children\ of\ v_2$
        $M[v_1, v_2] \leftarrow M[\ell_1, \ell_2] \cup M[r_1, r_2]$
        if $size(M[\ell_1, r_2]) + size(M[r_1, \ell_2]) > size(M[v_1, v_2])$ **then**
            $M[v_1, v_2] \leftarrow M[\ell_1, r_2] \cup M[r_1, \ell_2]$
        if $size(M[v_1, \ell_2]) > size(M[v_1, v_2])$ **then**
            $M[v_1, v_2] \leftarrow M[v_1, \ell_2]$
        if $size(M[v_1, r_2]) > size(M[v_1, v_2])$ **then**
            $M[v_1, v_2] \leftarrow M[v_1, r_2]$
        if $size(M[\ell_1, v_2]) > size(M[v_1, v_2])$ **then**
            $M[v_1, v_2] \leftarrow M[\ell_1, v_2]$
        if $size(M[r_1, v_2]) > size(M[v_1, v_2])$ **then**
            $M[v_1, v_2] \leftarrow M[r_1, v_2]$
return $M[root(T_1), root(T_2)]$

Example 7.8

A maximum agreement subtree of the fully resolved rooted phylogenetic trees with Newick string `(((A,B),(C,D)),(E,F));` (left) and with Newick string `(((A,B),(F,D)),(E,C));` (right) is the fully resolved rooted phylogenetic tree with Newick string `(((A,B),D),E);` (middle).

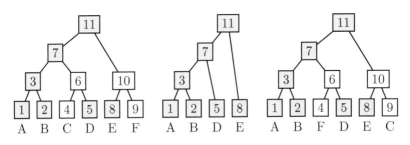

The maximum agreement subtree has four terminal nodes labeled A, B, D, and E, as shown in the last entry of the following dynamic programming table. Notice that there are two maximum agreement subtrees of the left subtree rooted at node 11 and the right subtree rooted at node 7, which are induced by the terminal nodes labeled A, B, D and by the terminal nodes labeled A, B, F (the latter shown in the dynamic programming table). Also, there are two maximum agreement subtrees of the left subtree rooted at node 7 and the right subtree rooted at node 11, which are induced by the terminal nodes labeled A, B, D (shown in the dynamic programming table) and by the terminal nodes labeled A, B, C.

	1	2	3	4	5	6	7	8	9	10	11
1	A		A				A				A
2		B	B				B				B
3	A	B	A,B				A,B				A,B
4									C	C	C
5					D	D	D				D
6					D	D	D		C	C	C,D
7	A	B	A,B		D	D	A,B,D		C	C	A,B,C
8								E		E	E
9				F	F	F					F
10				F	F	F		E		E	E,F
11	A	B	A,B	F	D	D,F	A,B,F	E	C	C,E	A,B,D,E

The previous algorithm for obtaining an actual maximum agreement subtree of two fully resolved rooted phylogenetic trees can easily be implemented by storing an array of terminal node labels at each position of the dynamic programming table, as shown in the following Perl script.

```perl
sub mast {
  my $tree1 = shift;
  my $tree2 = shift;
  my @m;
  for my $node1 (@{ $tree1->Bio::Tree::Compatible::
      postorder_traversal }) {
    my $n1 = $node1->internal_id;
    for my $node2 (@{ $tree2->Bio::Tree::Compatible::
        postorder_traversal }) {
      my $n2 = $node2->internal_id;
      my @mm = ();
      if ($node1->is_Leaf) {
        if ($node2->is_Leaf) {
          if ($node1->id eq $node2->id) {
```

```perl
          @mm = ($node1->id);
        }
    } else {
      my $label1 = $node1->id;
      if (grep { /$label1/ } map { $_->id } grep
          { $_->is_Leaf } $node2->
          get_all_Descendents) {
        @mm = ($label1);
      }
    }
  } else {
    if ($node2->is_Leaf) {
      my $label2 = $node2->id;
      if (grep { /$label2/ } map { $_->id } grep
          { $_->is_Leaf } $node1->
          get_all_Descendents) {
        @mm = ($label2);
      }
    } else {
      my ($l1,$r1) = map { $_->internal_id }
          $node1->each_Descendent;
      my ($l2,$r2) = map { $_->internal_id }
          $node2->each_Descendent;
      @mm = @{$m[$l1][$l2]};
      push @mm, @{$m[$r1][$r2]};
      if (scalar(@{$m[$l1][$r2]})+scalar(@{$m[$r1
          ][$l2]})>scalar(@mm)) {
        @mm = @{$m[$l1][$r2]};
        push @mm, @{$m[$r1][$l2]};
      }
      if (scalar(@{$m[$n1][$l2]})>scalar(@mm)) {
        @mm = @{$m[$n1][$l2]};
      }
      if (scalar(@{$m[$n1][$r2]})>scalar(@mm)) {
        @mm = @{$m[$n1][$r2]};
      }
      if (scalar(@{$m[$l1][$n2]})>scalar(@mm)) {
        @mm = @{$m[$l1][$n2]};
      }
      if (scalar(@{$m[$r1][$n2]})>scalar(@mm)) {
        @mm = @{$m[$r1][$n2]};
      }
    }
  }
}
$m[$n1][$n2] = \@mm;
```

```
      }
    }
    return \@m;
}
```

The previous algorithm for obtaining an actual maximum agreement subtree of two fully resolved rooted phylogenetic trees can also be implemented by
extending the previous R implementation, storing a vector of terminal node
labels at each position of the dynamic programming table, as shown in the
following R script.

```
> mast <- function (t1,t2) {
  po1 <- postorder(t1)
  po2 <- postorder(t2)
  m <- matrix(rep(list(),length(po1)),nrow=length(po1
    ),ncol=length(po2),dimnames=list(po1,po2))
  for (i in po1) {
    for (j in po2) {
      mm <- list()
      if (length(t1$edge[which(t1$edge[,1]==i),1])
        ==0) {
        if (length(t2$edge[which(t2$edge[,1]==j),1])
          ==0) {
          if (t1$tip.label[i]==t2$tip.label[j])
            mm <- c(t1$tip.label[i])
        } else {
          jj <- which(t2$tip.label==t1$tip.label[i])
          if (is.ancestor(t2,j,jj))
            mm <- c(t1$tip.label[i])
        }
      } else {
        if (length(t2$edge[which(t2$edge[,1]==j),1])
          ==0) {
          ii <- which(t1$tip.label==t2$tip.label[j])
          if (is.ancestor(t1,i,ii))
            mm <- c(t2$tip.label[j])
        } else {
          l1 <- children(t1,i)[1]
          r1 <- children(t1,i)[2]
          l2 <- children(t2,j)[1]
          r2 <- children(t2,j)[2]
          mm <- c(m[[l1,l2]],m[[r1,r2]])
          if ( length(m[[l1,r2]]) + length(m[[r1,l2
            ]]) > length(mm) )
            mm <- c(m[[l1,r2]],m[[r1,l2]])
          if ( length(m[[i,l2]]) > length(mm) )
```

```
           mm <- m[[i,12]]
           if ( length(m[[i,r2]]) > length(mm) )
              mm <- m[[i,r2]]
           if ( length(m[[l1,j]]) > length(mm) )
              mm <- m[[l1,j]]
           if ( length(m[[r1,j]]) > length(mm) )
              mm <- m[[r1,j]]
        }
     }
     m[[i,j]] <- mm
   }
 }
 unlist(m[[length(t1$tip.label)+1,length(t2$tip.
    label)+1]])
}

> t1 <- read.tree(text="(((A,B),(C,D)),(E,F));")
> t2 <- read.tree(text="(((A,B),(F,D)),(E,C));")
> mast(t1,t2)
[1] "a" "b" "d" "e"
```

7.2.2 Maximum Agreement of Unrooted Trees

A maximum agreement subtree of two fully resolved unrooted phylogenetic trees can be obtained by computing a maximum agreement subtree of the first tree rooted at an arbitrary node and the second tree rooted at each node in turn. The largest of them is a maximum agreement subtree of the unrooted phylogenetic trees.

7.3 Comparing Trees

The similarities and differences between two phylogenetic trees can be assessed by computing a distance measure between the two trees. The *triplets distance* is based on the subtrees induced by triplets of terminal nodes in two rooted phylogenetic trees, and the *quartets distance* is based on the subtrees induced by quartets of terminal nodes in two unrooted phylogenetic trees.

7.3.1 The Triplets Distance between Rooted Trees

The *triplets distance* is based on the subtrees induced by triplets of terminal nodes in the two trees under comparison. The sets of subtrees induced by triplets of terminal nodes reveal similarities and differences between two fully resolved rooted phylogenetic trees.

The triplets distance between two fully resolved rooted phylogenetic trees labeled over the same taxa is defined as the size of the symmetric difference of their sets of triplets, that is, the number of triplets in which the two phylogenetic trees differ.

Example 7.9

Consider again the fully resolved rooted phylogenetic trees with Newick string
`(((A,B),(C,D)),(E,F))`; and `(((A,B),(F,D)),(E,C))`; and $n = 6$ terminal nodes labeled A, B, C, D, E, F.

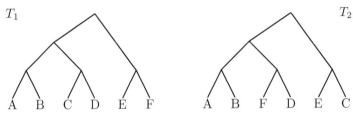

The sets of $\binom{6}{3} = 20$ triplets induced by the terminal nodes of each of them are given in the following table, where the triplets common to the two trees are marked with an asterisk.

	T_1	T_2	
*	`((A,B),C);`	`((A,B),C);`	*
*	`((A,B),D);`	`((A,B),D);`	*
*	`((A,B),E);`	`((A,B),E);`	*
*	`((A,B),F);`	`((A,B),F);`	*
	`((C,D),A);`	`((A,D),C);`	
	`((A,C),E);`	`((C,E),A);`	
	`((A,C),F);`	`((A,F),C);`	
*	`((A,D),E);`	`((A,D),E);`	*
	`((A,D),F);`	`((D,F),A);`	
	`((E,F),A);`	`((A,F),E);`	
	`((C,D),B);`	`((B,D),C);`	
	`((B,C),E);`	`((C,E),B);`	
	`((B,C),F);`	`((B,F),C);`	
*	`((B,D),E);`	`((B,D),E);`	*
	`((B,D),F);`	`((D,F),B);`	
	`((E,F),B);`	`((B,F),E);`	
	`((C,D),E);`	`((C,E),D);`	
	`((C,D),F);`	`((D,F),C);`	
	`((E,F),C);`	`((C,E),F);`	
	`((E,F),D);`	`((D,F),E);`	

The two phylogenetic trees differ in 14 of the 20 triplets and, thus, their triplets distance is 28.

The triplets distance between two phylogenetic trees can be computed by first obtaining the triplet induced by each set of three terminal node labels in each of the trees and then counting the number of triplets in which the two trees differ. In the following description, the two sets of $\binom{n}{3}$ triplets are obtained using the previous algorithm for finding the subtree induced by a triplet of terminal nodes upon each set of three terminal node labels in turn.

> **function** triplets_distance(T_1, T_2)
> $\quad L \leftarrow$ terminal node labels in T_1 and T_2
> $\quad n \leftarrow length(L)$
> $\quad d \leftarrow 0$
> $\quad$ **for** $i \leftarrow 1, \ldots, n$ **do**
> $\quad\quad$ **for** $j \leftarrow i+1, \ldots, n$ **do**
> $\quad\quad\quad$ **for** $k \leftarrow j+1, \ldots, n$ **do**
> $\quad\quad\quad\quad t_1 \leftarrow triplet(T_1, L[i], L[j], L[k])$
> $\quad\quad\quad\quad t_2 \leftarrow triplet(T_2, L[i], L[j], L[k])$
> $\quad\quad\quad\quad$ **if** $t_1 \neq t_2$ **then**
> $\quad\quad\quad\quad\quad d \leftarrow d+2$
> $\quad$ **return** d

The representation of phylogenetic trees in BioPerl does not include any method to compute the triplets distance between two fully resolved rooted phylogenetic trees with the same terminal node labels. However, the sets of triplets of terminal nodes can be computed using the `triplet` method and the previous algorithm can easily be implemented, as shown in the following Perl script.

```perl
sub triplets_distance {
  my $tree1 = shift;
  my $tree2 = shift;
  my @leaves = sort {$a->id cmp $b->id} $tree1->
    get_leaf_nodes;
  my @labels = map { $_->id } @leaves;
  my $dist = 0;
  for (my $i = 0; $i < @leaves; $i++) {
    for (my $j = $i+1; $j < @leaves; $j++) {
      for (my $k = $j+1; $k < @leaves; $k++) {
        my $t1 = triplet($tree1,$labels[$i],$labels[
          $j],$labels[$k]);
        my $t2 = triplet($tree2,$labels[$i],$labels[
          $j],$labels[$k]);
        $dist += 2 unless ($t1 eq $t2);
      }
    }
  }
}
```

```
      return $dist;
}
```

The representation of phylogenetic trees in R does not include any method to compute the triplets distance between two phylogenetic trees, either. However, the sets of triplets of terminal nodes can be computed using the `triplet` function and the previous algorithm can be implemented in a straightforward way, as illustrated by the following R script.

```
> triplets.distance <- function (t1,t2) {
  L <- sort(t1$tip.label)
  d <- 0
  for (i in L[1:(length(L)-2)]) {
    for (j in L[(match(i,L)+1):(length(L)-1)]) {
      for (k in L[(match(j,L)+1):length(L)]) {
        str1 <- triplet(t1,i,j,k)
        str2 <- triplet(t2,i,j,k)
        if (str1 != str2) { d <- d + 2 }
      }
    }
  }
  d
}

> t1 <- read.tree(text="(((A,B),(C,D)),(E,F));")
> t2 <- read.tree(text="(((A,B),(F,D)),(E,C));")
> triplets.distance(t1,t2)
[1] 28
```

7.3.2 The Quartets Distance between Unrooted Trees

The *quartets distance* is based on the subtrees induced by quartets of terminal nodes in the two trees under comparison. The sets of subtrees induced by quartets of terminal nodes reveal similarities and differences between two fully resolved unrooted phylogenetic trees.

The quartets distance between two fully resolved unrooted phylogenetic trees labeled over the same taxa is defined as the size of the symmetric difference of their sets of quartets, that is, the number of quartets in which the two phylogenetic trees differ.

Example 7.10

Consider the following fully resolved unrooted phylogenetic trees with Newick string `(((A,B),(C,D)),(E,F));` and `(((A,B),(F,D)),(E,C));` and $n = 6$ terminal nodes labeled A, B, C, D, E, F.

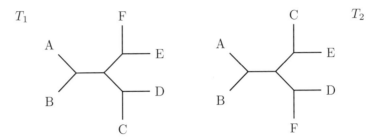

The sets of $\binom{6}{4} = 15$ quartets induced by the terminal nodes of each of them are given in the following table, where the quartets common to the two trees are marked with an asterisk.

	T_1	T_2	
*	((A,B),C,D);	((A,B),C,D);	*
*	((A,B),C,E);	((A,B),C,E);	*
*	((A,B),C,F);	((A,B),C,F);	*
*	((A,B),D,E);	((A,B),D,E);	*
*	((A,B),D,F);	((A,B),D,F);	*
*	((A,B),E,F);	((A,B),E,F);	*
	((A,E),C,D);	((A,D),C,E);	
	((A,F),C,D);	((A,C),D,F);	
	((A,C),E,F);	((A,F),C,E);	
	((A,D),E,F);	((A,E),D,F);	
	((B,E),C,D);	((B,D),C,E);	
	((B,F),C,D);	((B,C),D,F);	
	((B,C),E,F);	((B,F),C,E);	
	((B,D),E,F);	((B,E),D,F);	
	((C,D),E,F);	((C,E),D,F);	

The two phylogenetic trees differ in 9 of the 15 quartets and, thus, their quartets distance is 18.

The quartets distance between two phylogenetic trees can be computed by first obtaining the quartet induced by each set of four terminal node labels in each of the trees and then counting the number of quartets in which the two trees differ. In the following description, the two sets of $\binom{n}{4}$ quartets are obtained using the previous algorithm for finding the subtree induced by a quartet of terminal nodes upon each set of four terminal node labels in turn.

function quartets_distance(T_1, T_2)
    $L \leftarrow$ terminal node labels in T_1 and T_2
    $n \leftarrow length(L)$
    $d \leftarrow 0$
    for $i \leftarrow 1, \ldots, n$ **do**
        for $j \leftarrow i+1, \ldots, n$ **do**

```
        for k ← j + 1, ..., n do
            for ℓ ← k + 1, ..., n do
                q₁ ← quartet(T₁, L[i], L[j], L[k], L[ℓ])
                q₂ ← quartet(T₂, L[i], L[j], L[k], L[ℓ])
                if q₁ ≠ q₂ then
                    d ← d + 2
    return d
```

The representation of phylogenetic trees in BioPerl does not include any method to compute the quartets distance between two fully resolved unrooted phylogenetic trees with the same terminal node labels. However, the sets of quartets of terminal nodes can be computed using the `quartet` method and the previous algorithm can easily be implemented, as shown in the following Perl script.

```perl
sub quartets_distance {
  my $tree1 = shift;
  my $tree2 = shift;
  my @leaves = sort {$a->id cmp $b->id} $tree1->
    get_leaf_nodes;
  my @labels = map { $_->id } @leaves;
  my $dist = 0;
  for (my $i = 0; $i < @leaves; $i++) {
    for (my $j = $i+1; $j < @leaves; $j++) {
      for (my $k = $j+1; $k < @leaves; $k++) {
        for (my $l = $k+1; $l < @leaves; $l++) {
          my $q1 = quartet($tree1,$labels[$i],$labels
            [$j],$labels[$k],$labels[$l]);
          my $q2 = quartet($tree2,$labels[$i],$labels
            [$j],$labels[$k],$labels[$l]);
          if (partition_distance($q1,$q2) != 0) {
            $dist += 2;
          }
        }
      }
    }
  }
  return $dist;
}
```

The representation of phylogenetic trees in R does not include any method to compute the quartets distance between two phylogenetic trees, either. However, the sets of quartets of terminal nodes can be computed using the `quartet` function and the previous algorithm can be implemented in a straightforward way, as illustrated by the following R script.

```
quartets.distance <- function (t1,t2) {
  L <- sort(t1$tip.label)
  d <- 0
  for (i in L[1:(length(L)-3)]) {
    for (j in L[(match(i,L)+1):(length(L)-2)]) {
      for (k in L[(match(j,L)+1):(length(L)-1)]) {
        for (l in L[(match(k,L)+1):length(L)]) {
          q1 <- quartet(t1,i,j,k,l)
          q2 <- quartet(t2,i,j,k,l)
          if (dist.topo(q1,q2) != 0) { d <- d + 2 }
        }
      }
    }
  }
  d
}

> t1 <- read.tree(text="((A,B),(C,D),(E,F));")
> t2 <- read.tree(text="((A,B),(F,D),(E,C));")
> quartets.distance(t1,t2)
[1] 18
```

Bibliographic Notes

The triplets distance between fully resolved rooted phylogenetic trees was introduced by Critchlow et al. (1996).

The quartet distance was introduced in (Estabrook et al. 1985). The characterization of unrooted phylogenetic trees in term of their quartets was established in (Bandelt and Dress 1986). Algorithms for computing the quartets distance between fully resolved unrooted phylogenetic trees can be found in (Brodal et al. 2003; Bryant et al. 2000; Christiansen et al. 2006).

The maximum agreement subtree problem was introduced by Steel and Warnow (1993). Improved algorithms for computing a maximum agreement subtree between two fully resolved rooted phylogenetic trees can be found in (Amir and Keselman 1997; Cole et al. 2000; Goddard et al. 1994; Lee et al. 2005).

Part III

Graph Pattern Matching

Chapter 8

Graphs

Graphs, together with trees, count among the most useful mathematical abstractions and, at the same time, the most common combinatorial structures in computer science and computational biology. Basic notions underlying combinatorial algorithms on graphs, such as counting, generation, and traversal algorithms, as well as appropriate data structures for the representation of graphs, are the subject of this introductory chapter.

8.1 Graphs in Mathematics

The notion of graph most often found in discrete mathematics is that of an undirected graph, that is, a set (the nodes of the graph) equipped with a binary relation (the edges of the graph). Such a binary relation is symmetric in an undirected graph, and often it is also irreflexive (for graphs without self-loops).

Some applications of graphs in mathematics involve *labeled* graphs, where nodes and edges may have additional attributes such as, in the case of computational biology, gene names, protein names, taxa names and confidence values, evolutionary distances, or bootstrap values.

Example 8.1

The following four graphs are identical as unlabeled graphs, but they are all different labeled graphs.

8.1.1 Counting Labeled Graphs

Determining the number of possible graphs is an important problem in mathematics and computer science, and it becomes even more important in computational biology, where it is essential to the uniform generation of random graphs and to the validation of graphs produced by various phylogenetic reconstruction methods. Here, as in the case of trees, *counting* refers to determining the number of possible graphs that have certain properties, while *generation* is the process of obtaining the actual graphs with these properties such as, for instance, all labeled graphs without self-loops.

The number of possible labeled graphs increases very rapidly with the number of nodes, and for 10 nodes there are already more than 35 trillion labeled graphs.

There is $2^{1 \cdot 0/2} = 2^0 = 1$ way to arrange one labeled node A to make a graph, as illustrated by the following single labeled graph:

There are $2^{2 \cdot 1/2} = 2^1 = 2$ ways to arrange two labeled nodes A, B to make a graph, as illustrated by the following two labeled graphs:

Three labeled nodes A, B, C can be arranged in $2^{3 \cdot 2/2} = 2^3 = 8$ ways to make a graph, as illustrated by the following eight labeled graphs:

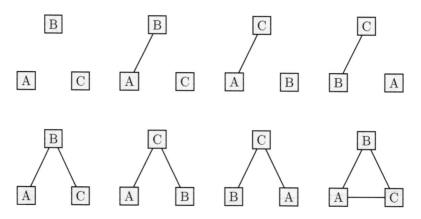

Similarly, four labeled nodes A, B, C, D can be arranged in $2^{4 \cdot 3/2} = 2^6 = 64$ ways to make a graph. In general, there are $2^{n(n-1)/2}$ different ways to arrange $n \geqslant 1$ labeled nodes to make a graph, because the edges of a graph with n nodes are a subset of the set of $\binom{n}{2} = n(n-1)/2$ pairs of nodes. The following R script computes the number of graphs with $1 \leqslant n \leqslant 9$ labeled nodes.

```
> t(sapply(1:9,function(n)c(n,2^choose(n,2))))
       [,1]         [,2]
[1,]    1            1
[2,]    2            2
[3,]    3            8
[4,]    4           64
[5,]    5         1024
[6,]    6        32768
[7,]    7      2097152
[8,]    8    268435456
[9,]    9  68719476736
```

8.2 Graphs in Computer Science

While the notion of graph most often found in discrete mathematics is that of an undirected graph, the notion of graph which is most useful in computer science is that of a directed graph, that is, a graph in which the edges are directed from a *source* node to a *target* node. In a directed acyclic graph, there is no path of directed edges starting and ending in the same node. In a rooted directed acyclic graph, there is a distinguished node, called the root of the graph, such that there is at least one directed path from the root to any node of the graph. The edges of the graph are directed away from the root.

Example 8.2
In the following rooted directed acyclic graph, there are paths from the root, A, to every node of the graph: A–B, A–B–C, A–D–C, A–D, A–B–E, A–B–C–F, A–D–C–F, and A–D–G.

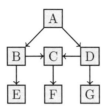

8.2.1 Traversing Directed Graphs

Most algorithms on graphs require a systematic method of visiting the nodes of a graph, and combinatorial pattern matching algorithms are no exception. The most common methods for exploring a graph are the *depth-first* and the *breadth-first* traversal.

In a depth-first traversal of a graph, also known as *depth-first search*, the nodes reachable by paths from a given initial node are visited before their adjacent nodes, and those non-visited nodes to which a visited node is adjacent are visited in left-to-right order.

In a breadth-first traversal of a graph, also known as *breadth-first search*, the nodes reachable by paths from a given initial node are visited in order of increasing distance from the initial node. The initial node is visited first, followed by the adjacent nodes in left-to-right order, then the non-visited nodes adjacent to them, also in left-to-right order, and so on.

Example 8.3
In a depth-first traversal of the following rooted directed acyclic graph, the nodes are visited in the order A, B, E, C, F, D, G. In a breadth-first traversal, the order in which the nodes are visited is A, B, D, E, C, G, F.

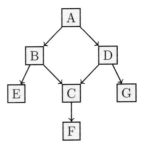

8.3 Graphs in Computational Biology

One of the notions of graph most often found in computational biology is that of a phylogenetic network, that is, an either unrooted or rooted directed acyclic graph whose terminal nodes are labeled by taxa names. However, directed graphs also arise as a mathematical model of metabolic pathways, and undirected graphs also arise as a mathematical model of RNA and protein structure, the regulatory interactions between genes, and the physical interactions between proteins.

A metabolic pathway is a complex network of biochemical reactions occurring within a cell. These biochemical reactions are activated or catalyzed by enzymes, and they often require additional cofactors to achieve their function of transforming substrate into product metabolites. The structure of a metabolic pathway can be represented as a directed graph of metabolites and biochemical reactions as nodes and directed edges from the substrate metabolites to the biochemical reaction, as well as from the biochemical reaction to the product metabolites. Such a graph is called *bipartite*, because the nodes can be partitioned in two subsets (metabolites and reactions) such that there are no edges between metabolites or between reactions. The edges are undirected if the biochemical reaction is reversible.

Example 8.4

The tricarboxylic acid cycle, also known as the TCA cycle, citric acic cycle, or Krebs cycle, is a cyclic metabolic pathway that oxidizes acetyl residues to carbon dioxide in a series of eight biochemical reactions. The following directed graph is an abstract representation of this metabolic pathway.

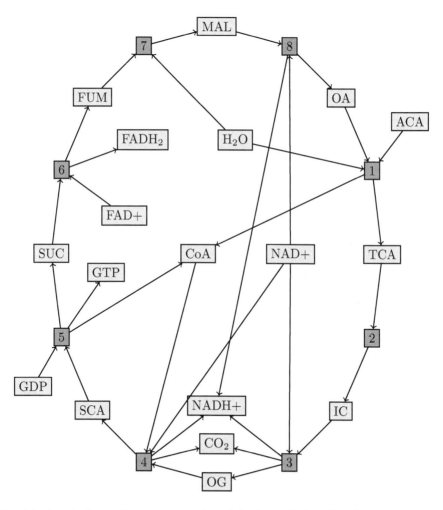

The biochemical reactions are numbered 1 through 8. The first reaction, activated by the enzyme *citrate synthase*, takes three substrates, OA (oxaloacetate), ACA (acetyl-CoA), and H_2O (water), and gives two products,

TCA (tricarboxylic acid) and CoA (coenzyme A). The second reaction, activated by the enzyme *aconitase hydratase*, takes a single substrate, TCA, and gives a single product, IC (isocitrate). The third reaction, activated by the enzyme *isocitrate*, takes two substrates, IC and NAD+ (nicotinamide adenine dinucleotide), and gives three products, OG (2-oxoglutarate), NADH+ (reduced nicotinamide adenine dinucleotide), and CO_2 (carbon dioxide). The fourth reaction, activated by the enzyme complex *2-oxoglutarate dehydrogenase*, takes three substrates, OG, NAD+, and CoA, and gives three products, SCA (succinyl coenzyme A), NADH+, and CO_2. The fifth reaction, activated by the enzyme *succinate-CoA ligase*, takes two substrates, SCA and GDP (guanosine diphosphate), and gives three products, SUC (succinate), GTP (guanosine triphosphate), and CoA. The sixth reaction, activated by the enzyme *succinate dehydrogenase*, takes two substrates, SUC and FAD+ (flavin adenine dinucleotide), and gives two products, FUM (fumarate) and $FADH_2$ (reduced flavin adenine dinucleotide). The seventh reaction, activated by the enzyme *fumarate hydratase*, takes two substrates, FUM and H_2O, and gives a single product, MAL (malate). Finally, the eighth biochemical reaction, activated by the enzyme *malate dehydrogenase*, takes two substrates, MAL and NAD+, and gives two products, OA and NADH+.

In a more abstract representation, a metabolic pathway can be seen as a directed graph of biochemical reactions as nodes and directed edges between reactions with some common metabolite as a product of the source reaction and a substrate of the target reaction.

Example 8.5
The following directed graph is a more abstract representation of the metabolic pathway of Example 8.4.

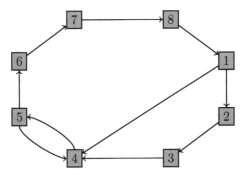

There is a directed edge from the first to the second biochemical reaction because they share the metabolite tricarboxylic acid as a product of the former and a substrate of the latter. There are also directed edges between the second and the third reactions (because they share the metabolite isocitrate), between the third and the fourth reactions (because they share the metabolite 2-oxoglutarate), between the fourth and the fifth reactions (because they share the metabolite succinyl coenzyme A), between the fifth and the sixth reactions (because they share the metabolite succinate), between the sixth and the seventh reactions (because they share the metabolite fumarate), between the seventh and the eighth reactions (because they share the metabolite malate), and between the eighth and the first reactions (because they share the metabolite oxaloacetate). There are also directed edges from the first and the fifth to the fourth biochemical reaction, because they share the metabolite coenzyme A as a product of the former ones and a substrate of the latter reaction.

A protein interaction network, on the other hand, is a complex network of associations between protein molecules. These associations are essential to various biological functions, including signal transduction (the process by which a signal or stimulus in a cell is converted into another, often involving a series of biochemical reactions inside the cell), the formation of protein complexes (a stable association of two or more proteins, in which one or more of the associated proteins are often activated or inhibited), transport (for instance, a protein carrying another protein between the cytoplasm and the nucleus of a cell), and the modification of proteins (such as the phosphorylation of a protein by a protein kinase). A protein interaction network can be represented as an undirected graph of proteins (or genes coding for proteins) as nodes and edges between interacting proteins.

Example 8.6
Protein interactions have been determined for several model organisms using a variety of experimental techniques, such as the high-throughput yeast two-hybrid method. The whole protein interaction network of the bacterium *Campylobacter jejuni* consists of 11,687 interactions among 1,654 proteins, and the chemotaxis signal transduction pathway (which is responsible for directional swimming in bacteria) of this organism has 23 interactions among 21 proteins. The following undirected graph is an abstract representation of the latter.

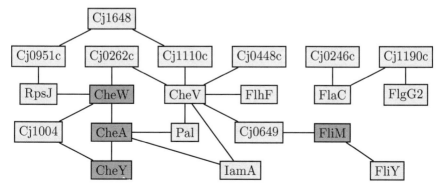

The nodes represent chemotaxis proteins (CheW, CheV, CheA, CheY); methyl-accepting chemotaxis proteins (Cj0951c, Cj0262c, Cj1110c, Cj0448c, Cj0246c, Cj1190c); flagellar proteins (FlhF, FlaC, FlgG2, FliM, FliY); and other proteins not related to motility (Cj0649, Cj1004, Cj1648, IamA, Pal, RpsJ). Nodes depicted in green (CheW, CheA, CheY, FliM) correspond to proteins in the canonical chemotaxis signal transduction pathway. The edges represent signal transduction interactions between proteins, such as complex formation between CheA and CheW, the phosphorylation of CheY by CheA, and the interaction of CheY and FliM to change the direction of flagellar rotation.

In the undirected graph representation of a protein interaction network, nodes stand for either proteins or genes coding for proteins, and there are undirected edges between interacting proteins or genes. A particular case of a protein interaction network is a transcription regulation network, which can be represented as a directed graph of transcription factors and genes as nodes and directed edges from transcription factors to the genes they transcribe.

Enough has been said about the representation of pathways of biochemical reactions and networks of interacting genes and proteins. The evolutionary relationships among a group of organisms are often illustrated by means of a phylogenetic tree, whose nodes represent taxonomic units (which can be species or taxa, higher or nested taxa, populations, individuals, or genes) and whose branches define the evolutionary relationships among the taxonomic units (where children nodes descend from their parents by mutation). However, there are evolutionary processes acting at the population level, such as recombination between genes, hybridization between lineages, and lateral gene transfer, that lead to reticulate relationships that can no longer be modeled by a phylogenetic tree. The modeling and explicit representation of reticulate evolutionary events turn phylogenetic trees into a particular form of directed acyclic graphs called phylogenetic networks.

A phylogenetic network is a directed acyclic graph whose terminal nodes are labeled by taxa names and whose internal nodes are either *tree nodes* (if they have only one parent) or *hybrid nodes* (if they have two or more parents). As in the case of phylogenetic trees, a phylogenetic network is *fully resolved* if every internal tree node in the network has two children and every hybrid node has two parents and a single child, which is often a tree node.

Example 8.7

The alcohol dehydrogenase enzyme is one of the most abundant proteins in *Drosophila melanogaster*, and it is encoded by a single gene. The alcohol dehydrogenase gene was studied by Kreitman (1983) on a sample of eleven species from five natural populations of *Drosophila melanogaster* (Af: Africa; Fl: Southern Florida; Fr: France; Ja: Japan; Wa: Seattle, Washington). The two most frequent forms of the alcohol dehydrogenase gene differ by a single nucleotide replacement, and they are denoted Slow (S) and Fast (F) because of their influence in the catalytic efficiency of the enzyme. The sampled sequences contain 44 polymorphic (segregating) sites, shown below, where the eleven cloned genes are numbered 1 (Wa-S), 2 (Fl-1S), 3 (Af-S), 4 (Fr-S), 5 (Fl-2S), 6 (Ja-S), 7 (Fl-F), 8 (Fr-F), 9 (Wa-F), 10 (Af-F), and 11 (Ja-F).

```
 1  CCGCAATAATGGCGCTACTCTCACAATAACCCACTAGACAGCCT
 2  CCCCAATATGGGCGCTACTTTCACAATAACCCACTAGACAGCCT
 3  CCGCAATATGGGCGCTACCCCCCGGAATCTCCACTAAACAGTCA
 4  CCGCAATATGGGCGCTGTCCCCCGGAATCTCCACTAAACTACCT
 5  CCGAGATAAGTCCGAGGTCCCCCGGAATCTCCACTAGCCAGCCT
 6  CCCCAATATGGGCGCGACCCCCCGGAATCTCTATTCACCAGCTT
 7  CCCCAATATGGGCGCGACCCCCCGGAATCTGTCTCCGCCAGCCT
 8  TGCAGATAAGTCGGCGACCCCCCGGAATCTGTCTCCGCGAGCCT
 9  TGCAGATAAGTCGGCGACCCCCCGGAATCTGTCTCCGCGAGCCT
10  TGCAGATAAGTCGGCGACCCCCCGGAATCTGTCTCCGCGAGCCT
11  TGCAGGGGAGGGCTCGACCCCACGGGATCTGTCTCCGCCAGCCT
```

Under the infinite sites assumption, by which mutations are rare enough to discard the possibility of more than one mutation to occur at the same site in a sample of sequences, no site of a sample can contain more than two different nucleotides. The most frequent nucleotide along a site is often taken as the base, with the least frequent nucleotide being taken as the mutant. The base and mutant nucleotides for each site of the previous sequences are as follows.

```
CCCCAATAAGGGCGCGACCCCCCGGAATCTCTATTCGCCAGCCT
TGGAGGGGTTTCGTATGTTTTAACAGTAACGCCCCAAAGTATTA
```

This allows for a binary representation of a sample of sequences, where the base nucleotide in each site is encoded as 0 and the mutant nucleotide is encoded as 1.

```
 1  00100000010000010010101110111101010101000000
 2  00000000100000010011101110111101010101000000
 3  00100000100000010000000000000001010111000101
 4  00100000100000011100000000000001010111011000
 5  00111000001100101100000000000001010100000000
 6  00000000100000000000000000000000010000010
 7  00000000100000000000000000000010101000000000
 8  11011000001110000000000000000010101000100000
 9  11011000001110000000000000000010101000100000
10  11011000001110000000000000000010101000100000
11  11011111000001000000010001000010101000000000
```

The evolutionary relationships among these eleven cloned genes cannot be modeled by a phylogenetic tree. In fact, any phylogenetic network with hybrid nodes representing recombination events explaining these evolutionary relationships must include at least 7 recombination events. One such possible explanation is the following fully resolved phylogenetic network, which has 11 leaves and exactly 7 hybrid nodes.

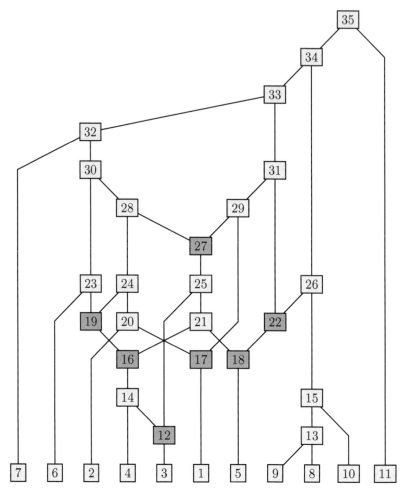

The edges are directed from top to bottom, and some of them represent mutation events. For instance, the edge from the node labeled 23 (which corresponds to the sequence 00000000100000000000000000000010101000000000) to the leaf node labeled 6 represents a single mutation of C to T at segregating site number 43, and the edge from the node labeled 21 (for the sequence 00100000100000011100000000000001010100000000) to the node labeled 16 (for the sequence 00100000100000011100000000000001010111000000) represents a mutation of G to A at segregating site number 37 and a mutation of C to A at segregating site number 38. The edges going into a hybrid node, on the other hand, represent recombination events. For instance, the edges from the nodes labeled 21 (which corresponds to the sequence 00100000100000011100000000000001010100000000) and 22 (which corresponds to the sequence 00111000001100000000000000000000010101000000000) to the node labeled 18 represent the single crossover recombination of a prefix of the latter (the first 16 segregating sites, sequence 0011100000110000)

with a suffix of the former (the last 28 segregating sites, that is, the sequence 1100000000000001010100000000) to give the 44 segregating sites at the node labeled 18 (sequence 00111000001100001100000000000001010100000000).

A phylogenetic network explaining the evolutionary relationships among a given set of taxonomic units can be very large indeed, as there is no upper bound on the number of hybrid nodes. There is, however, a lower bound on the number of recombination events needed to explain the evolutionary relationships among a set of organisms given their DNA or RNA sequences, and this lower bound is 7 for the phylogenetic network of Example 8.7.

While most phylogenetic reconstruction algorithms attempt to achieve the lower bound on the number of hybridization, recombination, or lateral gene transfer events, the resulting phylogenetic networks often lack topological properties that are essential to their further analysis. This is especially relevant to the comparative analysis of phylogenetic networks, which is the subject of the next chapter, and most distances and alignment algorithms impose some condition on phylogenetic network topology; for instance, that recombination or hybridization cycles be pairwise disjoint, that internal nodes have some child that is a tree node, or that hybrid nodes have some sibling that is a tree node.

A phylogenetic network is called *tree-sibling* if every hybrid node has at least one sibling that is a tree node. For instance, the phylogenetic network of Example 8.7 is not tree-sibling, because the hybrid nodes labeled 16 and 18 have each other as their only sibling. The biological meaning of the tree-sibling condition is that in each of the recombination or hybridization processes, at least one of the species involved in them also has some descendant through mutation.

A phylogenetic network is called *tree-child* if every internal node has at least one child that is a tree node. All tree-child networks are also tree-sibling; thus, the phylogenetic network of Example 8.7 is not tree-child, either. In fact, the nodes labeled 19, 21, 22, and 29 have all their children hybrid. The biological meaning of the tree-child condition is that every non-extant species has some descendant through mutation.

A recombination or hybridization cycle consists of two paths from some common ancestor to the two parents of a hybrid node, and a phylogenetic network is called a *galled-tree* if all recombination or hybridization cycles are pairwise disjoint. All galled-trees are also tree-child networks and, thus, the phylogenetic network of Example 8.7 is not a galled-tree, either. For instance, the recombination cycles that end up in the hybrid nodes labeled 18 and 19 are disjoint, but those ending up in the hybrid nodes labeled 17 and 27 share the hybrid node labeled 27 and the tree nodes labeled 28 and 29.

On the other hand, a phylogenetic network is *time-consistent* if there is a *temporal representation* of the network, that is, an assignment of times to the nodes of the network that strictly increases on tree edges (those edges whose head is a tree node) and remains the same on hybrid edges (whose head is a hybrid node). For instance, the phylogenetic network of Example 8.7 is not time-consistent, because the nodes labeled 18, 21, 22, 26, and 31 share the same time assignment (there are hybrid edges from the nodes labeled 26 and 31 to the node labeled 22, as well as from the nodes labeled 21 and 22 to the node labeled 18) and, thus, no time assignment to the nodes along the path from the node labeled 31 down to the node labeled 21 (that is, the nodes labeled 29, 27, and 25) that strictly increases on these tree edges is possible. The biological meaning of a temporal assignment is the time when certain species exist or when certain hybridization processes occur, because for these processes to take place, the species involved must coexist in time.

Example 8.8

The fully resolved phylogenetic network of Example 8.7 is not time-consistent. Let x be the time assigned to the node labeled 12. Since the time assignment must remain the same on hybrid edges, the nodes labeled 14 and 25 get assigned time x as well. Now, along the path between the parents of the former, the nodes labeled 21 and 16 must also have a time assignment strictly greater than x and, thus, the time assignment along the tree edge from the node labeled 16 to the node labeled 14 cannot be increasing.

In a similar way, let y be the time assigned to the node labeled 31. Since the time assignment must remain the same on hybrid edges, the nodes labeled 22 and 26, and also 18 and 21, get assigned time y as well. Now, the nodes labeled 29, 27, and 25 must have a time assignment strictly greater than y, so the time assignment along the tree edge from the node labeled 25 to the node labeled 21 cannot be increasing, either.

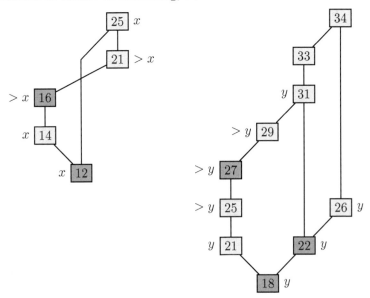

A phylogenetic network with the least possible number of hybrid nodes does not necessarily exhibit any of the topological properties of being tree-sibling, tree-child, a galled-tree, or time-consistent. However, under the hypothesis of more recombination events, the evolutionary history of the sample from Example 8.7 can be explained by a tree-sibling phylogenetic network.

Example 8.9

Another possible explanation of the evolutionary relationships among the eleven alcohol dehydrogenase genes from Example 8.7 is the following fully resolved tree-sibling phylogenetic network, which has 8 hybrid nodes.

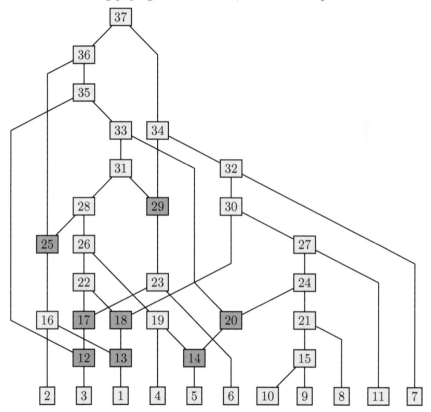

8.3.1   The eNewick Linear Representation

The eNewick format is an extension of the Newick format for representing phylogenetic trees, and it is quite convenient for representing phylogenetic networks since it makes it possible to describe a whole network in linear form in a unique way once the network is drawn or the ordering among parents and children nodes is fixed. The eNewick description of a network is a string of nested parentheses annotated with taxa names and possibly also with branch lengths or bootstrap values, with hybrid nodes appropriately tagged, as illustrated by the following simple example:

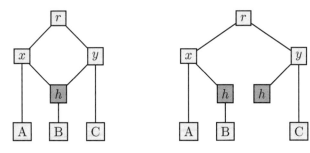

```
((A,(B)h#H1)x,(h#H1,C)y)r;
```

The eNewick description of a given network can be obtained by first splitting each hybrid node into as many copies as parents has the node, where the first such copy carries the children and the other copies have no children, and then obtaining the Newick description of the resulting tree. In this way, the leftmost occurrence of each hybrid node in an eNewick string corresponds to the full description of the network rooted at that node, and all labeled occurrences of a hybrid node in an eNewick string carry the same label.

A phylogenetic network can be recovered from an eNewick string by first recovering the tree and then identifying all copies of the same hybrid node, that is, identifying those nodes that are labeled as hybrid nodes and are tagged with the same identifier.

The reticulate evolutionary event represented by a hybrid node in a phylogenetic network can be a recombination between genes, a hybridization between lineages, or a lateral gene transfer. The unique representation of the latter as hybrid nodes requires encoding each gene transfer event as a hybrid edge. In the following example, a gene is transferred from the species corresponding to the node labeled B to the species corresponding to the node labeled C after the divergence of the species corresponding to the node labeled A from the species corresponding to the node labeled B. The eNewick string

```
((A,(B,(C)h#LGT1)y)x,h#LGT1)r;
```

describes such a phylogenetic network in a unique way.

Example 8.10

The representation of a lateral gene transfer event (left) as a hybrid edge in a phylogenetic network (right) is shown below.

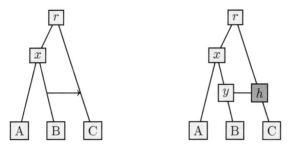

8.3.2 Counting Phylogenetic Networks

The number of possible phylogenetic networks increases much more rapidly with the number of terminal nodes than the number of possible phylogenetic trees. In fact, without any topological constraint such as being tree-sibling, tree-child, a galled-tree, or time-consistent, there is no upper bound on the size of a phylogenetic network with a given number of terminal nodes and, thus, there is no upper bound on the number of possible phylogenetic networks with a given number of terminal nodes, either.

Even with the constraint of having only one hybridization or recombination cycle, the number of possible fully resolved *unicyclic* phylogenetic networks increases more rapidly with the number of terminal nodes than the number of possible fully resolved phylogenetic trees, and for 10 terminal nodes there are already more than 58 million such unrooted phylogenetic networks: undirected acyclic graphs with the terminal nodes labeled by taxa names. Having exactly one cycle, these networks cannot have intersecting recombination or hybridization cycles and, thus, they are galled-trees.

There is $(2 \cdot 3 - 3 - 1)!/((3 - 3)!2^{3-3+1}) = 2!/(2^1) = 1$ way to connect three labeled nodes A, B, C to make an unrooted phylogenetic network with a single cycle of length three, as illustrated by the following single fully resolved phylogenetic network:

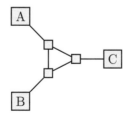

Four labeled nodes A, B, C, D can be connected in $(2 \cdot 4 - 3 - 1)!/((4 - 3)!2^{4-3+1}) = 4!/2^2 = 6$ different ways to make an unrooted phylogenetic network with a single cycle of length three, as illustrated by the following six fully resolved phylogenetic networks:

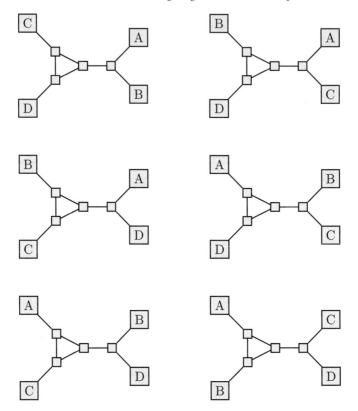

Four labeled nodes A, B, C, D can also be connected in $(2 \cdot 4 - 4 - 1)!/((4 - 4)!2^{4-4+1}) = 3!/2 = 3$ different ways to make an unrooted phylogenetic network with a single cycle of length four, as illustrated by the following three fully resolved phylogenetic networks:

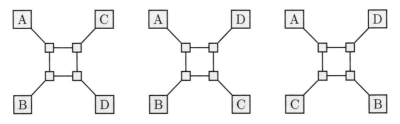

In general, the number $C(n, k)$ of fully resolved unrooted unicyclic phylogenetic networks with $n \geqslant 3$ terminal nodes and whose unique cycle has length $k \geqslant 3$ is $C(n, k) = (2n - k - 1)!/((n - k)!2^{n-k+1})$, and the number $C(n)$ of fully resolved unrooted unicyclic phylogenetic networks with $n \geqslant 3$ terminal nodes is $C(n) = (n-1)!2^{n-2} - (2n-2)!/((n-1)!2^{n-1})$. The following R script computes the number $C(n)$ of fully resolved unrooted unicyclic phylogenetic networks with $3 \leqslant n \leqslant 12$ terminal nodes.

```
> c <- function(n)(factorial(n-1)*2^(n-2)-factorial
  (2*n-2)/(factorial(n-1)*2^(n-1)))
> cbind(3:12,sapply(3:12,c))
        [,1]          [,2]
 [1,]    3             1
 [2,]    4             9
 [3,]    5            87
 [4,]    6           975
 [5,]    7         12645
 [6,]    8        187425
 [7,]    9       3133935
 [8,]   10      58437855
 [9,]   11    1203216525
[10,]   12   27125492625
```

Galled-trees with three labeled nodes are unicyclic phylogenetic networks, and there is $(2(3-2)-3+3\cdot1)!(3-2\cdot1-1)!2^{3-(3-2)-3\cdot1}/((3-2-3+2\cdot1)!(3-3\cdot1)!(1-1)!1!) = (2!0!2^{-1}/(0!0!0!1!)) = 1$ way to connect three labeled nodes to make an unrooted galled-tree. Four labeled nodes A,B,C,D can be connected in $(2(4-2)-3+3\cdot1)!(3-2\cdot1-1)!2^{3-(4-2)-3\cdot1}/((4-2-3+2\cdot1)!(3-3\cdot1)!(1-1)!1!) = 4!0!2^{-2}/(1!0!0!1!) = 6$ different ways to make an unrooted galled-tree with a single cycle of length three; they can be connected in $(2(4-2)-4+3\cdot1)!(4-2\cdot1-1)!2^{4-(4-2)-3\cdot1}/((4-2-4+2\cdot1)!(4-3\cdot1)!(1-1)!1!) = 3!1!2^{-1}/(0!1!0!1!) = 3$ different ways to make an unrooted galled-tree with a single cycle of length four; and they can be connected in $(2(4-2)-6+3\cdot2)!(6-2\cdot2-1)!2^{6-(4-2)-3\cdot2}/((4-2-6+2\cdot2)!(6-3\cdot2)!(2-1)!2!) = 4!1!2^{-2}/(0!0!1!2!) = 3$ different ways to make an unrooted galled-tree with two cycles of total length six, as illustrated by the following three galled-trees:

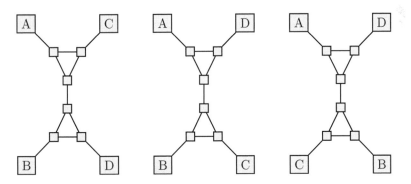

In general, the number $G(n,k,m)$ of fully resolved galled-trees with $n \geqslant 3$ terminal nodes, k cycles, and m edges across all the cycles is $G(n,k,m) = (2(n-2)-m+3k)!(m-2k-1)!2^{m-(n-2)-3k}/((n-2-m+2k)!(m-3k)!(k-1)!k!)$. The number $G(n)$ of fully resolved galled-trees with $n \geqslant 3$ terminal nodes is thus $G(n) = \sum_{k=1}^{n-2} \sum_{m=3k}^{n-2-2k} G(n,k,m)$. The following R script computes the number $G(n)$ of fully resolved galled-trees with $n \geqslant 3$ terminal nodes, k cycles, and m edges across all the cycles.

```
> gkm <- function(n,k,m)factorial(2*(n-2)-m+3*k)*
    factorial(m-2*k-1)*2^(m-(n-2)-3*k)/(factorial(n-2-
    m+2*k)*factorial(m-3*k)*factorial(k-1)*factorial(k
    ))
> g <- function(n)lapply(1:(n-2),function(k)lapply
    ((3*k):(n-2+2*k),function(m)gkm(n,k,m)))
> options("digits"=8)
> cbind(3:12,lapply(3:12,function(n)sum(unlist(g(n)))
    ))
        [,1] [,2]
 [1,]  3    1
 [2,]  4    12
 [3,]  5    177
 [4,]  6    3345
 [5,]  7    78795
 [6,]  8    2242485
 [7,]  9    75091905
 [8,]  10   2896454295
 [9,]  11   126536043375
[10,]  12   6176725787925
```

8.3.3 Generating Phylogenetic Networks

All phylogenetic networks cannot be generated without imposing any topological constraint such as being tree-sibling, tree-child, a galled-tree, or time-consistent, because, otherwise, there is no upper bound on network size and the number of possible phylogenetic networks also grows unbounded. However, under some of these constraints, a simple algorithm for their generation consists of taking each of the phylogenetic trees on the desired terminal nodes and adding one hybrid node to each of them in each possible way in turn, repeating this process until the produced networks do not satisfy the topological constraints. The procedure for adding a new hybrid node consists of splitting two edges with a new internal node along each of them and then adding a new edge from one of the new nodes to the other. The latter becomes the new hybrid node.

The same phylogenetic network may be obtained more than once with this algorithm, and many of the networks obtained with this algorithm may be discarded because of not being acyclic, fully resolved, tree-sibling, tree-child, galled-trees, or time-consistent. In any case, all tree-child phylogenetic networks and all tree-child, time-consistent phylogenetic networks with a given number of terminal nodes can be obtained using this algorithm.

All the fully resolved tree-child phylogenetic networks on $n \geqslant 2$ terminal nodes can be generated by taking each of the fully resolved rooted phylogenetic trees on n terminal nodes in turn and then, adding one hybrid node to each of them in each possible way in turn, repeating this process until the result is not a fully resolved tree-child phylogenetic network.

Example 8.11

The three fully resolved tree-child phylogenetic networks on two terminal nodes can be generated by taking the only fully resolved rooted phylogenetic tree topology on two terminal nodes,

and either adding an edge from a new tree node along the edge from the root to the node labeled A, to a new hybrid node along the edge from the root to the node labeled B, or vice versa, with the new hybrid node along the edge from the root to the node labeled A:

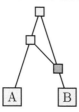 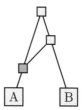

Each of these two tree-child phylogenetic networks with one hybrid node has 5 edges and, thus, there are $2 \cdot 5 \cdot (5-1) = 2 \cdot 20 = 40$ further possibilities for obtaining a phylogenetic network with two hybrid nodes, none of which is a fully resolved tree-child phylogenetic network.

In the same way, the six fully resolved time-consistent tree-child phylogenetic networks on 3 terminal nodes can be generated by taking each of the three fully resolved rooted phylogenetic tree topologies on three terminal nodes,

  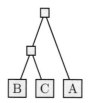

which have 4 branches, and adding a new hybrid node in each of the $3 \cdot 4 \cdot (4-1) = 3 \cdot 12 = 36$ possible ways. Only three of the resulting networks are fully resolved time-consistent tree-child phylogenetic networks, and each of them is obtained in two different ways:

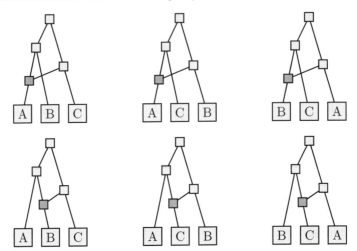

Each of the three fully resolved time-consistent tree-child phylogenetic networks with one hybrid node has 7 edges and, thus, there are $3 \cdot 7 \cdot (7 - 1) = 3 \cdot 42 = 126$ further possibilities for obtaining a phylogenetic network with two hybrid nodes, none of which is a fully resolved time-consistent tree-child phylogenetic network.

Such an algorithm for generating fully resolved rooted phylogenetic networks is implemented in the BioPerl modules for phylogenetic networks. The following Perl script uses them to generate the 66 fully resolved tree-child phylogenetic networks on 3 terminal nodes.

```perl
use Bio::PhyloNetwork;
use Bio::PhyloNetwork::Factory;

my $factory=Bio::PhyloNetwork::Factory->new(
  -leaves=>[qw(A B C)]
);

while (my $net=$factory->next_network()) {
  print $net->eNewick()."\n";
}
```

Running the previous Perl script produces the following output, which consists of the 66 fully resolved tree-child phylogenetic trees on three terminal nodes labeled A, B, C, in eNewick format:

```
((A,C)t3,B)t0;
((B)#H1,(#H1,(C,A)t3)T1)t0;
((((C)#H2,B)T2)#H1,((#H2,A)t3,#H1)T1)t0;
((B)#H1,(#H1,((C)#H2,(#H2,A)T2)t3)T1)t0;
((B)#H1,((#H1,(A,(C)#H2)t3)T1,#H2)T2)t0;
```

```
((((A)#H2,B)T2)#H1,((C,#H2)t3,#H1)T1)t0;
((B)#H1,(#H1,((A)#H2,(#H2,C)T2)t3)T1)t0;
((((B)#H1,(C,(A)#H2)t3)T1,#H2)T2,#H1)t0;
((A,((B)#H1,C)T1)t3,#H1)t0;
((((A)#H2,B)T2)#H1,((C,#H1)T1,#H2)t3)t0;
((B)#H1,((A)#H2,((#H1,C)T1,#H2)t3)T2)t0;
((B)#H1,((A,((#H1,C)T1)#H2)t3,#H2)T2)t0;
((C,((B)#H1,A)T1)t3,#H1)t0;
((((C)#H2,B)T2)#H1,((A,#H1)T1,#H2)t3)t0;
((B)#H1,((C)#H2,((#H1,A)T1,#H2)t3)T2)t0;
((B)#H1,((C,((#H1,A)T1)#H2)t3,#H2)T2)t0;
(((C,A)t3)#H1,(#H1,B)T1)t0;
((((C)#H2,(#H2,A)t3)T2)#H1,(B,#H1)T1)t0;
((((C)#H2,A)t3)#H1,((#H2,B)T2,#H1)T1)t0;
(((A,(C)#H2)t3)#H1,((#H1,B)T1,#H2)T2)t0;
((((A)#H2,(C,#H2)t3)T2)#H1,(B,#H1)T1)t0;
(((C,(A)#H2)t3)#H1,((#H2,B)T2,#H1)T1)t0;
(((((C,(A)#H2)t3)#H1,B)T1,#H2)T2,#H1)t0;
((A,(C)#H1)t3,(#H1,B)T1)t0;
(B,((C)#H1,(A,#H1)t3)T1)t0;
((C,(A)#H1)t3,(#H1,B)T1)t0;
(B,((A)#H1,(C,#H1)t3)T1)t0;
((B,C)t3,A)t0;
(((B,C)t3)#H1,(#H1,A)T1)t0;
((((B)#H2,(#H2,C)t3)T2)#H1,(A,#H1)T1)t0;
((((B)#H2,C)t3)#H1,(#H1,(#H2,A)T2)T1)t0;
(((C,(B)#H2)t3)#H1,((A,#H1)T1,#H2)T2)t0;
((((C)#H2,(#H2,B)T2)t3)#H1,(A,#H1)T1)t0;
((A)#H1,(#H1,(B,C)t3)T1)t0;
((((B)#H2,A)T2)#H1,((#H2,C)t3,#H1)T1)t0;
((A)#H1,(((B)#H2,(#H2,C)T2)t3,#H1)T1)t0;
((A)#H1,(((C,(B)#H2)t3,#H1)T1,#H2)T2)t0;
((A)#H1,(((C)#H2,(#H2,B)T2)t3,#H1)T1)t0;
((((A)#H1,(B,(C)#H2)t3)T1,#H2)T2,#H1)t0;
((C,((A)#H1,B)T1)t3,#H1)t0;
((A)#H1,((C)#H2,((#H1,B)T1,#H2)t3)T2)t0;
((A)#H1,((C,((#H1,B)T1)#H2)t3,#H2)T2)t0;
((B,((A)#H1,C)T1)t3,#H1)t0;
((((B)#H2,A)T2)#H1,(#H2,(C,#H1)T1)t3)t0;
((A)#H1,((B)#H2,((C,#H1)T1,#H2)t3)T2)t0;
((((C,(A)#H1)T1)#H2,(B,#H2)T2)t3,#H1)t0;
(A,((B)#H1,(C,#H1)t3)T1)t0;
((C,(B)#H1)t3,(#H1,A)T1)t0;
(A,((C)#H1,(B,#H1)t3)T1)t0;
((A,B)t0,C)t3;
```

```
(((B,A)t0)#H1,(#H1,C)T1)t3;
((((B)#H2,(#H2,A)t0)T2)#H1,(C,#H1)T1)t3;
((((A)#H2,(B,#H2)t0)T2)#H1,(C,#H1)T1)t3;
(C,((A)#H1,(B,#H1)t0)T1)t3;
((C)#H2,((((#H2,A)T2)#H1,B)t0,#H1)T1)t3;
((((A)#H1,((C)#H2,B)T2)t0,#H1)T1,#H2)t3;
((C)#H2,((A)#H1,(#H2,(#H1,B)t0)T2)T1)t3;
((C)#H2,(((B,(A)#H1)t0,#H1)T1,#H2)T2)t3;
((C)#H1,(#H1,(B,A)t0)T1)t3;
((C)#H1,(#H1,((B)#H2,(#H2,A)T2)t0)T1)t3;
((C)#H1,((#H1,(A,(B)#H2)t0)T1,#H2)T2)t3;
((B,((C)#H1,A)T1)t0,#H1)t3;
((C)#H1,((B)#H2,((#H1,A)T1,#H2)t0)T2)t3;
((A,((C)#H1,B)T1)t0,#H1)t3;
((C)#H1,((A,((#H1,B)T1)#H2)t0,#H2)T2)t3;
(C,((B)#H1,(A,#H1)t0)T1)t3;
```

8.3.4 Representing Graphs in Perl

There are many ways in which graphs can be represented in Perl and, as a matter of fact, many different Perl modules implementing various types of graphs are available for download from CPAN, the Comprehensive Perl Archive Network, at `http://www.cpan.org/`. Among them, let us focus on the BioPerl phylogenetic network representation, which is essentially an object-oriented representation of directed graphs that relies on the collection `Graph::Directed` of Perl modules for representing and manipulating graphs.

A phylogenetic network is represented in BioPerl as a `Bio::PhyloNetwork` object, whose nodes are represented as an array reference and whose edges are represented as an array of array references. The nodes can be accessed by means of such methods as `vertices` and `has_vertex`, and the edges can be accessed by means of the `edges` and `has_edge` methods.

For instance, the phylogenetic network with eNewick string `((A,(B)h#H1)x, (h#H1,C)y)r;` can be obtained by first creating an array of node labels, then creating an array of source and target node labels for the edges, creating a `Graph::Directed` object with the array of node labels as nodes and the array of node label arrays as edges, and, finally, creating a `Bio::PhyloNetwork` object with the latter as the underlying graph. This is all shown in the following Perl script.

```perl
use Bio::PhyloNetwork;

my $g = Graph::Directed->new(
  vertices=>[qw(A B C r x y h)],
  edges => [
    [qw(r x)], [qw(r y)], [qw(x h)], [qw(y h)],
```

```
    [qw(x A)], [qw(h B)], [qw(y C)]
  ]
);
```

```
my $net = Bio::PhyloNetwork->new(-graph => $g);
```

The same phylogenetic network can be obtained by adding the nodes and edges to an empty graph, as illustrated by the following Perl script.

```
use Bio::PhyloNetwork;

my $g = Graph::Directed->new;
$g->add_vertices(qw(A B C r x y h));
$g->add_edges(
  qw(r x), qw(r y), qw(x h), qw(y h),
  qw(x A), qw(h B), qw(y C)
);

my $net = Bio::PhyloNetwork->new(-graph => $g);
```

Since phylogenetic networks are connected graphs, though, the nodes do not need to be given explicitly and it suffices to give the edges, as shown in the following Perl script.

```
use Bio::PhyloNetwork;

my $g = Graph::Directed->new;
$g->add_edges(
  qw(r x), qw(r y), qw(x h), qw(y h),
  qw(x A), qw(h B), qw(y C)
);

my $net = Bio::PhyloNetwork->new(-graph => $g);
```

Moreover, the edges can be specified as an array of nodes, where the even elements are source nodes and the odd elements are the corresponding target nodes. This is shown in the following Perl script.

```
use Bio::PhyloNetwork;

my $g = Graph::Directed->new;
$g->add_edges(qw(r x r y x h y h x A h B y C));

my $net = Bio::PhyloNetwork->new(-graph => $g);
```

On the other hand, the phylogenetic network can be obtained straight from the array of source and target node of the edges, as illustrated by the following Perl script.

```
use Bio::PhyloNetwork;

my $net = Bio::PhyloNetwork->new(-edges => [
  qw(r x x A x h h B r y y h y C)
]);
```

Furthermore, a `Bio::PhyloNetwork` object can be also obtained from an eNewick string, as shown in the following Perl script.

```
my $net = Bio::PhyloNetwork->new(
  -eNewick => '((A,(B)h#H1)x,(h#H1,C)y)r;'
);
```

The representation of phylogenetic networks in BioPerl includes additional methods for performing various operations on networks and their nodes and edges; for instance, to access all the roots of a phylogenetic network,

```
my @roots = $net->roots;
my $rooted = scalar @roots == 1;
```

to access the terminal nodes of a phylogenetic network,

```
my @taxa = $net->leaves;
```

to access the tree and hybrid nodes of a phylogenetic network,

```
my @tree_nodes = $net->tree_nodes;
my @hybrid_nodes = $net->hybrid_nodes;
```

to access the tree and hybrid edges of a phylogenetic network,

```
my @tree_edges = $net->tree_edges;
my @hybrid_edges = $net->edges;
```

to obtain the directed acyclic graph underlying a phylogenetic network,

```
my $graph = $net->graph;
```

to test for time consistency and obtain a temporal representation of a time-consistent phylogenetic network,

```
if ( $net->is_time_consistent ) {
  my %time = $net->temporal_representation;
}
```

to test if a phylogenetic network is tree-child,

```
my $tree_child = $net->is_tree_child;
```

and to obtain all the phylogenetic trees contained in a phylogenetic network,

```
my @trees = $net->explode;
```

Phylogenetic networks can be displayed using BioPerl in a variety of ways, such as in eNewick format,

```
my $str = $net->eNewick;
```

and drawn as a layered directed acyclic graph, with tree nodes depicted as circles and hybrid nodes as rectangles, in several graphic formats, including Encapsulated PostScript (EPS), Graphics Interchange Format (GIF), Joint Photographic Experts Group (JPEG), and Scalable Vector Graphics (SVG).

```
use Bio::PhyloNetwork::GraphViz;

my $g = Bio::PhyloNetwork::GraphViz->new(
  -net => $net)
;

$g->as_ps("net.eps");
$g->as_gif("net.gif");
$g->as_jpeg("net.jpg");
$g->as_svg("net.svg");
```

8.3.5   Representing Graphs in R

There are also many ways in which graphs can be represented in R and, as a matter of fact, many different R contributed packages implementing various types of graphs are available for download from CRAN, the Comprehensive R Archive Network, at http://cran.r-project.org/. Among them, let us focus on the iGraph representation, which is essentially a vector-based representation of undirected and directed graphs and, in particular, phylogenetic networks.

A phylogenetic network in represented in the R package iGraph as a list of class `igraph` consisting of nine elements, including: the number of nodes in the network, a Boolean indicating whether the graph is directed or undirected and set to `TRUE` because phylogenetic networks are directed graphs, a numeric vector with the source node of each edge, and another numeric vector with the target node of each edge.

For instance, the phylogenetic network with eNewick string `((A,(B)h#H1)x, (h#H1,C)y)r;` has a root numbered 0 and labeled r; two internal tree nodes numbered 1, 2 and labeled x, y; one hybrid node numbered 3 and labeled h; three terminal nodes numbered 4, 5, 6 and labeled A, B, C; and edges 0–1, 0–2, 1–3, 1–4, 2–3, 2–6, 3–5. The following R script computes the representation of such a phylogenetic network, where the edges are specified as a vector of nodes, with the even elements standing for source nodes and the odd elements standing for the corresponding target nodes.

```
> library(igraph)
> net <- graph(c(0,1,0,2,1,4,1,3,2,3,2,6,3,5),n=7)
> net
Vertices: 7
```

```
Edges: 7
Directed: TRUE
Edges:

[0] 0 -> 1
[1] 0 -> 2
[2] 1 -> 4
[3] 1 -> 3
[4] 2 -> 3
[5] 2 -> 6
[6] 3 -> 5
```

Nodes are referred to by their numbers, although they can still have a label assigned to them.

```
> V(net)
Vertex sequence:
[1] 0 1 2 3 4 5 6
> V(net)$name <- c("r","x","y","h","A","B","C")
> net
Vertices: 7
Edges: 7
Directed: TRUE
Edges:

[0] r -> x
[1] r -> y
[2] x -> A
[3] x -> h
[4] y -> h
[5] y -> C
[6] h -> B
```

Edges make reference to their source and target nodes by label instead of number, though, when the nodes are labeled.

```
> V(net)
Vertex sequence:
[1] "r" "x" "y" "h" "A" "B" "C"
> E(net)
Edge sequence:

[0] r -> x
[1] r -> y
[2] x -> A
[3] x -> h
[4] y -> h
```

```
[5]  y -> C
[6]  h -> B
```

The number of nodes and edges of a phylogenetic network are also readily available.

```
> vcount(net)
[1] 7
> ecount(net)
[1] 7
```

Based on this vector representation, is is rather easy to code various operations on phylogenetic networks and their nodes and edges; for instance, to obtain the root node or nodes,

```
> V(net)[which(degree(net,mode="in")==0)-1]
Vertex sequence:
[1] "r"
```

to obtain the terminal nodes,

```
> V(net)[which(degree(net,mode="out")==0)-1]
Vertex sequence:
[1] "A" "B" "C"
```

to access the tree nodes and the hybrid nodes of the phylogenetic network,

```
> V(net)[which(degree(net,mode="in")<=1)-1]
Vertex sequence:
[1] "r" "x" "y" "A" "B" "C"
> V(net)[which(degree(net,mode="in")>1)-1]
Vertex sequence:
[1] "h"
```

to obtain the parent or parents of a node in the phylogenetic network,

```
> V(net)[nei(which(V(net)$name=="x")-1,"in")]
Vertex sequence:
[1] "r"
> V(net)[nei(which(V(net)$name=="h")-1,"in")]
Vertex sequence:
[1] "x" "y"
```

and to obtain the children of a node in the phylogenetic network,

```
> V(net)[nei(which(V(net)$name=="r")-1,"out")]
Vertex sequence:
[1] "x" "y"
> V(net)[nei(which(V(net)$name=="x")-1,"out")]
Vertex sequence:
[1] "h" "A"
```

Phylogenetic networks can also be displayed using R in a variety of ways, such as in edge list format,

```
> write.graph(net,file="net.txt",format="edgelist")
```

and drawn as a layered directed acyclic graph, with tree nodes depicted as circles and hybrid nodes as rectangles, in Adobe Portable Document Format (PDF),

```
> pdf(file="net.pdf")
> net$layout <- layout.reingold.tilford(net)
> V(net)$shape <- "circle"
> V(net)[which(degree(net,mode="in")>1)-1]$shape <- "
    square"
> plot.igraph(net,layout=cbind(net$layout[,1],-net$
    layout[,2]),vertex.shape=V(net)$shape,vertex.label
    =V(net)$name)
> dev.off()
```

among several other display options for undirected and directed graphs and, in particular, phylogenetic networks.

Bibliographic Notes

The number of possible, not necessarily connected, labeled graphs is given in (Sloane and Plouffe 1995, p. 19).

The depth-first traversal of undirected and directed graphs was first described in (Tarjan 1972). Breadth-first graph traversal was first described in (Lee 1961; Moore 1959). See also (Valiente 2002, ch. 5).

Graph-based models of metabolic pathways are reviewed in detail by Deville et al. (2003). The metabolic pathway of the tricarboxylic acid cycle is adapted from (Koolman and Roehm 2005, 136–139). See also (Nelson and Cox 2008, ch. 16).

The representation of protein interaction networks as undirected graphs underlies computational approaches to the prediction of protein function from protein interaction data (Pandey et al. 2008). The protein interaction network of *Campylobacter jejuni* was determined by Parrish et al. (2007).

The use of directed acyclic graphs to model reticulate evolution was first proposed in (Strimmer and Moulton 2000; Strimmer et al. 2001). Tree-sibling phylogenetic networks were introduced by Moret et al. (2004) and further studied by Cardona et al. (2008a;d). Tree-child phylogenetic networks were introduced by Cardona et al. (2009c). Galled-trees were introduced by Wang et al. (2001) and further studied by Gusfield et al. (2004a;b). The temporal representation of a phylogenetic network was first studied in (Baroni et al.

2006; Maddison 1997). Tree-sibling, tree-child, and galled-tree phylogenetic networks were further studied by Arenas et al. (2008) using computer simulation. The fully resolved phylogenetic networks explaining the evolutionary relationships among the eleven alcohol dehydrogenase genes of *Drosophila melanogaster* were reconstructed using the `beagle` tool (Lyngsø et al. 2005).

The eNewick linear representation for phylogenetic networks was first proposed in (Morin and Moret 2006) and further described by Cardona et al. (2008b;c). Further linear representations of graphs in computational biology include the notation for protein molecular structures developed by Levitt and Lifson (1969) and the SMILES (Simplified Molecular Input Line Entry System) notation for molecular structures of Weininger (1988).

The number of possible unicyclic phylogenetic networks and galled-trees as a function of the number of terminal nodes was determined by Semple and Steel (2006).

The representation of phylogenetic networks in BioPerl is described in more detail in (Cardona et al. 2008b), where the simple algorithm for generating fully resolved tree-child phylogenetic networks and the interface to the GraphViz toolset (`http://www.graphviz.org/`) are also discussed. The representation of undirected and directed graphs in the R package iGraph is described in (Csárdi and Nepusz 2006).

Chapter 9

Simple Pattern Matching in Graphs

Combinatorial pattern matching is the search for exact or approximate occurrences of a given pattern within a given text. When it comes to graphs in computational biology, both the pattern and the text are graphs and the pattern matching problem becomes one of finding the occurrences of a graph within another graph. For instance, scanning a metabolic pathway for the presence of a known pattern can help in finding conserved network motifs, and finding a phylogenetic network within another phylogenetic network can help in assessing their similarities and differences. This will be the subject of the next chapter.

A related pattern matching problem that arises in the analysis of graphs consists in finding simpler patterns, that is, paths and trees within a given graph. For instance, finding paths between two nodes of a graph is useful for computing distances in a graph and also for computing distances between two graphs, and finding the trees contained in a graph is also useful for computing distances between two graphs. This is the subject of this chapter.

9.1 Finding Paths in Graphs

Any two nodes may be connected by more than one path in a graph, even if no edge is to be traversed more than once in a path between the two nodes of the graph.

Example 9.1

In the following fully resolved phylogenetic network, there are two paths from the root to the terminal node labeled B, the paths r–x–h–B and r–y–h–B.

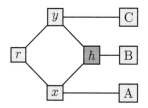

The *height* of a node in a phylogenetic network is the length of a longest path from the node to a terminal node, that is, the largest of the paths in the network from the node to the terminal nodes.

The paths from a given node to each of the terminal nodes of a phylogenetic network are the paths from the children of the node to the terminal nodes. Thus, the number of paths from the root to the terminal nodes of a phylogenetic network can be computed by order of increasing height of the nodes in the network, adding up the number of paths from the children to each of the terminal nodes. From a terminal node, there is one (trivial) path to itself and no path to any of the other terminal nodes.

Example 9.2
In the phylogenetic network of the previous example, the terminal nodes are at height 0, the hybrid node is at height 1, the internal tree nodes are at height 2, and the root is at height 3. There is one path from the root to each of the terminal nodes labeled A and C in this network, and there are two paths from the root to the terminal node labeled B, and this is all represented by the vector $(1, 2, 1)$ of path multiplicities.

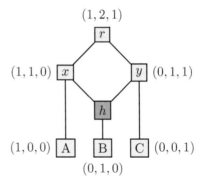

The children of the root have $(1, 1, 0)$ and $(0, 1, 1)$ as path multiplicity vectors, and $(1, 2, 1) = (1, 1, 0) + (0, 1, 1)$. The path multiplicity vector of the hybrid node is $(0, 1, 0)$, the same as the terminal node labeled B.

The height of the nodes in a phylogenetic network cannot be obtained by performing a depth-first or a breadth-first traversal of the network, though, and a *bottom-up traversal* is needed, from the terminal nodes up to the root of the network.

A bottom-up traversal of a phylogenetic network can be performed with the help of a queue of nodes, which holds the nodes waiting to be traversed. Initially, the queue contains the terminal nodes of the network, which have height 0. Every time a node is dequeued and visited, the heights of its parents are updated (by keeping the largest of their current height value and the height of the dequeued node plus 1), and the parents are enqueued as soon as all their children have been visited. The procedure finishes when the queue has been emptied and no nodes remain to be enqueued.

In the following description, the height $h(v)$ of each node v in a phylogenetic network N is computed during a bottom-up traversal of N, with the help of an (initially empty) queue Q of nodes.

---

function height(N)
    for all nodes v of N **do**
        $h(v) \leftarrow 0$
        if v is a terminal node **then**
            $enqueue(Q, v)$
    while Q is not empty **do**
        $v \leftarrow dequeue(Q)$
        mark node v as visited
        for all parents u of node v **do**
            $h(u) \leftarrow \max(h(u), h(v) + 1)$
            if all children of u are marked visited **then**
                $enqueue(Q, u)$
    return h

Example 9.3

The following table illustrates the step-by-step computation of node heights in the phylogenetic network of the previous example.

node	A	B	C	h	x	y	r
height after initial loop	0	0	0	0	0	0	0
height after dequeue of terminal node labeled 1	0	0	0	0	1	0	0
height after dequeue of terminal node labeled 2	0	0	0	1	1	0	0
height after dequeue of terminal node labeled 3	0	0	0	1	1	1	0
height after dequeue of hybrid node h	0	0	0	1	2	2	0
height after dequeue of internal tree node x	0	0	0	1	2	2	3
height after dequeue of internal tree node y	0	0	0	1	2	2	3
height after dequeue of root r	0	0	0	1	2	2	3

With the representation of phylogenetic networks in BioPerl, the height of the nodes is computed when constructing the `Bio::PhyloNetwork` object, and it is readily available by means of the `heights` method, as shown in the following Perl script.

```
use Bio::PhyloNetwork;

my $net = Bio::PhyloNetwork->new(
  -eNewick => '((A,(B)h#H1)x,(h#H1,C)y)r;'
);
```

```
my %h = $net->heights;
```

The representation of phylogenetic networks in R, on the other hand, does not include any method to compute the height of the nodes in a phylogenetic network. However, a `network.height` function can easily be defined using a vector to represent the queue of nodes waiting to be traversed, as illustrated by the following R script.

```
> library(igraph)
> net <- graph(c(0,1,0,2,1,4,1,3,2,3,2,6,3,5),n=7)
> V(net)$name <- c("r","x","y","h","A","B","C")

> network.height <- function (net) {
  V(net)$height <- 0
  V(net)$visited <- FALSE
  Q <- as.vector(V(net)[which(degree(net,mode="out")
     ==0)-1])
  while (length(Q)>0) {
     j <- Q[1]
     Q <- Q[-1]
     V(net)[j]$visited <- TRUE
     I <- as.vector(V(net)[nei(j,mode="in")])
     for (i in I) {
        V(net)[i]$height <- max(V(net)[i]$height,V(net)
           [j]$height+1)
        if (all(V(net)[nei(i,mode="out")]$visited)) Q
           <- c(Q,i)
     }
  }
  V(net)$height
}

> network.height(net)
[1] 3 2 2 1 0 0 0
```

9.1.1   Distances in Graphs

Recall that the path between any two given terminal nodes of a fully resolved rooted phylogenetic tree traverses the most recent common ancestor of the nodes in the tree. In a phylogenetic network, however, the path between any two terminal nodes need not be unique, because of the existence of hybrid nodes in the network.

Hybrid nodes make it necessary to distinguish between *strict* and *non-strict* descendants of a node in a phylogenetic network. For a strict descendant of a node, every path from the root to the descendant must contain the node, while for a non-strict descendant, there is at least one path from the root to the descendant containing the node and at least one path from the root to the descendant not containing it. Every node of a phylogenetic network is thus a (trivial) strict descendant of itself, and every strict ancestor of a node is connected by a path with every ancestor of the node.

A common semi-strict ancestor of two nodes in a phylogenetic network is a common ancestor of the nodes which is also a strict ancestor of at least one of them. The path between any two given terminal nodes of a fully resolved phylogenetic network is the shortest path that traverses the most recent common semi-strict ancestor of the nodes in the network, which always exists. Any other path between the two terminal nodes is not of much use when computing distances in phylogenetic networks.

Example 9.4

In the following phylogenetic network, the most recent common semi-strict ancestor of the terminal nodes labeled A and B is the node labeled x, because it is a strict ancestor of the terminal node labeled A and a (non-strict) ancestor of the terminal node labeled B, and none of its descendants is an ancestor of these two terminal nodes.

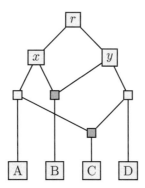

The most recent common semi-strict ancestor of the terminal nodes labeled B and C is the root r, because it is a strict ancestor of the two terminal nodes, and none of its descendants is also a strict ancestor of any of them. The internal tree nodes labeled x and y are indeed common ancestors of these two terminal nodes and their most recent common ancestors in terms of path length, but they are not strict ancestors of any of them.

The common ancestors are those nodes which the two terminal nodes can be reached from by directed paths and can thus be easily obtained by testing the existence of these paths in the directed acyclic graph representation of the phylogenetic network. Further, it can be decided if a common ancestor node is a strict ancestor of at least one of the two terminal nodes and thus a common semi-strict ancestor of them, by testing if at least one of the two terminal nodes is no longer reachable from the root after having removed from the phylogenetic network the common ancestor node.

In the following description, the common semi-strict ancestors of two terminal nodes v and w in a phylogenetic network N are collected in an (initially empty) queue Q of nodes.

> **function** CSA(N, v, w)
>   **for all** nodes u of N **do**
>   **if** v and w are reachable from u in N **then**
>   **if** $strict\_ancestor(N, u, v)$ **or** $strict\_ancestor(N, u, w)$ **then**
> $enqueue(Q, u)$
>   **return** Q

The test for a node u being a strict ancestor of a terminal node v involves removing u from a copy N' of the phylogenetic network N.

> **function** strict_ancestor(N, u, v)
> $N' \leftarrow N$
>   remove node u from N'
> $r \leftarrow root(N')$
>   **if** v is reachable from r in N' **then**
>   **return** *false*
>   **else**
>   **return** *true*

Now, the most recent common semi-strict ancestor of two terminal nodes is just the common semi-strict ancestor of the two nodes that has the smallest height in the directed acyclic graph representation of the phylogenetic network. In the following description, the node u of least height among all the common semi-strict ancestors of two terminal nodes v and w in a phylogenetic network N is selected.

> **function** LCSA(N, v, w)
> $Q \leftarrow CSA(N, v, w)$
> $u \leftarrow dequeue(Q)$
>   **while** Q is not empty **do**
> $x \leftarrow dequeue(Q)$
>   **if** $height(x) < height(u)$ **then**
> $u \leftarrow x$
>   **return** u

The representation of phylogenetic networks in Perl by means of directed acyclic graphs makes it possible to implement the previous algorithms in a straightforward way. In the following Perl script, the `is_strict_ancestor` test removes with the `delete_vertex` method the candidate ancestor node from a local copy of the directed acyclic graph, obtained with the `copy_graph` method; the root of the phylogenetic network is (the only) one of the `source_vertices` in the directed acyclic graph; and the node reachability test is performed by the `is_reachable` method of the `Graph::Directed` Perl module.

The CSA method uses both the `is_reachable` and the `is_strict_ancestor` tests, and the LCSA method uses the `heights` method of `Bio::PhyloNetwork`.

```perl
use Bio::PhyloNetwork;

sub is_strict_ancestor {
  my $net = shift;
  my $u = shift;
  my $v = shift;
  my $dag = $net->graph->copy_graph;

  my @roots = $dag->source_vertices;
  my $root = shift @roots;

  $dag->delete_vertex($u);
  return !$dag->is_reachable($root,$v);
}

sub CSA {
  my $net = shift;
  my $v = shift;
  my $w = shift;
  my $dag = $net->graph;

  my @csa = ();
  foreach my $u ( $net->nodes ) {
    if ( $dag->is_reachable($u,$v) and $dag->
        is_reachable($u,$w) ) { # common ancestor
      if (is_strict_ancestor($net,$u,$v) or
         is_strict_ancestor($net,$u,$w)) {
        push @csa, $u;
      }
    }
  }

  return \@csa;
}

sub LCSA {
```

```perl
my $net = shift;
my $v = shift;
my $w = shift;

my %height = $net->heights;
my @CSA = @{ CSA($net,$v,$w) };

my $lcsa = shift @CSA;
for my $node ( @CSA ) {
  if ( $height{$node} < $height{$lcsa} ) {
    $lcsa = $node;
  }
}

return $lcsa;
}
```

The representation of phylogenetic networks in R by means of directed acyclic graphs also makes it possible to implement the previous algorithms in a straightforward way. In the following R script, the shortest paths from the root to the descendant node are obtained using the get.all.shortest.paths function of the R package iGraph. Notice that the identifier of the descendant node in the representation of the phylogenetic network may change when deleting the ancestor node from the network.

```r
> library(igraph)
> net <- graph(c(0,1,0,2,1,4,1,3,2,3,2,6,3,5),n=7)
> V(net)$name <- c("r","x","y","h","A","B","C")

> is.strict.ancestor <- function (net,i,j) {
  if (length(get.all.shortest.paths(net,V(net)[i],V(
    net)[j],mode="out")) == 0) return(FALSE)
  r <- V(net)[which(degree(net,mode="in")==0)-1]
  if (i == r || i == j) return(TRUE)
  net <- delete.vertices(net,i)
  if (i < j) j <- j - 1 # account for deleted node
  length(get.all.shortest.paths(net,r,V(net)[j],mode=
    "out")) == 0
}

> V(net)
Vertex sequence:
[1] "r" "x" "y" "h" "A" "B" "C"

> is.strict.ancestor(net,1,4)
[1] TRUE
```

```
> is.strict.ancestor(net,1,5)
[1] FALSE
> is.strict.ancestor(net,1,6)
[1] FALSE
```

Based on the `is.strict.ancestor` function, it is straightforward to obtain the common semi-strict ancestors of two terminal nodes in a phylogenetic network, as illustrated by the following R script.

```
> is.ancestor <- function (net,i,j) {
  length(get.all.shortest.paths(net,V(net)[i],V(net)[
    j],mode="out")) != 0
}

> CSA <- function (net,i,j) {
  strict <- lapply(V(net),function (u) is.ancestor(
    net,u,i) && is.ancestor(net,u,j) && (is.strict.
    ancestor(net,u,i) || is.strict.ancestor(net,u,j)
    ))
  V(net)[unlist(strict)]
}

> V(net)
Vertex sequence:
[1] "r" "x" "y" "h" "A" "B" "C"

> CSA(net,4,5)
Vertex sequence:
[1] "r" "x"
> CSA(net,4,6)
Vertex sequence:
[1] "r"
> CSA(net,5,6)
Vertex sequence:
[1] "r" "y"
```

Then the most recent common semi-strict ancestor of two terminal nodes is just the common semi-strict ancestor of the two terminal nodes that has the smallest height, which can be obtained with help of the `network.height` function, as shown in the following R script.

```
> LCSA <- function (net,i,j) {
  csa <- match(CSA(net,i,j),V(net))
  V(net)[csa[which.min(network.height(net)[csa])]-1]
}

> V(net)
```

```
Vertex sequence:
[1] "r" "x" "y" "h" "A" "B" "C"

> LCSA(net,4,5)
Vertex sequence:
[1] "x"
> LCSA(net,4,6)
Vertex sequence:
[1] "r"
> LCSA(net,5,6)
Vertex sequence:
[1] "y"
```

9.1.2 The Path Multiplicity Distance between Graphs

The similarities and differences between two phylogenetic networks can be assessed by computing a distance measure between the two networks. The *path multiplicity distance* is based on the number of different paths from the internal nodes to each of the terminal nodes of the networks. While there is one (trivial) path from a terminal node to itself and no path to any other terminal node, the numbers of paths from the internal nodes to the terminal nodes reveal similarities and differences between two phylogenetic networks.

Example 9.5

The following fully resolved tree-child phylogenetic networks with eNewick string $(((A,(C)h1\#H1)x,(B)h2\#H2)y,(h2\#H2,(h1\#H1,D)z)w)r;$ (left) and $((A,(B,(C)h\#H)x)y,(h\#H,D)z)r;$ (right) differ in six path multiplicity vectors, and, thus, their path multiplicity distance is 6.

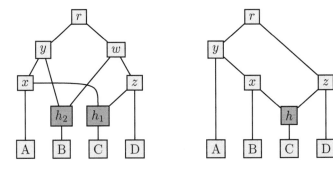

The path multiplicity vectors for the first phylogenetic network are given in the following table, with the nodes sorted by height.

node	height	vector
A	0	$(1,0,0,0)$
B	0	$(0,1,0,0)$
C	0	$(0,0,1,0)$
D	0	$(0,0,0,1)$
h_1	1	$(0,0,1,0)$
h_2	1	$(0,1,0,0)$
x	2	$(1,0,1,0)$
z	2	$(0,0,1,1)$
w	3	$(0,1,1,1)$
y	3	$(1,1,1,0)$
r	4	$(1,2,2,1)$

The path multiplicity vectors for the second phylogenetic network are given in the following table, also with the nodes sorted by height.

node	height	vector
A	0	$(1,0,0,0)$
B	0	$(0,1,0,0)$
C	0	$(0,0,1,0)$
D	0	$(0,0,0,1)$
h	1	$(0,0,1,0)$
x	2	$(0,1,1,0)$
z	2	$(0,0,1,1)$
y	3	$(1,1,1,0)$
r	4	$(1,1,2,1)$

The two networks differ in the path multiplicity vectors $(0,1,0,0)$, $(0,1,1,0)$, $(0,1,1,1)$, $(1,0,1,0)$, $(1,1,2,1)$, and $(1,2,2,1)$. Notice that the path multiplicity vector $(0,1,0,0)$ occurs twice in the first network but only once in the second network and, thus, it contributes $|2-1|=1$ to the symmetric difference of the multisets of path multiplicity vectors.

The path multiplicity distance between two phylogenetic networks labeled over the same taxa is defined as the size of the symmetric difference of their multisets of path multiplicity vectors, that is, the number of path multiplicity vectors in which the two phylogenetic networks differ. The symmetric difference applies to multisets rather than to sets, because path multiplicity vectors in a phylogenetic network are not necessarily unique. For instance, in a fully resolved phylogenetic network, a hybrid node and its single child share the same path multiplicity vector.

Now, since the paths from an internal node to the terminal nodes of a phylogenetic network are the paths from the children of the internal node to the terminal nodes, the vector of path multiplicities associated with each node of a phylogenetic network can be computed by performing a bottom-up traversal, from the terminal nodes up to the root of the network, adding the path multiplicity vectors of the children to obtain the path multiplicity vector of the parent node. The path multiplicity vector of the i-th terminal node has 1 in the i-th position and 0 everywhere else.

In the following description, the path multiplicity vector $\mu(v)$ of each node v in a phylogenetic network N is computed during a bottom-up traversal of N, with the help of an (initially empty) queue Q of nodes. The path multiplicity vector $\mu(v)$ of each child v of an internal node u is added in turn to the (initially all-zero) path multiplicity vector $\mu(u)$ of the parent node u.

---

procedure path_multiplicity(N, μ)
    for all nodes v of N **do**
        $\mu(v) \leftarrow (0, 0, \ldots, 0)$
        if v is a terminal node **then**
            $i \leftarrow$ rank of v in the terminal nodes of N
            $\mu(v)[i] \leftarrow 1$
            $enqueue(Q, v)$
    while Q is not empty **do**
        $v \leftarrow dequeue(Q)$
        mark node v as visited
        for all parents u of node v **do**
            $\mu(u) \leftarrow \mu(u) + \mu(v)$
            if all children of u are marked visited **then**
                $enqueue(Q, u)$

Example 9.6
Consider again the fully resolved tree-child phylogenetic network with eNewick string `(((A,(C)h1#H1)x,(B)h2#H2)y,(h2#H2,(h1#H1,D)z)w)r;` from the previous example.

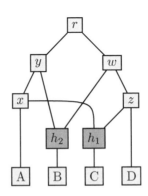

The following table illustrates the step-by-step computation of the path multiplicity vector for each of the internal nodes in the network.

dequeued node	height	computation of path multiplicity vector
A	0	$\mu(x) \leftarrow (0,0,0,0) + (1,0,0,0) = (1,0,0,0)$
B	0	$\mu(h_2) \leftarrow (0,0,0,0) + (0,1,0,0) = (0,1,0,0)$
C	0	$\mu(h_1) \leftarrow (0,0,0,0) + (0,0,1,0) = (0,0,1,0)$
D	0	$\mu(z) \leftarrow (0,0,0,0) + (0,0,0,1) = (0,0,0,1)$
h_1	1	$\mu(x) \leftarrow (1,0,0,0) + (0,0,1,0) = (1,0,1,0)$
		$\mu(z) \leftarrow (0,0,0,1) + (0,0,1,0) = (0,0,1,1)$
h_2	1	$\mu(w) \leftarrow (0,0,0,0) + (0,1,0,0) = (0,1,0,0)$
		$\mu(y) \leftarrow (0,0,0,0) + (0,1,0,0) = (0,1,0,0)$
x	2	$\mu(y) \leftarrow (0,1,0,0) + (1,0,1,0) = (1,1,1,0)$
z	2	$\mu(w) \leftarrow (0,1,0,0) + (0,0,1,1) = (0,1,1,1)$
w	3	$\mu(r) \leftarrow (0,0,0,0) + (0,1,1,1) = (0,1,1,1)$
y	3	$\mu(r) \leftarrow (0,1,1,1) + (1,1,1,0) = (1,2,2,1)$

With the representation of phylogenetic networks in BioPerl, the path multiplicities are computed when constructing the `Bio::PhyloNetwork` object, and they are readily available by means of the `mudata` method, as shown in the following Perl script.

```perl
use Bio::PhyloNetwork;

my $net = Bio::PhyloNetwork->new( -eNewick => '(((A,(
    C)h1#H1)x,(B)h2#H2)y,(h2#H2,(h1#H1,D)z)w)r;' );

my %mu = $net->mudata;

for my $node (sort keys %mu) {
  print "$node␣$mu{$node}\n";
}
```

Running the previous Perl script produces the following output:

```
#1 (0 0 1 0)
#2 (0 1 0 0)
A (1 0 0 0)
B (0 1 0 0)
C (0 0 1 0)
D (0 0 0 1)
r (1 2 2 1)
w (0 1 1 1)
x (1 0 1 0)
y (1 1 1 0)
z (0 0 1 1)
```

The representation of phylogenetic networks in R, on the other hand, does not include any method to compute the path multiplicity vectors of the nodes in a phylogenetic network. However, a `path.multiplicity` function can easily be defined using a vector to represent the queue of nodes waiting to be traversed and a matrix indexed by node names (rows) and terminal node names (columns) to represent the multiset of path multiplicity vectors, as illustrated by the following R script.

```
> library(igraph)
> el <- matrix(c("r","y","r","w","y","x","y","h2","w"
    ,"h2","w","z","x","A","x","h1","h2","B","z","h1","
    z","D","h1","C"),nc=2,byrow=TRUE)
> net <- graph.edgelist(el)

> path.multiplicity <- function (net) {
    leaves <- V(net)[which(degree(net,mode="out")==0)
        -1]
    mu <- matrix(0,nrow=vcount(net),ncol=length(leaves)
        ,dimnames=list(sort(V(net)$name),sort(leaves$
        name)))
    for (v in sort(V(net)[leaves]$name)) mu[v,v] <- 1
    V(net)$visited <- FALSE
    Q <- as.vector(V(net)[which(degree(net,mode="out")
        ==0)-1])
    while (length(Q)>0) {
        j <- Q[1]
        Q <- Q[-1]
        V(net)[j]$visited <- TRUE
        I <- as.vector(V(net)[nei(j,mode="in")])
        for (i in I) {
            mu[V(net)[i]$name,] <- mu[V(net)[i]$name,] + mu
                [V(net)[j]$name,]
            if (all(V(net)[nei(i,mode="out")]$visited)) Q
                <- c(Q,i)
        }
    }
    mu
}

> path.multiplicity(net)
    A B C D
A   1 0 0 0
B   0 1 0 0
C   0 0 1 0
D   0 0 0 1
h1  0 0 1 0
h2  0 1 0 0
```

r	1	2	2	1
w	0	1	1	1
x	1	0	1	0
y	1	1	1	0
z	0	0	1	1

The path multiplicity distance between two phylogenetic networks can be computed by counting the number of path multiplicity vectors shared by the two networks, during a simultaneous traversal of the sorted path multiplicity vectors of the two networks. In the following description, the matrices of path multiplicities are sorted by rows and then the simultaneous traversal is performed by advancing the row index to the path multiplicity matrix of the first network, the second network, or both, depending on the indexed path multiplicity vector of the first network being less than, greater than, or equal to the indexed path multiplicity vector of the second network. In the latter case, the number of common path multiplicity vectors is increased by one. Then the path multiplicity distance is the number of nodes in the two networks minus twice the number of path multiplicity vectors shared by the two networks.

```
function path_multiplicity_distance(N₁, N₂)
    path_multiplicity(N₁, μ₁)
    path_multiplicity(N₂, μ₂)
    sort μ₁ and μ₂
    n₁ ← number of nodes of N₁
    n₂ ← number of nodes of N₂
    i₁ ← 1
    i₂ ← 1
    c ← 0
    while i₁ ≤ n₁ and i₂ ≤ n₂ do
        if μ₁[i₁] < μ₂[i₂] then
            i₁ ← i₁ + 1
        else if μ₁[i₁] > μ₂[i₂] then
            i₂ ← i₂ + 1
        else
            i₁ ← i₁ + 1
            i₂ ← i₂ + 1
            c ← c + 1
    return n₁ + n₂ - 2 · c
```

Example 9.7
The fully resolved tree-child phylogenetic networks from the previous examples, with `(((A,(C)h1#H1)x,(B)h2#H2)y,(h2#H2,(h1#H1,D)z)w)r;` (left) and `((A,(B,(C)h#H)x)y,(h#H,D)z)r;` (right) as eNewick strings, have the

following sorted path multiplicity vectors.

$$i_1 \qquad\qquad\qquad i_2$$

D	1	0001	0001	1	D
C	2	0010	0010	2	C
h_1	3	0010	0010	3	h
z	4	0011	0011	4	z
B	5	0100	0100	5	B
h_2	6	0100	0110	6	x
w	7	0111	1000	7	A
A	8	1000	1110	8	y
x	9	1010	1121	9	r
y	10	1110			
r	11	1221			

There are 7 path multiplicity vectors shared by the two networks and, thus, their path multiplicity distance is $11 + 9 - 2 \cdot 7 = 6$.

The `mu_distance` method provided by the representation of phylogenetic networks in BioPerl allows one to compute the path multiplicity distance between two phylogenetic networks with the same terminal node labels, as shown in the following Perl script.

```
use Bio::PhyloNetwork;

my $net1 = Bio::PhyloNetwork->new( -eNewick => '((A
   ,((B,(C)h1#H1)w)h2#H2)x,(h1#H1,(h2#H2,D)y)z)r;' );
my $net2 = Bio::PhyloNetwork->new( -eNewick => '((A,(
   B,(C)h#H)x)y,(h#H,D)z)r;' );

my $dist = $net1->mu_distance($net2);
```

The representation of phylogenetic networks in R, on the other hand, does not include any method to compute the path multiplicity distance between two phylogenetic networks. However, the matrices of path multiplicities can be computed using the `path.multiplicity` function and then sorted by rows, and the simultaneous traversal algorithm can be implemented in a straightforward way by defining a `lex.cmp` function for the lexicographical comparison of two path multiplicity vectors, as illustrated by the following R script.

```
> library(igraph)
> el <- matrix(c("r","y","r","w","y","x","y","h2","w"
   ,"h2","w","z","x","A","x","h1","h2","B","z","h1","
   z","D","h1","C"),nc=2,byrow=TRUE)
```

```
> net1 <- graph.edgelist(el)

> el <- matrix(c("r","z","r","y","y","A","y","x","z",
    "h","z","D","x","B","x","h","h","C"),nc=2,byrow=
    TRUE)
> net2 <- graph.edgelist(el)

> lex.cmp <- function(vec1,vec2) {
  index <- which.min(vec1 == vec2)
  as.numeric(sign(vec1[index] - vec2[index]))
}

> path.multiplicity.distance <- function (net1,net2)
    {
  mu1 <- path.multiplicity(net1)
  mu1 <- mu1[do.call("order",lapply(1:ncol(mu1),
      function (j) mu1[,j])),]
  mu2 <- path.multiplicity(net2)
  mu2 <- mu2[do.call("order",lapply(1:ncol(mu2),
      function (j) mu2[,j])),]
  i1 <- 1
  i2 <- 1
  c <- 0
  while (i1 <= nrow(mu1) && i2 <= nrow(mu2)) {
    lex <- lex.cmp(mu1[i1,],mu2[i2,])
    if ( lex == -1 ) {
      i1 <- i1 + 1
    } else if ( lex == 1 ) {
      i2 <- i2 + 1
    } else {
      i1 <- i1 + 1
      i2 <- i2 + 1
      c <- c + 1
    }
  }
  nrow(mu1)+nrow(mu2)-2*c
}

> path.multiplicity.distance(net1,net2)
[1] 6
```

The path multiplicity distance is a metric on the space of all tree-child phylogenetic networks, as well as on the space of all fully resolved time-consistent tree-sibling phylogenetic networks, and it generalizes the partition distance between rooted phylogenetic trees.

9.1.3   The Tripartition Distance between Graphs

The *tripartition distance* is based on the partition of the taxa into strict descendants, non-strict descendants, and non-descendants induced by each node in the two phylogenetic networks under comparison. While terminal nodes are their only (strict) descendants, the tripartitions induced by the internal nodes reveal similarities and differences between two phylogenetic networks.

Example 9.8
Consider again the fully resolved tree-child phylogenetic network with eNewick string `(((A,(C)h1#H1)x,(B)h2#H2)y,(h2#H2,(h1#H1,D)z)w)r;` (left) and with eNewick string `((A,(B,(C)h#H)x)y,(h#H,D)z)r;` (right) from the previous examples.

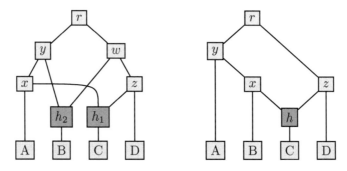

The tripartition vectors for the first phylogenetic network are given in the following table, with the nodes sorted by height, where A denotes a strict descendant, B denotes a non-strict descendant, and C denotes a non-descendant.

node	height	tripartition
A	0	(A, C, C, C)
B	0	(C, A, C, C)
C	0	(C, C, A, C)
D	0	(C, C, C, A)
h_1	1	(C, C, A, C)
h_2	1	(C, A, C, C)
x	2	(A, C, B, C)
z	2	(C, C, B, A)
w	3	(C, B, B, A)
y	3	(A, B, B, C)
r	4	(A, A, A, A)

The tripartition vectors for the second phylogenetic network are given in the following table, also with the nodes sorted by height.

node	height	tripartition
A	0	(A, C, C, C)
B	0	(C, A, C, C)
C	0	(C, C, A, C)
D	0	(C, C, C, A)
h	1	(C, C, A, C)
x	2	(C, A, B, C)
z	2	(C, C, B, A)
y	3	(A, A, B, C)
r	4	(A, A, A, A)

The two phylogenetic networks differ in the tripartition vectors (A, A, B, C), (A, B, B, C), (A, C, B, C), (C, A, B, C), (C, A, C, C), (C, B, B, A) and, thus, their tripartition distance is 6.

The tripartition distance between two phylogenetic networks labeled over the same taxa is defined as the size of the symmetric difference of their multisets of tripartition vectors, that is, the number of tripartition vectors in which the two phylogenetic networks differ. The symmetric difference applies to multisets rather than to sets, because tripartition vectors in a phylogenetic network are not necessarily unique. For instance, a hybrid node and its single child share the same tripartition vector in a fully resolved network.

Now, the multiset of tripartition vectors of a phylogenetic network can be obtained by testing for each node of the network if it is a strict ancestor, a non-strict ancestor, or not an ancestor at all of each of the terminal nodes in turn. In the following description, the tripartition vector $\theta(v)$ of each node v in a phylogenetic network N is computed by testing if the node v is a strict ancestor, a non-strict ancestor, or not an ancestor of each terminal node w in the network N.

procedure tripartition(N, θ)
    for all nodes v of N **do**
        for all terminal nodes w of N **do**
            if w is reachable from v in N **then**
                if $strict\_ancestor(N, v, w)$ **then**
                    $\theta[v][w] \leftarrow A$
                else
                    $\theta[v][w] \leftarrow B$
            else
                $\theta[v][w] \leftarrow C$

With the representation of phylogenetic networks in BioPerl, the tripartition vector of each node in a network is computed when constructing the corresponding `Bio::PhyloNetwork` object, and it is readily available by means of the `tripartitions` method, as shown in the following Perl script.

```
use Bio::PhyloNetwork;

my $net = Bio::PhyloNetwork->new( -eNewick => '(((A,(
    C)h1#H1)x,(B)h2#H2)y,(h2#H2,(h1#H1,D)z)w)r;' );

my %theta = $net->tripartitions;

map { print "$_ $theta{$_}\n" } sort keys %theta;
```

Running the previous Perl script produces the following output:

```
#1  CCAC
#2  CACC
A   ACCC
B   CACC
C   CCAC
D   CCCA
r   AAAA
w   CBBA
x   ACBC
y   ABBC
z   CCBA
```

The representation of phylogenetic networks in R, on the other hand, does not include any method to compute the tripartition vectors of the nodes in a phylogenetic network. However, a `tripartitions` function can easily be defined using a matrix indexed by node names (rows) and terminal node names (columns) to represent the multiset of tripartition vectors, as illustrated by the following R script.

```
> library(igraph)
> el <- matrix(c("r","y","r","w","y","x","y","h2","w"
    ,"h2","w","z","x","A","x","h1","h2","B","z","h1","
    z","D","h1","C"),nc=2,byrow=TRUE)
> net <- graph.edgelist(el)

> tripartitions <- function (net) {
  leaves <- V(net)[which(degree(net,mode="out")==0)
      -1]
  theta <- matrix("",nrow=vcount(net),ncol=length(
      leaves),dimnames=list(sort(V(net)$name),sort(
      leaves$name)))
  for (i in V(net)) {
    for (j in leaves) {
      if ( is.ancestor(net,i,j) ) {
        if ( is.strict.ancestor(net,i,j) ) {
          theta[V(net)[i]$name,V(net)[j]$name] <- "A"
```

```
         } else {
             theta[V(net)[i]$name,V(net)[j]$name] <- "B"
         }
       } else {
           theta[V(net)[i]$name,V(net)[j]$name] <- "C"
       }
     }
   }
   theta
}

> tripartitions(net)
    A   B   C   D
A  "A" "C" "C" "C"
B  "C" "A" "C" "C"
C  "C" "C" "A" "C"
D  "C" "C" "C" "A"
h1 "C" "C" "A" "C"
h2 "C" "A" "C" "C"
r  "A" "A" "A" "A"
w  "C" "B" "B" "A"
x  "A" "C" "B" "C"
y  "A" "B" "B" "C"
z  "C" "C" "B" "A"
```

As in the case of the path multiplicity distance, the tripartition distance between two phylogenetic networks can be computed by counting the number of tripartition vectors shared by the two networks during a simultaneous traversal of the sorted tripartition vectors of the two networks. In the following description, the matrices of tripartitions are sorted by rows and then the simultaneous traversal is performed by advancing the row index to the tripartitions matrix of the first network, the second network, or both, depending on the indexed tripartition vector of the first network being less than, greater than, or equal to the indexed tripartition vector of the second network. In the latter case, the number of common tripartition vectors is increased by one. Then the tripartition distance is the number of nodes in the two networks minus twice the number of tripartition vectors shared by the two networks.

function tripartition_distance(N_1, N_2)
 $tripartition(N_1, \theta_1)$
 $tripartition(N_2, \theta_2)$
 sort θ_1 and θ_2
 $n_1 \leftarrow$ number of nodes of N_1
 $n_2 \leftarrow$ number of nodes of N_2
 $i_1 \leftarrow 1$

$$i_2 \leftarrow 1$$
$$c \leftarrow 0$$
while $i_1 \leqslant n_1$ **and** $i_2 \leqslant n_2$ **do**
    if $\theta_1[i_1] < \theta_2[i_2]$ **then**
        $i_1 \leftarrow i_1 + 1$
    else if $\theta_1[i_1] > \theta_2[i_2]$ **then**
        $i_2 \leftarrow i_2 + 1$
    else
        $i_1 \leftarrow i_1 + 1$
        $i_2 \leftarrow i_2 + 1$
        $c \leftarrow c + 1$
return $n_1 + n_2 - 2 \cdot c$

The representation of phylogenetic networks in BioPerl does not include any method to compute the tripartition distance between two phylogenetic networks. However, the matrices of tripartition vectors can be computed using the `tripartitions` method and then sorted by rows, and the simultaneous traversal algorithm over two phylogenetic networks with the same terminal node labels can be implemented in a straightforward way, as illustrated by the following Perl script.

```perl
use Bio::PhyloNetwork;

sub tripartition_distance {
  my $net1 = shift;
  my $net2 = shift;

  $net1->compute_tripartitions() unless defined $net1
      ->{tripartitions};
  $net2->compute_tripartitions() unless defined $net2
      ->{tripartitions};

  my @tri1 = sort map {$net1->{tripartitions}->{$_}}
      $net1->nodes;
  my @tri2 = sort map {$net2->{tripartitions}->{$_}}
      $net2->nodes;

  my $i1 = 0;
  my $i2 = 0;
  my $c = 0;
  while ($i1 < scalar @tri1 && $i2 < scalar @tri2) {
    if ($tri1[$i1] lt $tri2[$i2]) {
      $i1++;
    } elsif ($tri1[$i1] gt $tri2[$i2]) {
      $i2++;
```

```
    } else {
      $i1++;
      $i2++;
      $c++;
    }
  }

  return scalar @tri1+scalar @tri2-2*$c;
}
```

The representation of phylogenetic networks in R does not include any method to compute the tripartition distance between two phylogenetic networks, either. However, the matrices of tripartition vectors can be computed using the `tripartitions` function and then sorted by rows, and the simultaneous traversal algorithm over two phylogenetic networks with the same terminal node labels can be implemented in a straightforward way by defining a `lex.cmp.char` function for the lexicographical comparison of two tripartition vectors, as illustrated by the following R script.

```
> library(igraph)
> el <- matrix(c("r","y","r","w","y","x","y","h2","w"
    ,"h2","w","z","x","A","x","h1","h2","B","z","h1","
    z","D","h1","C"),nc=2,byrow=TRUE)
> net1 <- graph.edgelist(el)

> el <- matrix(c("r","z","r","y","y","A","y","x","z",
    "h","z","D","x","B","x","h","h","C"),nc=2,byrow=
    TRUE)
> net2 <- graph.edgelist(el)

> lex.cmp.char <- function (vec1,vec2) {
  for (j in 1:length(vec1)) {
    if (vec1[j] < vec2[j]) { return(-1) }
    if (vec1[j] > vec2[j]) { return(1) }
  }
  return(0)
}

> tripartition.distance <- function (net1,net2) {
  theta1 <- tripartitions(net1)
  theta1 <- theta1[do.call("order",lapply(1:ncol(
      theta1),function (j) theta1[,j])),]
  theta2 <- tripartitions(net2)
  theta2 <- theta2[do.call("order",lapply(1:ncol(
      theta2),function (j) theta2[,j])),]
  i1 <- 1
```

```
i2 <- 1
c <- 0
while (i1 <= nrow(theta1) && i2 <= nrow(theta2)) {
    lex <- lex.cmp.char(theta1[i1,],theta2[i2,])
    if ( lex == -1 ) {
        i1 <- i1 + 1
    } else if ( lex == 1 ) {
        i2 <- i2 + 1
    } else {
        i1 <- i1 + 1
        i2 <- i2 + 1
        c <- c + 1
    }
}
nrow(theta1)+nrow(theta2)-2*c
}

> tripartition.distance(net1,net2)
[1] 6
```

The tripartition distance is a metric on the space of all time-consistent tree-child phylogenetic networks, and it also generalizes the partition distance between rooted phylogenetic trees.

9.1.4 The Nodal Distance between Graphs

The *nodal distance* is based on the shortest paths between terminal nodes in the two phylogenetic networks under comparison. The matrices of distances between each pair of terminal nodes and their most recent common semi-strict ancestors reveal similarities and differences between two phylogenetic networks.

Example 9.9

Consider again the two fully resolved tree-child phylogenetic networks with the eNewick string (((A,(C)h1#H1)x,(B)h2#H2)y,(h2#H2,(h1#H1,D)z)w)r; and ((A,(B,(C)h#H)x)y,(h#H,D)z)r; from the previous examples. The most recent common semi-strict ancestors of each pair of terminal nodes in the first phylogenetic network are given in the following table.

	A	B	C	D
A	A	y	x	r
B	y	B	r	w
C	x	r	C	z
D	r	w	z	D

The most recent common semi-strict ancestors of each pair of terminal nodes in the second phylogenetic network are given in the following table.

	A	B	C	D
A	A	y	y	r
B	y	B	x	r
C	y	x	C	z
D	r	r	z	D

The shortest path from the most recent common semi-strict ancestor of the terminal nodes labeled A and B to the terminal node labeled B has length 2 in the two networks, while the shortest path to the terminal node labeled A has length 2 in the first network and length 2 in the second network.

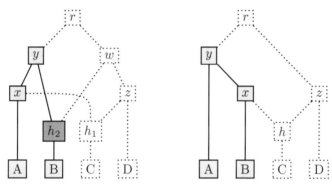

The absolute differences between the corresponding shortest path length matrices are as follows.

$$\left| \begin{pmatrix} 0 & 2 & 1 & 3 \\ 2 & 0 & 3 & 2 \\ 2 & 4 & 0 & 2 \\ 3 & 2 & 1 & 0 \end{pmatrix} - \begin{pmatrix} 0 & 1 & 1 & 2 \\ 2 & 0 & 1 & 3 \\ 3 & 2 & 0 & 2 \\ 2 & 2 & 1 & 0 \end{pmatrix} \right| = \begin{pmatrix} 0 & 1 & 0 & 1 \\ 0 & 0 & 2 & 1 \\ 1 & 2 & 0 & 0 \\ 1 & 0 & 0 & 0 \end{pmatrix}$$

The nodal distance between the two phylogenetic networks is thus 9.

The nodal distance between two phylogenetic networks labeled over the same taxa is defined as the sum of the absolute differences of distance between each pair of terminal nodes and their most recent common semi-strict ancestor in the two networks. Therefore, the nodal distance can be obtained by computing the distance between each pair of terminal nodes and their most recent common semi-strict ancestor in each of the two networks and then computing the absolute difference between the two matrices of nodal distances.

In the following description, the most recent common semi-strict ancestor of each pair of terminal nodes in a network is computed only once, in order to obtain the two nodal distances between them.

```
function nodal distance(N₁, N₂)
    L ← terminal node labels in N₁ and N₂
    n ← length(L)
    d ← 0
    for i ← 1, …, n do
        i₁ ← terminal node of N₁ labeled L[i]
        i₂ ← terminal node of N₂ labeled L[i]
        for j ← i + 1, …, n do
            j₁ ← terminal node of N₁ labeled L[j]
            j₂ ← terminal node of N₂ labeled L[j]
            ℓ₁ ← LCSA(N₁, i₁, j₁)
            ℓ₂ ← LCSA(N₂, i₂, j₂)
            d₁ ← distance(N₁, ℓ₁, i₁)
            d₂ ← distance(N₂, ℓ₂, i₂)
            d ← d + |d₁ − d₂|
            d₁ ← distance(N₁, ℓ₁, j₁)
            d₂ ← distance(N₂, ℓ₂, j₂)
            d ← d + |d₁ − d₂|
    return d
```

The representation of phylogenetic networks in BioPerl does not include any method to compute the nodal distance between two phylogenetic networks. However, using the `SP_Dijkstra` method of the `Graph::Directed` Perl module to obtain the shortest paths between two terminal nodes and their most recent common semi-strict ancestor, the previous algorithm can be implemented in a straightforward way, as illustrated by the following Perl script.

```
sub nodal_distance {
  my $net1 = shift;
  my $net2 = shift;
  my @L = $net1->leaves;
  my $dist = 0;
  for (my $i = 0; $i < @L; $i++) {
    for (my $j = $i+1; $j < @L; $j++) {
      my $lcsa1 = LCSA($net1,$L[$i],$L[$j]);
      my $lcsa2 = LCSA($net2,$L[$i],$L[$j]);
      my @p1 = $net1->graph->SP_Dijkstra($lcsa1,$L[$i
          ]);
      my @p2 = $net2->graph->SP_Dijkstra($lcsa2,$L[$i
          ]);
      my $d1 = @p1 - 1;
      my $d2 = @p2 - 1;
      $dist += abs($d1 - $d2);
      @p1 = $net1->graph->SP_Dijkstra($lcsa1,$L[$j]);
```

```
        @p2 = $net2->graph->SP_Dijkstra($lcsa2,$L[$j]);
        $d1 = @p1 - 1;
        $d2 = @p2 - 1;
        $dist += abs($d1 - $d2);
    }
  }
  return $dist;
}
```

The representation of phylogenetic networks in R does not include any method to compute the nodal distance between two phylogenetic networks, either. However, using the `get.shortest.paths` function of the R package iGraph to obtain the shortest paths between two terminal nodes and their most recent common semi-strict ancestor, the previous algorithm can also be implemented in a straightforward way, as illustrated by the following R script.

```
> library(igraph)
> el <- matrix(c("r","y","r","w","y","x","y","h2","w"
  ,"h2","w","z","x","A","x","h1","h2","B","z","h1","
  z","D","h1","C"),nc=2,byrow=TRUE)
> net1 <- graph.edgelist(el)

> el <- matrix(c("r","z","r","y","y","A","y","x","z",
  "h","z","D","x","B","x","h","h","C"),nc=2,byrow=
  TRUE)
> net2 <- graph.edgelist(el)

> nodal.distance <- function (net1,net2) {
  l1 <- V(net1)[which(degree(net1,mode="out")==0)-1]
  n1 <- matrix(0,nrow=length(l1),ncol=length(l1),
    dimnames=list(sort(l1$name),sort(l1$name)))
  l2 <- V(net2)[which(degree(net2,mode="out")==0)-1]
  n2 <- matrix(0,nrow=length(l2),ncol=length(l2),
    dimnames=list(sort(l2$name),sort(l2$name)))
  for (i in l1) {
    for (j in c(l1)[match(i,l1):length(l1)]) {
      lcsa <- LCSA(net1,i,j)
      paths <- get.shortest.paths(net1,lcsa,i,mode="
        out")
      n1[V(net1)[i]$name,V(net1)[j]$name] <- length(
        paths[[1]])-1 # number of edges
      paths <- get.shortest.paths(net1,lcsa,j,mode="
        out")
      n1[V(net1)[j]$name,V(net1)[i]$name] <- length(
        paths[[1]])-1 # number of edges
    }
```

```
  }
  for (i in 12) {
    for (j in c(12)[match(i,12):length(12)]) {
      lcsa <- LCSA(net2,i,j)
      paths <- get.shortest.paths(net2,lcsa,i,mode="
        out")
      n2[V(net2)[i]$name,V(net2)[j]$name] <- length(
        paths[[1]])-1 # number of edges
      paths <- get.shortest.paths(net2,lcsa,j,mode="
        out")
      n2[V(net2)[j]$name,V(net2)[i]$name] <- length(
        paths[[1]])-1 # number of edges
    }
  }
  sum(abs(n1-n2))
}

> nodal.distance(net1,net2)
[1] 9
```

The nodal distance is a metric on the space of all time-consistent tree-child phylogenetic networks, and it generalizes the nodal distance between rooted phylogenetic trees.

9.2 Finding Trees in Graphs

Finding the trees contained in a graph is useful for computing distances between two graphs. A fully resolved phylogenetic network with n terminal nodes and m hybrid nodes contains 2^m phylogenetic trees, each of them with n terminal nodes, which result from cutting one of the two edges coming into each hybrid node and then contracting any elementary paths.

Example 9.10
The fully resolved tree-child phylogenetic network with the eNewick string ((A,((B,(C)h1#H1)w)h2#H2)x,(h1#H1,(h2#H2,D)y)z)r; from the previous examples, which has $m = 2$ hybrid nodes, contains $2^m = 4$ phylogenetic trees.

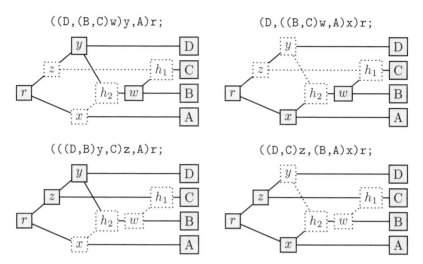

The 2^m phylogenetic trees contained in a phylogenetic network with m hybrid nodes can be obtained by removing each of the edges coming into each of the hybrid nodes in turn and contracting any elementary paths whenever there are no hybrid nodes left.

In the following description, the removed edges are added back to the network N after all the trees not containing the edge have been generated, in order to avoid making a local copy of the phylogenetic network at each stage, and the result is collected in an (initially empty) set T of phylogenetic trees.

> **procedure** explode(N, T)
>   **if** N has no hybrid nodes **then**
>   contract any elementary paths in N
> $T \leftarrow T \cup N$
>   **else**
> $v \leftarrow$ an hybrid node of N
>   **for all** parents u of node v **do**
>   delete edge (u, v) from N
>   *explode*(N, T)
>   add edge (u, v) to N

The representation of phylogenetic networks in BioPerl includes a method **explode** to obtain all the phylogenetic trees contained in a phylogenetic network, as illustrated by the following Perl script.

```
use Bio::PhyloNetwork;

my @trees = $net->explode;

my $output = new Bio::TreeIO('-format' => 'newick');
```

```
for my $tree (@trees) {
  $output->write_tree($tree);
}
```

The actual Perl code of the `explode` method in BioPerl is as follows.

```
sub explode_rec {
  my ($self,$trees) = @_;
  my @h = $self->hybrid_nodes;
  if (scalar @h) {
    my $v = shift @h;
    for my $u ($self->{graph}->predecessors($v)) {
      $self->{graph}->delete_edge($u,$v);
      $self->explode_rec($trees);
      $self->{graph}->add_edge($u,$v);
    }
  } else {
    my $io = IO::String->new($self->eNewick);
    my $treeio = Bio::TreeIO->new(-format => 'newick'
      , -fh => $io);
    my $tree = $treeio->next_tree;
    $tree->contract_linear_paths;
    push @{$trees}, $tree;
  }
}

sub explode {
  my ($self) = @_;
  my @trees;
  $self->explode_rec(\@trees);
  return @trees;
}
```

The representation of phylogenetic networks in R, on the other hand, does not include any method to obtain the trees contained in a network, although the representation of phylogenetic networks by means of directed acyclic graphs makes it possible to implement the previous algorithm in a straightforward way.

Recall that the Newick description of a phylogenetic tree can be obtained by traversing the tree in postorder and writing down the name or label of the node when visiting a terminal node, a left parenthesis (preceded by a comma unless the node is the first child of its parent) when visiting a non-terminal node for the first time, and a right parenthesis followed by the name or label of the node (if any) when visiting a non-terminal node for the second time, that is, after having visited all its descendants, where the name of a node is preceded by a comma unless it is the first child of its parent. This algorithm can be implemented as shown in the following R script.

```
> network.to.newick <- function (net) {
  r <- V(net)[which(degree(net,mode="in")==0)-1]
  newick <- obtain.newick(net,r,c())
  paste(c(newick,";"),collapse="")
}

> obtain.newick <- function (net,v,newick) {
  if (length(V(net)[nei(v,"out")]) != 0) {
    if ( length(V(net)[nei(v,"in")]) != 0 && match(v,
        V(net)[nei(V(net)[nei(v,"in")],"out")]) != 1 )
      newick <- c(newick,",")
    newick <- c(newick,"(")
  }
  for (w in V(net)[nei(v,"out")]) {
    newick <- obtain.newick(net,w,newick)
  }
  if (length(V(net)[nei(v,"out")]) != 0)
    newick <- c(newick,")")
  if ( length(V(net)[nei(v,"in")]) != 0 && length(V(
      net)[nei(v,"out")]) == 0 && match(v,V(net)[nei(V
      (net)[nei(v,"in")],"out")]) != 1 )
    newick <- c(newick,",")
  c(newick,V(net)[v]$name)
}
```

Now, the algorithm for obtaining all the phylogenetic trees contained in a phylogenetic network can be implemented by removing from the directed acyclic graph representing the network each of the incoming edges of each hybrid node in turn and then contracting any elementary paths during a postorder traversal of the tree. In the following R script, those nodes and edges to be deleted in order to contract any elementary paths are first marked with the $del attribute by the **postorder.traversal** function, and then they are all removed from the directed acyclic graph by the **network.to.tree** function to produce a tree, which is output in Newick format by the previous **network.to.newick** function.

```
> library(igraph)
> el <- matrix(c("r","x","r","z","z","h1","z","y","y"
  ,"h2","y","D","x","A","x","h2","h2","w","w","B","w
  ","h1","h1","C"),nc=2,byrow=TRUE)
> net <- graph.edgelist(el)

> network.to.tree <- function (net) {
  V(net)$del <- FALSE
  E(net)$del <- FALSE
  r <- V(net)[which(degree(net,mode="in")==0)-1]
```

```
  net <- postorder.traversal(net,r)
  net <- delete.vertices(net,V(net)[del])
  net <- delete.edges(net,E(net)[del])
  network.to.newick(net)
}

> postorder.traversal <- function (net,v) {
  for (w in V(net)[nei(v,"out")]) {
    net <- postorder.traversal(net,w)
  }
  VW <- E(net)[from == v & del==FALSE]
  if (length(c(VW)) == 1) {
    u <- V(net)[nei(v,"in")]
    w <- get.edge(net,VW)[2]
    E(net,path=c(V(net)[u],V(net)[v]))$del <- TRUE
    E(net,path=c(V(net)[v],V(net)[w]))$del <- TRUE
    V(net)[v]$del <- TRUE
    net <- add.edges(net,c(V(net)[u],V(net)[w]))
    E(net,path=c(V(net)[u],V(net)[w]))$del <- FALSE
  }
  net
}

> explode <- function (net) {
  trees <- c()
  H <- V(net)[which(degree(net,mode="in")>1)-1]
  if (length(H) == 0) {
    trees <- c(trees,network.to.tree(net))
  } else {
    v <- c(H)[1]
    for (u in V(net)[nei(v,"in")]) {
      new <- delete.edges(net,E(net,P=c(u,v)))
      trees <- c(trees,explode(new))
    }
  }
  trees
}

> options(width="40")
> explode(net)
[1] "((D,(B,C)w)y,A)r;"
[2] "((A,(B,C)w)x,D)r;"
[3] "(((D,B)y,C)z,A)r;"
[4] "((A,B)x,(D,C)z)r;"
```

9.2.1 The Statistical Error between Graphs

The path multiplicity distance, tripartition distance, and nodal distance can be computed between any two fully resolved phylogenetic networks labeled over the same taxa, but they only yield a metric when the networks exhibit some of the topological properties of being tree-sibling, tree-child, or time-consistent. An alternative method for the assessment of similarities and differences between two phylogenetic networks that lack the necessary topological properties consists of computing the statistical error between the phylogenetic trees contained in the two networks.

The false negative rate is the fraction of trees contained in the first network that are not contained in the second network. Conversely, the false positive rate is the fraction of trees contained in the second network that are not contained in the first network. The error rate between two phylogenetic networks is just the average of their false negative and false positive rates.

Example 9.11

Consider again the two fully resolved tree-child phylogenetic networks with the eNewick string `((A,((B,(C)h1#H1)w)h2#H2)x,(h1#H1,(h2#H2,D)y)z)r;` and `((A,(B,(C)h#H)x)y,(h#H,D)z)r;` from the previous examples. The first network has $m = 2$ hybrid nodes and, thus, it contains the following $2^m = 4$ phylogenetic trees.

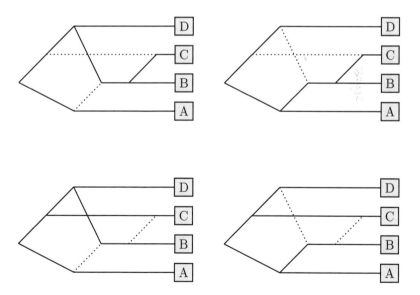

The second network only has $m = 1$ hybrid nodes and, thus, it contains the following $2^m = 2$ phylogenetic trees.

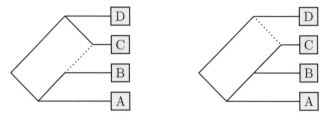

Two of the four trees in the first network are not contained in the second network (false negatives), while none of the two trees in the second network is not contained in the first network (false positives). Therefore, the error rate between the two networks is $(2/4 + 0/2)/2 = 0.25$.

Using the `explode` method of the representation of phylogenetic networks in BioPerl, the computation of the error rate between two phylogenetic networks can be implemented in a straightforward way, where identical trees in the two networks are identified as those at partition distance zero by using the `partition_distance` method, as shown in the following Perl script.

```perl
sub error_ratio {
  my $net1 = shift;
  my $net2 = shift;
  my @trees1 = $net1->explode;
  my @trees2 = $net2->explode;
  my $FN = 0;
  my $FP = 0;
  for my $tree1 (@trees1) {
    my $found = 0; # false
    for my $tree2 (@trees2) {
      if (partition_distance($tree1,$tree2) == 0) {
        $found = 1; # true
      }
    }
    if (!$found) { $FN++; }
  }
  for my $tree2 (@trees2) {
    my $found = 0; # false
    for my $tree1 (@trees1) {
      if (partition_distance($tree1,$tree2) == 0) {
        $found = 1; # true
      }
    }
    if (!$found) { $FP++; }
  }
  return (($FN/@trees1)+($FP/@trees2))/2;
}
```

Using the previous R implementation of the `explode` function to obtain the trees contained in a network, the computation of the error rate between two phylogenetic networks can be also implemented in a straightforward way, where identical trees in the two networks are identified as those at partition distance zero by using the `dist.topo` function, as illustrated by the following R script.

```
> library(ape)
> library(igraph)

> el <- matrix(c("r","x","r","z","z","h1","z","y","y"
  ,"h2","y","D","x","A","x","h2","h2","w","w","B","w
  ","h1","h1","C"),nc=2,byrow=TRUE)
> net1 <- graph.edgelist(el)

> el <- matrix(c("r","z","r","y","y","A","y","x","z",
  "h","z","D","x","B","x","h","h","C"),nc=2,byrow=
  TRUE)
> net2 <- graph.edgelist(el)

> error.rate <- function (net1,net2) {
  trees1 <- explode(net1)
  trees2 <- explode(net2)
  FN <- 0
  FP <- 0
  for (t1 in trees1) {
    tree1 <- read.tree(text=t1)
    found <- FALSE
    for (t2 in trees2) {
      tree2 <- read.tree(text=t2)
      if (dist.topo(tree1,tree2) == 0) {
        found <- TRUE
        break
      }
    }
    if (!found) FN <- FN+1
  }
  for (t2 in trees2) {
    tree2 <- read.tree(text=t2)
    found <- FALSE
    for (t1 in trees1) {
      tree1 <- read.tree(text=t1)
      if (dist.topo(tree1,tree2) == 0) {
        found <- TRUE
        break
```

```
        }
      }
    if (!found) FP <- FP+1
  }
  (FN/length(trees1)+FP/length(trees2))/2
}

> options(width="40")
> explode(net1)
[1] "((D,(B,C)w)y,A)r;"
[2] "((A,(B,C)w)x,D)r;"
[3] "(((D,B)y,C)z,A)r;"
[4] "((A,B)x,(D,C)z)r;"
> explode(net2)
[1] "((A,(B,C)x)y,D)r;"
[2] "((D,C)z,(A,B)y)r;"
> error.rate(net1,net2)
[1] 0.25
```

Bibliographic Notes

The most recent common semi-strict ancestor of two terminal nodes in a phylogenetic network was shown to play the role of the most recent common ancestor in a phylogenetic tree in (Cardona et al. 2009a;b).

The path multiplicity distance between tree-child phylogenetic networks was introduced by Cardona et al. (2009c), and further studied in (Cardona et al. 2008a;d) for fully resolved time-consistent tree-sibling phylogenetic networks.

The tripartition distance between time-consistent tree-sibling phylogenetic networks was introduced by Moret et al. (2004) and further studied by Cardona et al. (2008d) for fully resolved time-consistent tree-sibling phylogenetic networks and by Cardona et al. (2009c) for time-consistent tree-child phylogenetic networks.

The nodal distance between time-consistent tree-child phylogenetic networks was introduced in (Cardona et al. 2009b).

The use of statistical error for the assessment of similarities and differences between phylogenetic networks was advocated by Moret et al. (2004) and Woolley et al. (2008), although it fails to be a metric on the space of all phylogenetic networks.

Chapter 10

General Pattern Matching in Graphs

Combinatorial pattern matching is the search for exact or approximate occurrences of a given pattern within a given text. When it comes to graphs in computational biology, both the pattern and the text are graphs and the pattern matching problem becomes one of finding the occurrences of a graph within another graph. For instance, scanning a metabolic pathway for the presence of a known pattern can help in finding conserved network motifs, and finding a phylogenetic network within another phylogenetic network can help in assessing their similarities and differences. This is the subject of this chapter.

10.1 Finding Subgraphs

There are several ways in which a graph can be contained in another graph. In the most general sense, a subgraph of a given (unrooted or rooted) phylogenetic network is a connected subgraph of the network, while in the case of rooted phylogenetic networks, a distinction can be made between *top-down* and *bottom-up* subgraphs.

A bottom-up subgraph of a given rooted phylogenetic network is the whole subgraph rooted at some node of the network, and a connected subgraph of a rooted phylogenetic network is called a top-down subgraph if the parents of all nodes in the subgraph (up to, and including, the most recent common semi-strict ancestor of all nodes in the subgraph) also belong to the subgraph. Further, the subgraph of an (unrooted or rooted) network induced by a set of terminal nodes is the unique top-down subgraph that contains the set of terminal nodes but does not include any other top-down subgraph of the given network with these terminal nodes, where elementary paths (paths of two or more edges without internal branching) are contracted to single edges in a subgraph.

In the following example, the elementary path of two edges between the root and node y, as well as the elementary path of two edges between hybrid node h_2 and the terminal node labeled B, is contracted to single edges in the subgraph induced by the terminal nodes labeled A, B, and D.

Example 10.1

In the following fully resolved rooted phylogenetic network, with eNewick string `((A,((B,(C)h1#H1)w)h2#H2)x,(h1#H1,(h2#H2,D)y)z)r;` and terminal nodes labeled A, B, C, D, a top-down subgraph is shown highlighted (left), and the bottom-up subgraph rooted at the most recent common semi-strict ancestor of the terminal nodes labeled A, B, and C is also shown (middle), together with the subgraph induced by the terminal nodes labeled A, B, and D (right).

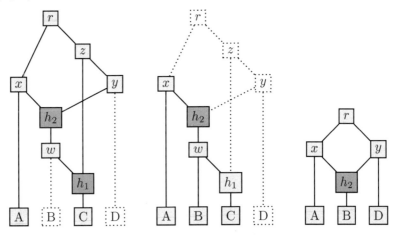

Example 10.2

In the following fully resolved unrooted phylogenetic network (left), the subgraph induced by the terminal nodes labeled A, B, C, and D is shown highlighted before contraction of elementary paths. The subgraph resulting from contraction of elementary paths is shown to the right.

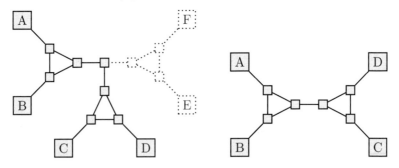

10.1.1 Finding Subgraphs Induced by Triplets

Finding subgraphs induced by triplets of terminal nodes in rooted phylogenetic networks is an interesting problem, because a rooted phylogenetic network can be reconstructed (although not in a unique way) from the set of all its triplet topologies. Given a triplet of terminal nodes, there are only six possible induced subgraphs if the phylogenetic network is fully resolved and time consistent.

Example 10.3
There are six fully resolved time-consistent phylogenetic networks on three terminal nodes.

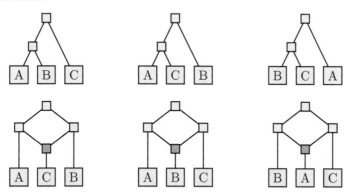

However, if the phylogenetic network is not time consistent, there is an infinite number of possible subgraphs induced by a triplet of terminal nodes, even if the phylogenetic network is fully resolved.

Example 10.4
There is an infinite number of fully resolved phylogenetic networks on three terminal nodes.

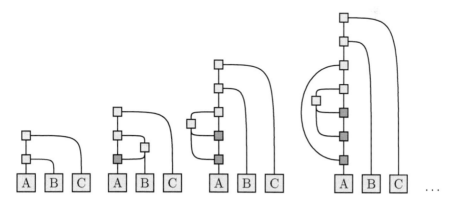

The subgraph induced by a triplet of terminal nodes of a fully resolved time-consistent phylogenetic network can be obtained by finding the most recent common semi-strict ancestor of the three terminal nodes and then removing all those nodes which are not in a path from one of the three terminal nodes to the most recent common semi-strict ancestor, contracting also any elementary paths. However, a more efficient algorithm consists of first finding the most recent common ancestor of each pair of terminal nodes from the given triplet and then building the induced subgraph by distinguishing among the six possible topologies on the basis of the relationship among the most recent common semi-strict ancestors. In the following description, the six possible triplet topologies are distinguished by

- $LCSA(i,j)$ is not an ancestor of $LCSA(i,k)$, $LCSA(i,k)$ is an ancestor of $LCSA(i,j)$, and $LCSA(j,k)$ is an ancestor of $LCSA(i,j)$

- $LCSA(i,j)$ is an ancestor of $LCSA(i,k)$, $LCSA(i,k)$ is not an ancestor of $LCSA(i,j)$, and $LCSA(j,k)$ is an ancestor of $LCSA(i,j)$

- $LCSA(i,j)$ is an ancestor of $LCSA(i,k)$, $LCSA(i,k)$ is an ancestor of $LCSA(i,j)$, and $LCSA(j,k)$ is not an ancestor of $LCSA(i,j)$

- $LCSA(i,j)$ is an ancestor of $LCSA(i,k)$ but neither $LCSA(i,k)$ nor $LCSA(j,k)$ is an ancestor of $LCSA(i,j)$

- $LCSA(i,j)$ is not an ancestor of $LCSA(i,k)$, $LCSA(i,k)$ is an ancestor of $LCSA(i,j)$, and $LCSA(j,k)$ is not an ancestor of $LCSA(i,j)$

- $LCSA(i,j)$ is not an ancestor of $LCSA(i,k)$, $LCSA(i,k)$ is not an ancestor of $LCSA(i,j)$, and $LCSA(j,k)$ is an ancestor of $LCSA(i,j)$

and their eNewick string is output.

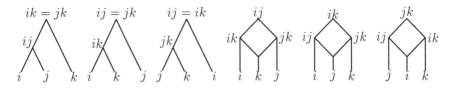

```
function triplet(N, i, j, k)
    ij ← LCSA(N, i, j)
    ik ← LCSA(N, i, k)
    jk ← LCSA(N, j, k)
    if ij is an ancestor of ik in N then
        if ik is an ancestor of ij in N then
            return ((j, k), i);
        else
            if jk is an ancestor of ij in N then
                return ((i, k), j);
            else
                return ((i, (k)#H1), (#H1, j));
    else
```

if ik is an ancestor of ij in N **then**
    if jk is an ancestor of ij in N **then**
        return $((i,j),k)$;
    else
        return $((i,(j)\#H1),(\#H1,k))$;
  else
    return $((j,(i)\#H1),(\#H1,k))$;

Example 10.5

The subgraphs induced by triplets of terminal nodes of the phylogenetic network with eNewick string `((A,(B,(C)h#H)x)y,(h#H,D)z)r;`, which is fully resolved and time consistent, are shown next.

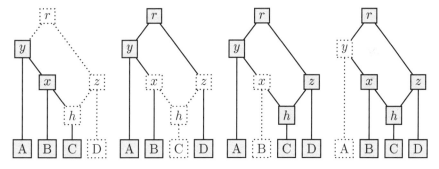

In the subgraph induced by the terminal nodes labeled A, B, C, node $y = LCSA(A,B) = LCSA(A,C)$ is an ancestor of node $x = LCSA(B,C)$. In the subgraph induced by the terminal nodes labeled A, B, D, node $r = LCSA(A,D) = LCSA(B,D)$ is an ancestor of $y = LCSA(A,B)$. In the subgraph induced by the terminal nodes labeled A, C, D, node $r = LCSA(A,D)$ is an ancestor of both $y = LCSA(A,C)$ and $z = LCSA(C,D)$. Finally, in the subgraph induced by the terminal nodes labeled B, C, D, node $r = LCSA(B,D)$ is an ancestor of both $x = LCSA(B,C)$ and $z = LCSA(C,D)$.

The representation of phylogenetic networks in BioPerl does not include any method to compute subgraphs of a time-consistent phylogenetic network induced by triplets of terminal nodes. Nevertheless, the subgraph of a fully resolved time-consistent phylogenetic tree induced by a triplet of terminal nodes can be easily obtained by using the `LCSA` method to find the most recent common semi-strict ancestor of the terminal nodes in the given triplet, as illustrated by the following Perl script.

```
sub triplet {
  my ($net,@triple) = @_;
  my ($i,$j,$k) = @triple;
```

```perl
my @L = $net->leaves;
my $ij = LCSA($net,$L[$i],$L[$j]);
my $ik = LCSA($net,$L[$i],$L[$k]);
my $jk = LCSA($net,$L[$j],$L[$k]);

my $t;
if ($net->graph->SP_Dijkstra($ij,$ik)) {
  if ($net->graph->SP_Dijkstra($ik,$ij)) {
    $t = "(($L[$j],$L[$k]),$L[$i]);";
  } else {
    if ($net->graph->SP_Dijkstra($jk,$ij)) {
      $t = "(($L[$i],$L[$k]),$L[$j]);";
    } else {
      $t = "(($L[$i],($L[$k])#H1),(#H1,$L[$j]));";
    }
  }
} else {
  if ($net->graph->SP_Dijkstra($ik,$ij)) {
    if ($net->graph->SP_Dijkstra($jk,$ij)) {
      $t = "(($L[$i],$L[$j]),$L[$k]);";
    } else {
      $t = "(($L[$i],($L[$j])#H1),(#H1,$L[$k]));";
    }
  } else {
    $t = "(($L[$j],($L[$i])#H1),(#H1,$L[$k]));";
  }
}
return $t;
}
```

The representation of phylogenetic networks in R does not include any method to compute subgraphs of a phylogenetic network induced by triplets of terminal nodes, either. However, a `triplet` function can easily be defined using the `LCSA` function, as illustrated by the following R script.

```r
> library(igraph)

> triplet <- function (net,i,j,k) {
    ij <- LCSA(net,i,j)
    ik <- LCSA(net,i,k)
    jk <- LCSA(net,j,k)

    ij.ancestor.of.ik <- length(get.all.shortest.
        paths(net,ij,ik,mode="out")) != 0
    ik.ancestor.of.ij <- length(get.all.shortest.
        paths(net,ik,ij,mode="out")) != 0
```

```
      ij.ancestor.of.jk <- length(get.all.shortest.
         paths(net,ij,jk,mode="out")) != 0
      jk.ancestor.of.ij <- length(get.all.shortest.
         paths(net,jk,ij,mode="out")) != 0

    if (ij.ancestor.of.ik)
      if (ik.ancestor.of.ij)
        paste("((",V(net)[j]$name,",",V(net)[k]$name,
           ")",",",V(net)[i]$name,");",sep="")
      else
        if (jk.ancestor.of.ij)
          paste("((",V(net)[i]$name,",",V(net)[k]$
             name,")",",",V(net)[j]$name,");",sep="")
        else
          paste("((",V(net)[i]$name,",(",V(net)[k]$
             name,"))#H,(#H,",V(net)[j]$name,"));",
             sep="")
    else
      if (ik.ancestor.of.ij)
        if (jk.ancestor.of.ij)
          paste("((",V(net)[i]$name,",",V(net)[j]$
             name,")",",",V(net)[k]$name,");",sep="")
        else
          paste("((",V(net)[i]$name,",(",V(net)[j]$
             name,"))#H,(#H,",V(net)[k]$name,"));",
             sep="")
      else
        paste("((",V(net)[j]$name,",(",V(net)[i]$name
           ,"))#H,(#H,",V(net)[k]$name,"));",sep="")
}

> el <- matrix(c("r","z","r","y","y","A","y","x","z",
   "h","z","D","x","B","x","h","h","C"),nc=2,byrow=
   TRUE)
> net <- graph.edgelist(el)

> triplet(net,3,7,8)
[1] "((B,C),A);"
> triplet(net,3,7,6)
[1] "((A,B),D);"
> triplet(net,3,8,6)
[1] "((A,(C))#H,(#H,D));"
> triplet(net,7,8,6)
[1] "((B,(C))#H,(#H,D));"
```

Finding subtrees induced by triplets of terminal nodes in rooted phylogenetic networks is also an interesting problem, because a rooted phylogenetic
network can be reconstructed (although, again, not in a unique way) from a
dense set of triplet topologies, that is, from a set of triplet topologies containing at least one triplet for each three terminal nodes. Given a triplet of
terminal nodes, there are only three possible induced subtrees if the phylogenetic network is fully resolved, as already shown in the previous example.
However, a phylogenetic network can contain more than one subtree induced
by the same triplet of terminal nodes.

Example 10.6

The fully resolved time-consistent phylogenetic network with eNewick string
`((A,(B,(C)h#H)x)y,(h#H,D)z)r;` has the following six subtrees induced by
triplets of terminal nodes.

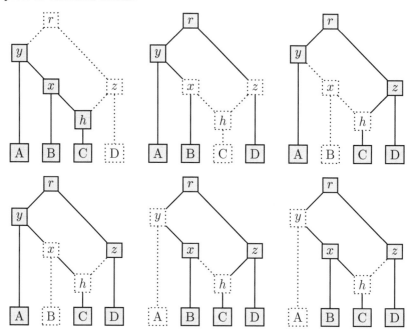

The subtrees induced by a triplet of terminal nodes of a fully resolved
time-consistent phylogenetic network can be obtained by finding the subgraph
induced by the triplet of terminal nodes using the previous algorithm and
then finding the trees contained therein using the *explode* algorithm from the
previous chapter.

Finding subtrees induced by the most recent common semi-strict ancestors
of triplets of terminal nodes in rooted phylogenetic networks is another interesting problem, because a rooted phylogenetic network can be reconstructed
in a unique way from the set of all such triplet topologies if the network is

tree-child and also time consistent. Given a triplet of terminal nodes, there are only five possible subgraphs induced by their most recent common semi-strict ancestors if the time-consistent tree-child phylogenetic network is fully resolved, and all these triplets are trees.

Example 10.7

There are five possible triplets induced by the recent common semi-strict ancestors of any three terminal nodes in a fully resolved rooted time-consistent tree-child phylogenetic network.

These five trees can be distinguished in several ways; for instance, by means of their adjacency matrices.

	$[i,j]$	$[i,k]$	$[j,k]$		$[i,j]$	$[i,k]$	$[j,k]$		$[i,j]$	$[i,k]$	$[j,k]$		$[i,j]$	$[i,k]$	$[j,k]$		$[i,j]$	$[i,k]$	$[j,k]$
$[i,j]$	1	1	1	$[i,j]$	1	1	1	$[i,j]$	1	1	0	$[i,j]$	1	1	1	$[i,j]$	1	0	0
$[i,k]$	1	1	1	$[i,k]$	1	1	1	$[i,k]$	1	1	0	$[i,k]$	0	1	1	$[i,k]$	1	1	1
$[j,k]$	1	1	1	$[j,k]$	0	0	1	$[j,k]$	1	1	1	$[j,k]$	0	0	1	$[j,k]$	0	0	1

Example 10.8

In the fully resolved time-consistent phylogenetic network with eNewick string ((A,(B,(C)h#H)x)y,(h#H,D)z)r;, the subtree induced by the most recent common semi-strict ancestors of the terminal nodes labeled A, B, C, node $y = LCSA(A, B) = LCSA(A, C)$ is an ancestor of node $x = LCSA(B, C)$. In the subtree induced by the terminal nodes labeled A, B, D, node $r = LCSA(A, D) = LCSA(B, D)$ is an ancestor of $y = LCSA(A, B)$. In the subtree induced by the terminal nodes labeled A, C, D, node $r = LCSA(A, D)$ is an ancestor of both $y = LCSA(A, C)$ and $z = LCSA(C, D)$. Finally, in the subtree induced by the terminal nodes labeled B, C, D, node $r = LCSA(B, D)$ is an ancestor of both $x = LCSA(B, C)$ and $z = LCSA(C, D)$.

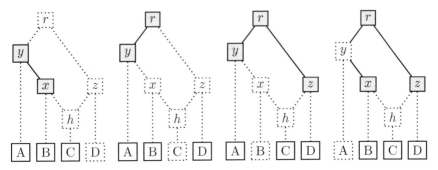

The topological relationships among the most recent common semi-strict ancestors are encoded in the following adjacency matrices.

	$[A,B]$	$[A,C]$	$[B,C]$
$[A,B]$	1	1	1
$[A,C]$	1	1	1
$[B,C]$	0	0	1

	$[A,B]$	$[A,D]$	$[B,D]$
$[A,B]$	1	0	0
$[A,D]$	1	1	1
$[B,D]$	1	1	1

	$[A,C]$	$[A,D]$	$[C,D]$
$[A,C]$	1	0	0
$[A,D]$	1	1	1
$[C,D]$	0	0	1

	$[B,C]$	$[B,D]$	$[C,D]$
$[B,C]$	1	0	0
$[B,D]$	1	1	1
$[C,D]$	0	0	1

The subtree induced by a triplet of terminal nodes of a fully resolved phylogenetic network can be obtained by first finding the most recent common semi-strict ancestor of each pair of terminal nodes from the given triplet and then building the induced subtree by distinguishing among the five possible topologies on the basis of the relationship among the most recent common semi-strict ancestors. In the following description, such a relationship is encoded in an adjacency matrix R.

```
function triplet(N, i, j, k)
    ij ← LCSA(N, i, j)
    ik ← LCSA(N, i, k)
    jk ← LCSA(N, j, k)
    R[ij, ij] ← R[ik, ik] ← R[jk, jk] ← true
    R[ij, ik] ← ij is an ancestor of ik in N
    R[ij, jk] ← ij is an ancestor of jk in N
    R[ik, ij] ← ik is an ancestor of ij in N
    R[ik, jk] ← ik is an ancestor of jk in N
    R[jk, ij] ← jk is an ancestor of ij in N
    R[jk, ik] ← jk is an ancestor of ik in N
    return R
```

The representation of phylogenetic networks in BioPerl does not include any method to compute subtrees of a phylogenetic network induced by the most recent common semi-strict ancestors of terminal nodes. Nevertheless, the subtree of a fully resolved time-consistent phylogenetic tree induced by the most recent common semi-strict ancestors of a triplet of terminal nodes can be easily obtained by using the LCSA method to find the most recent common semi-strict ancestor of the terminal nodes in the given triplet, as illustrated by the following Perl script, where the adjacency matrix of the induced subtree is encoded in a vector in row order and later packed into a string.

```perl
sub triplet {
  my ($net,@triple) = @_;
  my ($i,$j,$k) = @triple;

  my @L = $net->leaves;
  my $ij = LCSA($net,$L[$i],$L[$j]);
  my $ik = LCSA($net,$L[$i],$L[$k]);
  my $jk = LCSA($net,$L[$j],$L[$k]);

  my @R;
  $R[0] = 1;
  $R[1] = $net->graph->SP_Dijkstra($ij,$ik) ? 1 : 0;
  $R[2] = $net->graph->SP_Dijkstra($ij,$jk) ? 1 : 0;
  $R[3] = $net->graph->SP_Dijkstra($ik,$ij) ? 1 : 0;
  $R[4] = 1;
  $R[5] = $net->graph->SP_Dijkstra($ik,$jk) ? 1 : 0;
  $R[6] = $net->graph->SP_Dijkstra($jk,$ij) ? 1 : 0;
  $R[7] = $net->graph->SP_Dijkstra($jk,$ik) ? 1 : 0;
  $R[8] = 1;

  return join "", @R;
}
```

The representation of phylogenetic networks in R does not include any method to compute subtrees of a phylogenetic network induced by the most recent common semi-strict ancestors of a triplet of terminal nodes, either. However, a triplet function can easily be defined using the LCSA function, as illustrated by the following R script.

```r
> library(igraph)

> triplet <- function (net,i,j,k) {
  ij <- LCSA(net,i,j)
  ik <- LCSA(net,i,k)
  jk <- LCSA(net,j,k)

  L <- c(V(net)[i]$name,V(net)[j]$name,V(net)[k]$name
    )
  R <- matrix(rep(NA,9),nrow=3,dimnames=list(L,L))
```

```
  R[1,1] <- TRUE
  R[1,2] <- length(get.all.shortest.paths(net,ij,ik,
    mode="out")) != 0
  R[1,3] <- length(get.all.shortest.paths(net,ij,jk,
    mode="out")) != 0
  R[2,1] <- length(get.all.shortest.paths(net,ik,ij,
    mode="out")) != 0
  R[2,2] <- TRUE
  R[2,3] <- length(get.all.shortest.paths(net,ik,jk,
    mode="out")) != 0
  R[3,1] <- length(get.all.shortest.paths(net,jk,ij,
    mode="out")) != 0
  R[3,2] <- length(get.all.shortest.paths(net,jk,ik,
    mode="out")) != 0
  R[3,3] <- TRUE

  R
}

> el <- matrix(c("r","z","r","y","y","A","y","x","z",
   "h","z","D","x","B","x","h","h","C"),nc=2,byrow=
   TRUE)
> net <- graph.edgelist(el)

> V(net)
Vertex sequence:
[1] "r" "z" "y" "A" "x" "h" "D" "B" "C"

> triplet(net,3,7,8)
      A     B    C
A  TRUE  TRUE TRUE
B  TRUE  TRUE TRUE
C FALSE FALSE TRUE
> triplet(net,3,7,6)
     A     B     D
A TRUE FALSE FALSE
B TRUE  TRUE  TRUE
D TRUE  TRUE  TRUE
> triplet(net,3,8,6)
      A     C     D
A  TRUE FALSE FALSE
C  TRUE  TRUE  TRUE
D FALSE FALSE  TRUE
> triplet(net,7,8,6)
      B     C     D
```

B	TRUE	FALSE	FALSE
C	TRUE	TRUE	TRUE
D	FALSE	FALSE	TRUE

10.2 Finding Common Subgraphs

Subgraphs shared by two graphs reveal information common to the two graphs. As there are several ways in which a graph can be contained in another graph, common subgraphs can be connected, bottom-up, top-down, induced by a set of terminal nodes, or induced by the most recent common semi-strict ancestors of a set of terminal nodes. Further, in order to reveal the most of their shared information, it is interesting to find common subgraphs of largest size between two given graphs.

10.2.1 Maximum Agreement of Rooted Networks

Two (unrooted or rooted) phylogenetic networks are said to agree on a set of terminal nodes if they have isomorphic bottom-up subgraphs with that set of terminal nodes, and a maximum agreement subgraph of two phylogenetic networks is a common bottom-up subgraph with the largest possible number of terminal nodes.

Example 10.9

Consider again the fully resolved phylogenetic networks with eNewick string $((A,((B,(C)h1\#H1)w)h2\#H2)x,(h1\#H1,(h2\#H2,D)y)z)r$; and with eNewick string $((A,(B,(C)h\#H)x)y,(h\#H,D)z)r$; from previous examples. The two networks agree on the set of terminal nodes labeled A, B, C, and this is the largest set of terminal nodes on which they agree, as shown below. In fact, the subgraph of the first phylogenetic network rooted at node x has three terminal nodes labeled A, B, C and is isomorphic (after contraction of elementary paths) to the subgraph of the second phylogenetic network rooted at node y. On the other hand, the subgraphs of the first phylogenetic network rooted at nodes y and z also have three terminal nodes, labeled B, C, D, but no bottom-up subgraph of the second phylogenetic network has these terminal nodes. Thus, the maximum agreement subgraph of the two networks has three terminal nodes.

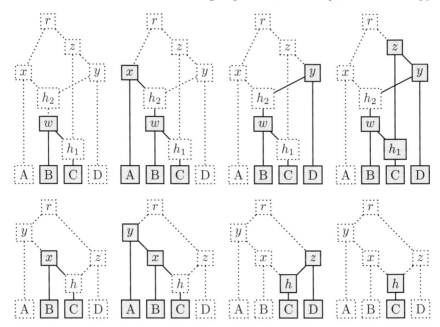

Computing the maximum agreement of two fully resolved rooted phylogenetic networks requires obtaining the subgraph rooted at each internal node
in each of the two networks and also testing each subgraph of one network
and each subgraph of the other network for isomorphism.

The subgraph rooted at an internal node of a phylogenetic network can be
obtained by adding all the descendant nodes of the given internal node to
an empty graph and then adding a directed edge between each pair of nodes
of the subgraph that are connected in the original graph. In the following
description, a directed edge (v, w) of the original graph N is added to the
subgraph N' only if both v and w are also nodes of N', and any elementary
paths are contracted by removing each elementary node v right after adding
a directed edge from the only parent u to the only child w of node v.

```
function bottom up subgraph(N, v)
    N' ← ∅
    for each node w of N do
        if v is an ancestor of w in N then
            add w to N'
    for each edge (v, w) of N do
        if v and w are nodes of N' then
            add edge (v, w) to N'

    while N' has some elementary node do
        v := an elementary node of N'
        u := parent of v in N'
```

$w :=$ child of v in N'
add edge (u, w) to N'
delete node v from N'
return N'

The representation of phylogenetic networks in BioPerl does not include any method to compute the subgraph rooted at an internal node in the directed acyclic graph representation of a phylogenetic network. However, the descendant nodes of the given node $node in the directed acyclic graph $graph can be easily obtained with the help of a queue @Q of nodes, and the previous algorithm can be implemented as shown in the following Perl script.

```perl
sub bottom_up_subgraph {
  my $graph = shift;
  my $node = shift;
  my $sub = Graph->new;
  my @Q = ($node);
  while (@Q) {
    $node = shift @Q;
    $sub->add_vertex($node);
    push @Q, $graph->successors($node);
  }
  for my $edge ($graph->edges) {
    if ($sub->has_vertex(@$edge[0]) &&
        $sub->has_vertex(@$edge[1])) {
      $sub->add_edge(@$edge[0],@$edge[1]);
    }
  }
  while (my @V = grep {
      $sub->in_degree($_)==1 &&
      $sub->out_degree($_)==1 }
      $sub->interior_vertices) {
    my $v = shift @V;
    my @U = $sub->predecessors($v);
    my $u = shift @U;
    my @W = $sub->successors($v);
    my $w = shift @W;
    $sub->add_edge($u,$w);
    $sub->delete_vertex($v);
  }
  return $sub;
}
```

The representation of phylogenetic networks in R does not include any method to compute the subgraph rooted at an internal node in the directed acyclic graph representation of a phylogenetic network, either. However, the previous algorithm can be easily implemented using the `subgraph` function of the R package iGraph to obtain the subgraph induced by a set of nodes, as illustrated by the following R script.

```
bottom.up.subgraph <- function (net,i) {
  Q <- c(i)
  S <- c()
  while (length(Q)>0) {
    i <- Q[1]
    Q <- Q[-1]
    S <- c(S,i)
    Q <- c(Q,as.vector(V(net)[nei(i,mode="out")]))
  }
  g <- subgraph(net,unique(S))
  while (length(V(g)[which(degree(g,mode="in")==1 &
      degree(g,mode="out")==1)-1])>0) {
    D <- V(g)[which(degree(g,mode="in")==1 &
      degree(g,mode="out")==1)-1]
    v <- head(c(D),n=1) # elementary node
    u <- V(g)[nei(v,"in")]
    w <- V(g)[nei(v,"out")]
    g <- add.edges(g,c(V(g)[u],V(g)[w]))
    g <- delete.vertices(g,V(g)[v])
  }
  g
}
```

Computing the maximum agreement of two phylogenetic networks also requires testing each subgraph of one network and each subgraph of the other network for isomorphism. Testing isomorphism of directed acyclic graphs is as hard as testing graph isomorphism in general, but in the case of phylogenetic networks, the set of descendant terminal node labels of each node, known as the *cluster map* of the network, allows for a significant reduction in the number of possible isomorphic mappings of the nodes of one network to the nodes of the other network, since in an isomorphic mapping, the nodes of one network can only be mapped to *equivalent* nodes of the other network, that is, to nodes with the same cluster of descendant node labels.

Mapping nodes in one network only to equivalent nodes in the other network is the basis of a fast backtracking algorithm for testing isomorphism of phylogenetic networks.

Example 10.10

The following fully resolved tree-child phylogenetic network with eNewick strings `(((A,(C)h1#H1)x,(B)h2#H2)y,(h2#H2,(h1#H1,D)z)w)r;` (left) and

`((A,((B,(C)h1#H1)w)h2#H2)x,(h1#H1,(h2#H2,D)y)z)r;` (right) have the same number of nodes, the same number of directed edges, the same terminal node labels, the same number of tree nodes, and the same number of hybrid nodes.

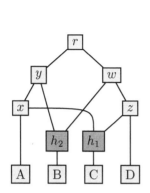 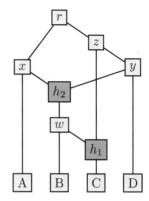

However, they differ in their cluster maps, and there are nodes in one network with no equivalent node in the other network, while there are nodes in one network with more than one equivalent node in the other network.

node	cluster	equivalent	node	cluster	equivalent
r	A, B, C, D	r	r	A, B, C, D	r
w	B, C, D	y, z	w	B, C	—
x	A, C	—	x	A, B, C	y
y	A, B, C	x	y	B, C, D	w
z	C, D	—	z	B, C, D	w
h_1	C	h_1, C	h_1	C	h_1, C
h_2	B	B	h_2	B, C	—
A	A	A	A	A	A
B	B	B	B	B	h_2, B
C	C	h_1, C	C	C	h_1, C
D	D	D	D	D	D

The backtracking test of two phylogenetic networks, or two bottom-up subgraphs of phylogenetic networks, for isomorphism involves computing the set of equivalent nodes in the second network for each node of the first network and then trying to extend an empty mapping of nodes of the first network to equivalent nodes of the second network in all possible ways, until either an isomorphic mapping is found (in which case the two networks are isomorphic) or no isomorphic mapping can be found (because the networks are not isomorphic).

In the following description, N_1 and N_2 are assumed to have the same number of nodes and directed edges, the nodes v_2 of N_2 that are equivalent to each node v_1 of N_1 are collected in $X[v_1]$, each node v_1 is mapped to an equivalent node $M[v_1] \in X[v_1]$, and the Boolean variable *found* indicates whether or not the phylogenetic networks have been found to be isomorphic.

function graph isomorphic(N_1, N_2)
    $C_1 \leftarrow$ cluster map of N_1
    $C_2' \leftarrow$ cluster map of N_2
    for each node v_1 of N_1 **do**
        $X[v_1] \leftarrow \emptyset$
        for each node v_2 of N_2 **do**
            if $C_1[v_1] = C_2[v_2]$ **then**
                $X[v_1] \leftarrow X[v_1] \cup \{v_2\}$
    $v_1 \leftarrow$ undefined
    $M \leftarrow \emptyset$
    $V_1 \leftarrow$ nodes of N_1 in some fixed order
    found $\leftarrow$ false
    backtrack$(N_1, N_2, X, v_1, M, V_1, found)$
    return *found*

The mapping M of the nodes of N_1 to the nodes of N_2, initially empty, is extended in all possible ways by mapping each node v_1 to each equivalent node $v_2 \in X[v_1]$ in turn, while ensuring that the resulting mapping M is valid, until all the nodes of N_1 have been mapped to the nodes of N_2. In the following description, M is extended according to a fixed, but arbitrary, order V_1 of the nodes of N_1. The extension of M by mapping v_1 to $M[v_1]$ is valid if, for each other node w_1 of N_1 already mapped to a node $M[w_1]$ of N_2, either v_1 is adjacent to w_1 in N_1 and $M[v_1]$ is adjacent to $M[w_1]$ in N_2, or v_1 is not adjacent to w_1 in N_1 and $M[v_1]$ is not adjacent to $M[w_1]$ in N_2. Also, M is an isomorphic mapping of N_1 to N_2 if all the nodes of N_1 have been mapped to the nodes of N_2, that is, if V_1 is empty, in which case *found* is set to true.

procedure backtrack$(N_1, N_2, X, v_1, M, V_1, found)$
    if not *found* and M is valid **then**
        if V_1 is empty **then**
            found $\leftarrow$ true
        else
            $v_1 \leftarrow$ first node in V_1
            $V_1 \leftarrow V_1 \setminus \{v_1\}$
            for each node v_2 in $X[v_1]$ **do**
                $M[v_1] \leftarrow v_2$
                backtrack$(N_1, N_2, X, v_1, M, V_1, found)$
                $M[v_1] \leftarrow$ undefined

The representation of phylogenetic networks in Perl by means of directed acyclic graphs makes it possible to implement the previous backtracking algorithm in a straightforward way. In the following Perl script, the cluster equality test is performed by the is_LequivalentR method of the List::Compare Perl module, and references to the hash %X of equivalent nodes, to the hash %M of mapped nodes, to the array @V of remaining nodes to be mapped, and to the variable $found are passed together with the variable $node1 to the backtrack subroutine as arguments.

The input directed acyclic graphs $g1 and $g2 are assumed to have the same number of nodes and directed edges, and the could_be_isomorphic method of the Graph::Directed Perl module returns true if they have the same number of nodes and edges and also the same distribution of node in-degrees and out-degrees, that is, if the fully resolved rooted phylogenetic networks have the same number of tree nodes and hybrid nodes.

```
sub graph_isomorphic {
  my $g1 = shift;
  my $g2 = shift;
  my $found = 0;
  if ($g1->could_be_isomorphic($g2)) {
    my $C1 = cluster_map($g1);
    my $C2 = cluster_map($g2);
    my %X;
    for my $node1 ($g1->vertices) {
      for my $node2 ($g2->vertices) {
        my $lc = List::Compare->new(@{$C1}{$node1},
          @{$C2}{$node2});
        if ($lc->is_LequivalentR) {
          push @{$X{$node1}}, $node2;
        }
      }
    }
    my $node1 = undef;
    my %M;
    my @V1 = sort $g1->vertices;
    backtrack($g1,$g2,\%X,$node1,\%M,\@V1,\$found);
  }
  return $found;
}
```

In the backtrack subroutine, validity of the extension of %M by mapping $node1 to $M{node1} is determined by checking that $prev1 is adjacent to $node1 in $g1 if and only if $M{$node1} is adjacent to $M{$prev1} in $g2, for all nodes $prev1 already mapped. Notice that a reference to a scalar value is dereferenced by either enclosing the reference inside ${} or just adding a second dollar sign in front of the variable name.

```
sub backtrack {
```

```perl
my ($g1,$g2,$X,$node1,$M,$V1,$found) = @_;
my %X = %{$X};
my %M = %{$M};
my @V1 = @{$V1};
unless ($$found) {
  for my $prev1 (keys %M) {
    if ($g1->has_edge($prev1,$node1) &&
        !$g2->has_edge($M{$prev1},$M{$node1}) ||
        $g1->has_edge($node1,$prev1) &&
        !$g2->has_edge($M{$node1},$M{$prev1}) ||
        !$g1->has_edge($prev1,$node1) &&
        $g2->has_edge($M{$prev1},$M{$node1}) ||
        !$g1->has_edge($node1,$prev1) &&
        $g2->has_edge($M{$node1},$M{$prev1})) {
      return;
    }
  }
  if (scalar @V1) {
    $node1 = shift @V1;
    for my $node2 (@{$X{$node1}}) {
      $M{$node1} = $node2;
      backtrack($g1,$g2,$X,$node1,\%M,\@V1,$found);
      $M{$node1} = undef;
    }
  } else {
    $$found = 1;
  }
}
}
```

The cluster map of a phylogenetic network can be obtained by collecting in cluster $C[v]$ each terminal node w that is a descendant of node v in the network.

function cluster map(N)
    for each node v of N **do**
        $C[v] \leftarrow \emptyset$
        for each terminal node w of N **do**
            if v is an ancestor of w in N **then**
                $C[v] \leftarrow C[v] \cup \{w\}$
    return C

This algorithm for computing the cluster map of a phylogenetic network can also be easily implemented in Perl, as shown in the following script.

```perl
sub cluster_map {
```

```perl
  my $graph = shift;
  my %C;
  for my $node ($graph->vertices) {
    for my $leaf ($graph->successorless_vertices) {
      if ($graph->is_reachable($node,$leaf)) {
        push @{$C{$node}}, $leaf;
      }
    }
  }
  return \%C;
}
```

The representation of phylogenetic networks in R, on the other hand, has a function `graph.isomorphic` for testing isomorphism of two directed graphs or two undirected graphs.

Now, the maximum agreement of two fully resolved rooted phylogenetic networks is computed by testing the subgraph rooted at each internal node of one network and the subgraph rooted at each internal node of the other network for isomorphism. In the following description, a common bottom-up subgraph C of two fully resolved rooted phylogenetic networks N_1 and N_2 with the largest set L_1 of terminal nodes is obtained.

> **function** maximum agreement subgraph(N_1, N_2)
> $C \leftarrow$ empty graph
>   size $\leftarrow 0$
>   **for** each node v_1 of N_1 **do**
> $G_1 \leftarrow$ bottom up subgraph(N_1, v_1)
>   **for** each node v_2 of N_2 **do**
> $G_2 \leftarrow$ bottom up subgraph(N_2, v_2)
>   **if** graph isomorphic(G_1, G_2) **then**
> $L_1 \leftarrow$ terminal nodes in G_1
>   **if** length$(L_1) >$ size **then**
> $C \leftarrow G_1$
>   size $\leftarrow$ length(L_1)
>   **return** C

The representation of phylogenetic networks in Perl by means of directed acyclic graphs also makes it easy to implement the previous algorithm. In the following Perl script, the terminal nodes of the bottom-up subgraph rooted at a node of a network are obtained by means of the `successorless_vertices` method of the `Graph::Directed` Perl module.

```perl
sub max_agreement_subgraph {
  my $net1 = shift;
  my $net2 = shift;
  my $graph1 = $net1->graph;
```

```perl
my $graph2 = $net2->graph;
my $common = Graph->new;
my $size = 0;
for my $node1 ($graph1->interior_vertices) {
  my $sub1 = subgraph($graph1,$node1);
  for my $node2 ($graph2->interior_vertices) {
    my $sub2 = subgraph($graph2,$node2);
    if (graph_isomorphic($sub1,$sub2)) {
      my $leaves1 = $sub1->successorless_vertices;
      if ($leaves1 > $size) {
        $common = $sub1;
        $size = $leaves1;
      }
    }
  }
}
return $common;
}
```

The representation of phylogenetic networks in R does not include any method to compute the maximum agreement of two phylogenetic networks, either. However, the previous algorithm can be easily implemented using the bottom.up.subgraph function and the graph.isomorphic function defined above, as illustrated by the following R script.

```r
max.agreement.subgraph <- function (net1,net2) {
  max <- graph.empty()
  max.size <- 0
  for (i in V(net1)) {
    g1 <- bottom.up.subgraph(net1,i)
    for (j in V(net2)) {
      g2 <- bottom.up.subgraph(net2,j)
      if (graph.isomorphic(g1,g2)) {
        L <- V(g1)[which(degree(g1,mode="out")==0)-1]
        if (length(L)>max.size) {
          max <- g1
          max.size <- length(L)
        }
      }
    }
  }
  max
}
```

The maximum agreement subgraph of the two fully resolved phylogenetic networks from the previous example has, indeed, three terminal nodes.

```
> library(igraph)
> el <- matrix(c("r","x","r","z","z","h1","z","y","y"
  ,"h2","y","D","x","A","x","h2","h2","w","w","B","w
  ","h1","h1","C"),nc=2,byrow=TRUE)
> net1 <- graph.edgelist(el)
> el <- matrix(c("r","z","r","y","y","A","y","x","z",
  "h","z","D","x","B","x","h","h","C"),nc=2,byrow=
  TRUE)
> net2 <- graph.edgelist(el)
> max.agreement.subgraph(net1,net2)
Vertices: 5
Edges: 4
Directed: TRUE
Edges:

[0] x -> A
[1] w -> B
[2] w -> C
[3] x -> w
```

10.3 Comparing Graphs

The similarities and differences between two phylogenetic networks can be assessed by computing a distance measure between the two phylogenetic networks. The *triplets distance* is based on the subtrees induced by the most recent common semi-strict ancestors of triplets of terminal nodes in two rooted phylogenetic networks.

10.3.1 The Triplets Distance between Graphs

The *triplets distance* is based on the subtrees induced by the most recent common semi-strict ancestors of triplets of terminal nodes in the two networks under comparison. The sets of subtrees induced by the most recent common semi-strict ancestors of triplets of terminal nodes reveal similarities and differences between two fully resolved rooted phylogenetic networks.

The triplets distance between two fully resolved phylogenetic networks is defined as the size of the symmetric difference of their sets of subtrees induced by the most recent common semi-strict ancestors of triplets, that is, the number of induced subtrees in which the two phylogenetic networks differ.

Example 10.11

Consider once more the fully resolved tree-child phylogenetic networks with the eNewick string `(((A,(C)h1#H1)x,(B)h2#H2)y,(h2#H2,(h1#H1,D)z)w)r;` and `((A,(B,(C)h#H)x)y,(h#H,D)z)r;` from previous examples. They have $n = 4$ terminal nodes labeled A, B, C, D.

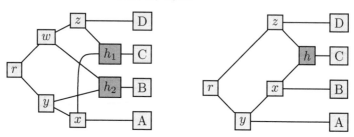

The $\binom{4}{3} = 4$ subtrees induced by the most recent common semi-strict ancestors of triplets of terminal nodes in the first phylogenetic network are given in the following table.

	[A,B]	[A,C]	[B,C]		[A,B]	[A,D]	[B,D]		[A,C]	[A,D]	[C,D]		[B,C]	[B,D]	[C,D]
[A,B]	1	1	0	[A,B]	1	0	0	[A,C]	1	0	0	[B,C]	1	1	1
[A,C]	0	1	0	[A,D]	1	1	1	[A,D]	1	1	1	[B,D]	0	1	1
[B,C]	1	1	1	[B,D]	0	0	1	[C,D]	0	0	1	[C,D]	0	0	1

The subtrees induced by the most recent common semi-strict ancestors of triplets of terminal nodes in the second phylogenetic network are as follows.

	[A,B]	[A,C]	[B,C]		[A,B]	[A,D]	[B,D]		[A,C]	[A,D]	[C,D]		[B,C]	[B,D]	[C,D]
[A,B]	1	1	1	[A,B]	1	0	0	[A,C]	1	0	0	[B,C]	1	0	0
[A,C]	1	1	1	[A,D]	1	1	1	[A,D]	1	1	1	[B,D]	1	1	1
[B,C]	0	0	1	[B,D]	1	1	1	[C,D]	0	0	1	[C,D]	0	0	1

The two phylogenetic networks differ in 3 of the 4 induced subtrees and, thus, their triplets distance is 6.

The triplets distance between two phylogenetic networks can be computed by first obtaining the subtree induced by the most recent common semi-strict ancestors of each set of three terminal nodes in each of the networks and then counting the number of induced subtrees in which the two networks differ. In the following description, the two sets of $\binom{n}{3}$ subtrees are obtained using the previous algorithm for finding the subgraph induced by the most recent common semi-strict ancestors of a triplet of terminal nodes, upon each set of three terminal node labels in turn.

```
function triplets_distance(N₁, N₂)
    L ← terminal node labels in N₁ and N₂
    n ← length(L)
    d ← 0
    for i ← 1, …, n do
        for j ← i + 1, …, n do
            for k ← j + 1, …, n do
                R₁ ← triplet(N₁, L[i], L[j], L[k])
                R₂ ← triplet(N₂, L[i], L[j], L[k])
                if R₁ ≠ R₂ then
                    d ← d + 2
    return d
```

The representation of phylogenetic networks in BioPerl does not include any method to compute the triplets distance between two fully resolved rooted phylogenetic networks with the same terminal node labels. However, the sets of subtrees induced by the most recent common semi-strict ancestors of triplets of terminal nodes can be computed using the `triplet` method and the previous algorithm can easily be implemented, as shown in the following Perl script.

```perl
sub triplets_distance {
  my $net1 = shift;
  my $net2 = shift;
  my $comp = Array::Compare->new;
  my @L = $net1->leaves;
  my $dist = 0;
  for (my $i = 0; $i < @L; $i++) {
    for (my $j = $i+1; $j < @L; $j++) {
      for (my $k = $j+1; $k < @L; $k++) {
        my $t1 = triplet($net1,$i,$j,$k);
        my $t2 = triplet($net2,$i,$j,$k);
        $dist += 2 unless ($t1 eq $t2);
      }
    }
  }
  return $dist;
}
```

The representation of phylogenetic trees in R does not include any method to compute the triplets distance between two phylogenetic networks, either. However, the sets of subtrees induced by the most recent common semi-strict ancestors of triplets of terminal nodes can be computed using the `triplet` function and the previous algorithm can be implemented in a straightforward way, as illustrated by the following R script.

```
triplets.distance <- function (net1,net2) {
  L <- sort(V(net1)[which(degree(net1,mode="out")==0)
    -1]$name)
  d <- 0
  for (i in L[1:(length(L)-2)]) {
    i1 <- V(net1)[V(net1)$name == i]
    i2 <- V(net2)[V(net2)$name == i]
    for (j in L[(match(i,L)+1):(length(L)-1)]) {
      j1 <- V(net1)[V(net1)$name == j]
      j2 <- V(net2)[V(net2)$name == j]
      for (k in L[(match(j,L)+1):length(L)]) {
        k1 <- V(net1)[V(net1)$name == k]
        k2 <- V(net2)[V(net2)$name == k]
        t1 <- triplet(net1,i1,j1,k1)
        t2 <- triplet(net2,i2,j2,k2)
        if (!isTRUE(all.equal(t1,t2))) { d <- d + 2 }
      }
    }
  }
  d
}

> el <- matrix(c("r","y","r","w","y","x","y","h2","w"
  ,"h2","w","z","x","A","x","h1","h2","B","z","h1","
  z","D","h1","C"),nc=2,byrow=TRUE)
> net1 <- graph.edgelist(el)

> el <- matrix(c("r","z","r","y","y","A","y","x","z",
  "h","z","D","x","B","x","h","h","C"),nc=2,byrow=
  TRUE)
> net2 <- graph.edgelist(el)

> triplets.distance(net1,net2)
[1] 6
```

The triplets distance is a metric on the space of all time-consistent tree-child phylogenetic networks, and it generalizes the triplets distance between rooted phylogenetic trees.

Bibliographic Notes

The problem of reconstructing a galled-tree from a set of rooted triplets of terminal nodes was studied in (Jansson and Sung 2006; Jansson et al. 2006). See also (He et al. 2006).

The maximum agreement of galled-trees was introduced by Choy et al. (2005) and further studied in (Jansson and Sung 2008). Bottom-up subgraphs and graph isomorphism algorithms are discussed in more detail in (Valiente 2002). See also (Read and Corneil 1977; Gati 1979).

The triplets distance between phylogenetic networks was introduced by Cardona et al. (2009b).

Appendix A

Elements of Perl

A brief introduction to Perl is given in this appendix by way of sample scripts that solve a simple computational biology problem using different methods. These scripts are then dissected in order to explain basic aspects of the Perl language, followed by an overview of more advanced aspects of the language, not covered in the sample scripts. This is all summarized for convenience in a Perl quick reference card.

A.1 Perl Scripts

Perl is an interpreted scripting language. A Perl program is a *script* containing a series of instructions, which are interpreted when the program is run instead of being compiled first into machine instructions and then assembled or linked into an executable program, thus avoiding the need for separate compilation and linking.

There are Perl distributions available for almost every computing platform, and free distributions can be downloaded from `http://www.perl.org/`. The actual mechanism of running a script using a Perl interpreter depends on the particular operating system, the common denominator being the Unix or Linux command line. Assuming one of the Perl scripts shown further below was already written using a text editor and stored in a file named `sample.pl` (where `pl` is the standard file extension for Perl scripts), the following command invokes the Perl interpreter on the sample script:

```
$ perl sample.pl
```

The simple computational biology problem at hand consists of translating to protein a messenger RNA sequence stored (in 5′ to 3′ direction) in a text file. Recall that the primary structure of a protein can be represented as a sequence over the alphabet of amino acids A (alanine, Ala), R (arginine, Arg), N (asparagine, Asn), D (aspartate, Asp), C (cysteine, Cys), E (glutamate, Glu), Q (glutamine, Gln), G (glycine, Gly), H (histidine, His), I (isoleucine, Ile), L (leucine, Leu), K (lysine, Lys), M (methionine, Met), F (phenylalanine, Phe), P (proline, Pro), S (serine, Ser), T (threonine, Thr), W (tryptophan, Trp), Y (tyrosine, Tyr), and V (valine, Val).

A codon of three nucleotides is translated into a single amino acid within a protein, with translation beginning with a start codon (AUG) and ending with a stop codon (UAA, UAG, or UGA). The $4^3 = 64$ different nucleotide triplets code for 20 amino acids, one translation start signal (methionine, one of these amino acids) and three translation stop signals, with some redundancies. The genetic code defines a mapping between codons and amino acids, and despite variations in the genetic code across species, there is a standard genetic code common to most species, shown in the following circular table.

¿ sapply(1:16,function(x)x*45/2) [1] 22.5 45.0 67.5 90.0 112.5 135.0 157.5 180.0 202.5 225.0 247.5 270.0 292.5 315.0 337.5 360.0

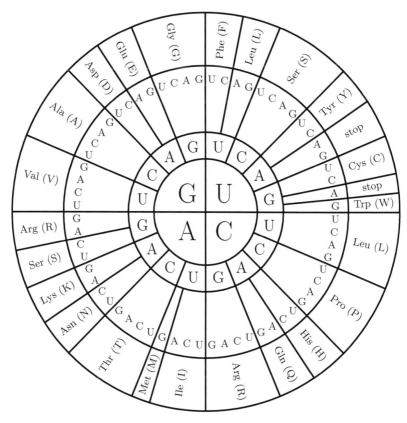

A codon is looked up in the circular genetic code table by starting with the first nucleotide at the center of the table, following with the second nucleotide at the inner circle, and finishing with the third nucleotide at the outer circle. For instance, UAC and UAU both code for Tyr (Y).

Let us assume first the input messenger RNA sequence is already stored in a variable $rna and, thus, readily available for translation to protein. The translation method will consist of skipping any nucleotides before the first start codon and then translating the rest of the sequence to protein using the standard genetic code, until the first stop codon. As an example, translating sequence GUCGCCAUGAUGGUGGUUAUUAUACCGUCAAGGACUGUGUGACUA to protein involves skipping GUCGCC and then translating AUGAUGGUGGUUAUUAUACCG UCAAGGACUGUGUGA to MVVIIPSRTV, because AUG codes for Met (M), GUG and GUU code for Val (V), AUU and AUA code for Ile (I), CCG for Pro (P), UCA for Ser (S), AGG for Arg (R), and ACU codes for Thr (T).

First Script

In a Perl script, the messenger RNA sequence stored in variable $rna can be translated to a protein sequence, stored in a variable $protein, and output to the console window, as shown in the following script.

```
#!/usr/bin/perl -w
use strict;

my $rna = "GCCAAUGACUAAGGCCUAAAGA";
my $protein = rna_to_protein($rna);
print "$rna translates to $protein\n";
```

Well, this is not quite the case. The Perl interpreter cannot run this short script because rna_to_protein has not been defined yet. Nevertheless, let us dissect this script in order to discuss some basic elements of Perl programming.

The Perl script contains a few header lines followed by some instructions to translate the $rna sequence and output the resulting $protein sequence. The first header line, #!/usr/bin/perl -w, points to the Perl interpreter and turns warnings on by setting the -w command line switch. This is perhaps the most important tool for writing good Perl code, because turning warnings on will make the Perl interpreter report on various things that are almost always sources of error in a Perl script. Further, the second header line is an instruction to make the Perl interpreter enforce good programming style by reporting on, among other things, any missing variable declarations.

The first line after the header declares the scalar variable $rna and makes the character string "GCCAAUGACUAAGGCCUAAAGA" the value stored in the memory location referenced by that variable. In other words, it assigns the character string to the variable, and a variable can be assigned different values at the same or different places in a Perl script. Unlike variables, the character string is a constant, whose value cannot change unless the Perl script is changed.

Variables are thus named references to memory locations, and there are scalar, array, and hash variables in Perl. Scalar variables hold scalar values such as numbers and character strings, and their names must begin with a dollar sign, such as $rna. Array and hash variables, on the other hand, hold lists of values, and their names must begin with either an at sign, such as @rna, for array variables or a percent sign, such as %rna, for hash variables.

Almost any name can follow the dollar, at, or percent sign in a variable name, as long as it consists of uppercase or lowercase letters, digits, or underscores and begins with a letter, with a few exceptions. Some variable names have a special meaning, such as $_, @_, @ARGV, and $1 through $9, which cannot be used to declare new variables. Besides, variable names are case sensitive and, thus, $rna, $Rna, and $RNA all reference different memory locations when used in the same Perl script.

An initial value can be assigned to a variable when declaring it,

```
my $rna = "GCCAAUGACUAAGGCCUAAAGA";
```

or the variable can be declared first and later have a value assigned,

```
my $rna;
...
$rna = "GCCAAUGACUAAGGCCUAAAGA";
```

In any case, a value is assigned to a variable in an assignment instruction or statement that resembles an equation, with the variable name to the left and the value to the right of the equal sign. Like most Perl statements, the assignment ends with a semicolon.

The declaration of a variable consists of the variable name preceded by `my` and followed by a semicolon. These variables are called *lexical*, and their scope is local to the block where they are declared, something that will be made clear further below, unlike global variables, which do not need to be declared and have the whole script as scope. Global variables are frequent sources of error and, thus, they are forbidden by the `use strict;` header line.

More than one variable can be declared in a single instruction, with the variable names enclosed in parentheses and separated by commas. Thus,

```
my $dna;
my $rna;
my $protein;
```

and

```
my ($dna, $rna, $protein);
```

both declare the same three variables.

In the next line of the Perl script, the `rna_to_protein` subroutine is invoked upon the value of the variable $rna, and the resulting value is assigned to the variable $protein. Before discussing subroutines, let us see how the protein sequence stored in variable $protein finds its way to the console window.

A `print` statement outputs the value of one or more constants or variables to the console window, also known as *standard output* in Unix and Linux. The value returned by a subroutine invocation can also be output in this way. In a print statement, the values to be output are separated by commas, as in

```
print $rna," translates to ",$protein,"\n";
```

or

```
print $rna,"␣translates␣to␣",rna_to_protein($rna),
"\n";
```

which both output the value of the variable `$rna` followed by the string constant " `translates to` " followed by either the value of the variable `$protein` or the result of the subroutine invocation `rna_to_protein($rna)`, followed also by the special string constant `"\n"` that indicates the end of an output line on the console window, and ending with a semicolon.

String constants must be enclosed in either double or single quotes in Perl, and a string enclosed in double quotes is subject to *variable interpolation*, the substitution of the value for any variable embedded in the string. In the last line of the Perl script, the values of the variables `$rna` and `$protein` are interpolated in a string constant,

```
print "$rna␣translates␣to␣$protein\n";
```

and invoking the Perl interpreter on the sample script will result in the following output,

```
GUCGCCAUGAUGGUGGUUAUUAUACCGUCAAGGACUGUGUGACUA trans
lates to MVVIIPSRTV
```

where this long output line might be broken at a different position, depending on the actual console window size.

Let us come back to discussing subroutines now. As in any programming language, there are various functions readily available in Perl, which take one or more arguments as input and give a single result as output. Examples include numeric functions such as `abs` (absolute value), `exp` (raise to a power), `int` (integer part of a real number), `log` (natural logarithm), and `sqrt` (square root), and string functions such as `lc` (convert to lowercase), `up` (uppercase), and `length` (number of characters), among many others.

Subroutines are, like functions, named pieces of code that perform a specific task and are evaluated when invoked upon particular argument values. Unlike functions, however, subroutines need not return a value, although they often do, and the `rna_to_protein` subroutine will take a string (a messenger RNA sequence) as input and return a string (a protein sequence) as output.

A subroutine begins with `sub` followed by the subroutine name and by the subroutine code enclosed in braces, such as in

```
sub rna_to_protein {
    ...
}
```

In this subroutine, the input messenger RNA sequence, stored in a string variable `$rna`, will be translated to protein and stored in a string variable `$protein` by first skipping any nucleotides before the first start codon and then translating the rest of the sequence to protein using the standard genetic code, until the first stop codon. The translation of each codon to amino acid will be done in turn in a `codon_to_amino_acid` subroutine, to be defined later. The code of the subroutine at hand is as follows.

```
sub rna_to_protein {
  my $rna = shift;
  my $protein;
  my $i = 0;
  while ($i < length($rna) - 2 &&
      substr($rna,$i,3) ne "AUG") { # start codon
    $i++;
  }
  $i += 3; # skip the start codon
  while ($i < length($rna) - 2 &&
      substr($rna,$i,3) ne "UAA" &&
      substr($rna,$i,3) ne "UAG" &&
      substr($rna,$i,3) ne "UGA") {
    my $codon = substr($rna,$i,3);
    $protein .= codon_to_amino_acid($codon);
    $i += 3;
  }
  return $protein;
}
```

Values such as the actual sequence stored in the string variable `$rna` can be passed to a subroutine as arguments, by listing them right after the subroutine name. Within the subroutine, these values are accessed as a special temporary array named `@_` either all together or one by one.

Recall that array variables hold lists of values. For instance, an array variable named `@stop` can be declared and have the three stop codons assigned as a list of values,

```
my @stop = ("UAA", "UAG", "UGA");
```

or it can be declared first and then have each of the three stop codons stored at a different position in the array, starting with position 0,

```
my @stop;
$stop[0] = "UAA";
$stop[1] = "UAG";
$stop[2] = "UGA";
```

or have each of the three stop codons pushed into the array,

```
my @stop;
push @stop, "UAA", "UAG", "UGA";
```

or have each of the stop codons pushed in reverse order into the array,

```
my @stop;
unshift @stop, "UGA", "UAG", "UAA";
```

While the value of the array variable is the whole list, individual elements can be accessed by position, as in the previous assignment instructions, or by either shifting off the first value in the array,

```perl
my $first_stop_codon = shift @stop;
```

or popping the last value in the array,

```perl
my $last_stop_codon = pop @stop;
```

The value of the array variable is the list of elements when evaluated in *list context*, that is, when used in a list valued expression. However, in *scalar context*, when used in a scalar valued expression, the value of the array variable is the number of elements in the array,

```perl
my $length = @stop;
```

In fact, every expression is interpreted in either list or scalar context, and this is a feature almost unique to the Perl language.

Now, within the `rna_to_protein` subroutine, the string value passed as an argument can be accessed by either assigning the whole temporary array to a list of variables, which consists of just one variable in this case,

```perl
sub rna_to_protein {
  my ($rna) = @_;
  ...
}
```

or shifting off the only element in the temporary array, which is the default argument for the `shift` function,

```perl
sub rna_to_protein {
  my $rna = shift;
  ...
}
```

With the string value passed as an argument and stored in the `$rna` variable, the subroutine translates it to protein and returns the string value of the `$protein` variable. The first step consists of skipping any nucleotides before the first start codon, and the scalar variable `$i` will hold the initial position in the `$rna` string of the first start codon, which is position 6 after skipping six nucleotides in the following sample sequence:

```
GUCGCCAUGAUGGUGGUUAUUAUACCGUCAAGGACUGUGUGACUA
.UCGCCAUGAUGGUGGUUAUUAUACCGUCAAGGACUGUGUGACUA
..CGCCAUGAUGGUGGUUAUUAUACCGUCAAGGACUGUGUGACUA
...GCCAUGAUGGUGGUUAUUAUACCGUCAAGGACUGUGUGACUA
....CCAUGAUGGUGGUUAUUAUACCGUCAAGGACUGUGUGACUA
.....CAUGAUGGUGGUUAUUAUACCGUCAAGGACUGUGUGACUA
......AUGAUGGUGGUUAUUAUACCGUCAAGGACUGUGUGACUA
0123456789...
```

The position $i of the first start codon in the $rna string can be found by starting with $i = 0 and increasing the value of $i as many times as needed, until the codon substr($rna,$i,3) in string $rna that begins at position $i and has length 3 is a start codon. This is achieved by continuing to increase $i by one, where $i++ is shorthand for $i = $i + 1, while substr($rna,$i,3) is not a start codon, that is, until substr($rna,$i,3) is a start codon:

```
my $i = 0;
while (substr($rna,$i,3) ne "AUG") { # start codon
    $i++;
}
```

Comments can be placed starting with a hash symbol either at the end of a line or on one or more separate lines, to make the code easier to understand. The substring function substr returns a substring of a given string starting at a given position and of a given length, and, thus, substr($rna,$i,3) is the substring of $rna starting at position $i and of length 3, that is, the codon starting at position $i of string $rna.

Scalar values can be compared by means of various operators, and the result is always a Boolean value. The equality test, == for numeric and eq for string values, returns true if the values are equal and false otherwise. Testing for the opposite, not equal, is done with != for numeric and ne for string values, which return true if the values are different and false if they are equal.

Testing if two numeric values are greater than, or greater than or equal to, each other is done with > and >=, respectively, and testing if two string values are greater than, or greater than or equal to, each other in lexicographical order is done with gt and ge, respectively. Similarly, testing if two numeric values are less than, or less than or equal to, each other is done with < and <=, respectively, and with lt and le for string values, in lexicographical order.

On the other hand, the signed inequality test, <=> for numeric and cmp for string values, returns −1 if the first value is less than the second value, 0 if the values are equal, and 1 if the second value is greater than the first value. Since any numeric value other than 0 qualifies as truth in Perl, the signed inequality test returns true (either −1 or 1) if the values are different and false (0) otherwise. Also, any string value other than the empty string qualifies as truth in Perl, and the Boolean comparison operators return the scalar value 1 (as a number or as a string, depending on the context) for true and the empty string for false.

There are several ways of looping across a block of Perl code. In a *for* loop, a block of code is executed a specific number of times: a variable controlling the loop is set to an initial value, a test expression is evaluated true or false on each repetition of the loop, and the control variable is reset in some way at the end of each loop repetition. For instance, by setting the control variable $i to the initial value $i = 0, executing the block of code as long as the test expression $i < @stop evaluates to true, and increasing the control variable with `$i++` at the end of each repetition, the block of code is executed three times, the length of the `@stop` array:

```perl
my @stop = ("UAA", "UAG", "UGA");
for (my $i = 0; $i < @stop; $i++) {
  print "$stop[$i]\n";
}
```

In a *for each* loop, a block of code is executed once upon each of the values in a given list, without need for the explicit position of the values in the list. For instance, by looping over the list stored in the `@stop` array, the block of code is executed once for each of the three `$codon` values:

```perl
foreach my $codon (@stop) {
  print "$codon\n";
}
```

In a *while* loop, a block of code is executed while a specific condition is met: if the expression controlling the loop evaluates to true, the block of code is executed once and then it continues to execute in a loop until the expression evaluates to false, but if the expression initially evaluates to false, the block of code is not executed at all. For instance, by setting a variable to the initial value `$i = 0` before the loop and then executing the block of code as long as the test expression `$i < @stop` evaluates to true, where the variable is increased with `$i++` within the block of code, the block of code is also executed a number of times equal to the length of the `@stop` array:

```perl
my $i = 0;
while ($i < @stop) {
  print "$stop[$i]\n";
  $i++;
}
```

Finally, in a *repeat* loop, a block of code is executed until a specific condition is met: the block of code is executed once and then it continues to execute in a loop until the expression evaluates to true. Unlike the while loop, the block of code is always executed at least once in a repeat loop. For instance, by setting a variable to the initial value `$i = 0` before the loop and then executing the block of code until the test expression `$i >= @stop` evaluates to true, where the variable is increased with `$i++` within the block of code, the block of code is also executed a number of times equal to the length of the `@stop` array:

```perl
my $i = 0;
do {
  print "$stop[$i]\n";
  $i++;
} until ($i >= @stop);
```

Back to the `rna_to_protein` subroutine, the previous while loop for skipping any nucleotides before the first start codon was correct only if the `$rna` string contained at least one start codon. Otherwise, the loop would eventually fall off beyond the end of the string, and `substr($rna,$i,3)` would then report a *substr outside of string* error. In order to skip nucleotides within the `$rna` string only, an additional condition is needed in the expression controlling the loop: the position `$i` must be less than or equal to `length($rna)` - 3, the starting position of the last codon in the string. This condition is equivalent to `$i < length($rna)` - 2, thus leading to the following code:

```
my $i = 0;
while ($i < length($rna) - 2 &&
    substr($rna,$i,3) ne "AUG") { # start codon
  $i++;
}
```

In the expression controlling this while loop, two expressions are combined by means of the Boolean `&&` (and) binary operator, which returns true if the two expressions evaluate to true and returns false otherwise, that is, if at least one of the expressions evaluates to false. Further, the Boolean `||` (or) binary operator returns true if at least one of the expressions evaluates to true and returns false otherwise, that is, if the two expressions evaluate to false, and the Boolean `!` (not) unary operator returns true if the expression evaluates to false and returns false otherwise, that is, if the expression evaluates to true.

The order of the expressions matters here, because they are always evaluated from the left to the right. Therefore, if the test for a start codon is put before the test for the end of the string,

```
my $i = 0;
while (substr($rna,$i,3) ne "AUG") &&
    $i < length($rna) - 2 { # start of last codon
  $i++;
}
```

the *substr outside of string* error will still be avoided, but in the case in which the `$rna` string does not contain any start codon, the loop will indeed fall off the end of the string, and `substr($rna,$i,3)` will then return only two nucleotides, instead of a codon of three nucleotides:

```
. . . . . . . . . . . . . . . . . . . . . . . . . . . . . . . . . . . . . . . . . . . . . . . . UGACUA
. . . . . . . . . . . . . . . . . . . . . . . . . . . . . . . . . . . . . . . . . . . . . . . . . GACUA
. . . . . . . . . . . . . . . . . . . . . . . . . . . . . . . . . . . . . . . . . . . . . . . . . . ACUA
. . . . . . . . . . . . . . . . . . . . . . . . . . . . . . . . . . . . . . . . . . . . . . . . . . . CUA
. . . . . . . . . . . . . . . . . . . . . . . . . . . . . . . . . . . . . . . . . . . . . . . . . . . . UA
```

Once the nucleotides before the first start codon have been skipped, the start codon itself has to be skipped as well,

```
$i += 3; # skip the start codon
```

where `$i += 3` is just shorthand for `$i = $i + 3` to skip three positions in the string. The second step consists of translating to protein the rest of the string, until the first stop codon. In the second while loop of the subroutine, each `$codon` is translated to amino acid, the amino acid is added to the end of the `$protein`, the codon is skipped, and the loop is repeated as long as there are still enough nucleotides in the string and the codon is not any of the stop codons.

```
while ($i < length($rna) - 2 &&
    substr($rna,$i,3) ne "UAA" &&
    substr($rna,$i,3) ne "UAG" &&
    substr($rna,$i,3) ne "UGA") {
  my $codon = substr($rna,$i,3);
  $protein .= codon_to_amino_acid($codon);
  $i += 3;
}
```

The dot assignment `$protein .= codon_to_amino_acid($codon)` is just shorthand for `$protein = $protein.codon_to_amino_acid($codon)`, where the dot operator denotes string concatenation. During the execution of this second while loop, the `$protein` string grows from an initial empty string to the 10 amino acids `MVVIIPSRTV` coded by the 30 nucleotides `AUGGUGGUUAUUAUA` `CCGUCAAGGACUGUG`, as the 10 codons are translated to protein and added one after the other to the end of the string:

```
..AUGAUGGUGGUUAUUAUACCGUCAAGGACUGUGUGA..
.....AUGGUGGUUAUUAUACCGUCAAGGACUGUGUGA..   M
........GUGGUUAUUAUACCGUCAAGGACUGUGUGA..   MV
...........GUUAUUAUACCGUCAAGGACUGUGUGA..   MVV
..............AUUAUACCGUCAAGGACUGUGUGA..   MVVI
.................AUACCGUCAAGGACUGUGUGA..   MVVII
....................CCGUCAAGGACUGUGUGA..   MVVIIP
.......................UCAAGGACUGUGUGA..   MVVIIPS
..........................AGGACUGUGUGA..   MVVIIPSR
.............................ACUGUGUGA..   MVVIIPSRT
................................GUGUGA..   MVVIIPSRTV
```

The empty string is the default initial value when a string variable such as `$protein` is declared without an explicit value assignment, and the final value of the `$protein` variable is the string returned by the subroutine:

```
sub rna_to_protein {
  my $rna = shift;
  my $protein;
  ...
  return $protein;
}
```

In the absence of an explicit `return` instruction, the subroutine would return the value of the last expression that was evaluated, which in this case is the empty string, when `substr($rna,$i,3)` is equal to one of the stop codons.

Now, a codon can be translated to protein by looking it up in the genetic code table, and the 64 entries of the genetic code table can be encoded in a series of 64 tests within the `codon_to_amino_acid` subroutine. Reading the circular genetic code table in clockwise order, for instance, leads to the following subroutine code:

```
sub codon_to_amino_acid {
  my $codon = shift;
  if    ($codon eq "UUU") { return "F" } # Phe
  elsif ($codon eq "UUC") { return "F" } # Phe
  elsif ($codon eq "UUA") { return "L" } # Leu
  elsif ($codon eq "UUG") { return "L" } # Leu
  elsif ($codon eq "UCU") { return "S" } # Ser
  elsif ($codon eq "UCC") { return "S" } # Ser
  elsif ($codon eq "UCA") { return "S" } # Ser
  elsif ($codon eq "UCG") { return "S" } # Ser
  elsif ($codon eq "UAU") { return "Y" } # Tyr
  elsif ($codon eq "UAC") { return "Y" } # Tyr
  elsif ($codon eq "UAA") { return "-" } # stop
  elsif ($codon eq "UAG") { return "-" } # stop
  elsif ($codon eq "UGU") { return "C" } # Cys
  elsif ($codon eq "UGC") { return "C" } # Cys
  elsif ($codon eq "UGA") { return "-" } # stop
  elsif ($codon eq "UGG") { return "W" } # Trp
  elsif ($codon eq "CUU") { return "L" } # Leu
  elsif ($codon eq "CUC") { return "L" } # Leu
  elsif ($codon eq "CUA") { return "L" } # Leu
  elsif ($codon eq "CUG") { return "L" } # Leu
  elsif ($codon eq "CCU") { return "P" } # Pro
  elsif ($codon eq "CCC") { return "P" } # Pro
  elsif ($codon eq "CCA") { return "P" } # Pro
  elsif ($codon eq "CCG") { return "P" } # Pro
  elsif ($codon eq "CAU") { return "H" } # His
  elsif ($codon eq "CAC") { return "H" } # His
  elsif ($codon eq "CAA") { return "Q" } # Gln
  elsif ($codon eq "CAG") { return "Q" } # Gln
  elsif ($codon eq "CGU") { return "R" } # Arg
  elsif ($codon eq "CGC") { return "R" } # Arg
  elsif ($codon eq "CGA") { return "R" } # Arg
  elsif ($codon eq "CGG") { return "R" } # Arg
  elsif ($codon eq "AUU") { return "I" } # Ile
  elsif ($codon eq "AUC") { return "I" } # Ile
```

```
    elsif ($codon eq "AUA") { return "I" } # Ile
    elsif ($codon eq "AUG") { return "M" } # Met
    elsif ($codon eq "ACU") { return "T" } # Thr
    elsif ($codon eq "ACC") { return "T" } # Thr
    elsif ($codon eq "ACA") { return "T" } # Thr
    elsif ($codon eq "ACG") { return "T" } # Thr
    elsif ($codon eq "AAU") { return "N" } # Asn
    elsif ($codon eq "AAC") { return "N" } # Asn
    elsif ($codon eq "AAA") { return "K" } # Lys
    elsif ($codon eq "AAG") { return "K" } # Lys
    elsif ($codon eq "AGU") { return "S" } # Ser
    elsif ($codon eq "AGC") { return "S" } # Ser
    elsif ($codon eq "AGA") { return "R" } # Arg
    elsif ($codon eq "AGG") { return "R" } # Arg
    elsif ($codon eq "GUU") { return "V" } # Val
    elsif ($codon eq "GUC") { return "V" } # Val
    elsif ($codon eq "GUA") { return "V" } # Val
    elsif ($codon eq "GUG") { return "V" } # Val
    elsif ($codon eq "GCU") { return "A" } # Ala
    elsif ($codon eq "GCC") { return "A" } # Ala
    elsif ($codon eq "GCA") { return "A" } # Ala
    elsif ($codon eq "GCG") { return "A" } # Ala
    elsif ($codon eq "GAU") { return "D" } # Asp
    elsif ($codon eq "GAC") { return "D" } # Asp
    elsif ($codon eq "GAA") { return "E" } # Glu
    elsif ($codon eq "GAG") { return "E" } # Glu
    elsif ($codon eq "GGU") { return "G" } # Gly
    elsif ($codon eq "GGC") { return "G" } # Gly
    elsif ($codon eq "GGA") { return "G" } # Gly
    elsif ($codon eq "GGG") { return "G" } # Gly
    else { return "*" }
}
```

This long instruction is called a *conditional* and allows for the conditional execution of a block of code, depending on whether or not a given expression evaluates to true. In the simplest form of the conditional, a block of code is executed if an expression evaluates to true:

```
if ($codon eq "AUG") {
  print "stop␣codon\n";
}
```

Another common form of the conditional provides for the alternative execution of two blocks of code, depending on the outcome of the evaluation of a given expression. A block of code is executed if an expression evaluates to true. Otherwise, if the expression evaluates to false, another block of code is executed:

```
if ($codon eq "AUG") {
  print "stop codon\n";
} else {
  $i += 3;
}
```

The whole Perl script, stored in the `sample.pl` file, will look as follows.

```
#!/usr/bin/perl -w
use strict;

my $rna = "GCCAAUGACUAAGGCCUAAAGA";
my $protein = rna_to_protein($rna);
print "$rna translates to $protein\n";

sub rna_to_protein {
  ...
}

sub codon_to_amino_acid {
  ...
}
```

Second Script

The previous Perl script would be more flexible if the `$rna` string could be input from the console, instead of being hardwired in the script, since this would allow for using exactly the same script over and over again, to translate to protein any given messenger RNA sequence.

A Perl script can be invoked upon particular argument values, much like a subroutine. For example, the following command invokes the Perl interpreter on the sample script with a short string as first argument:

```
$ perl sample.pl AUGGUGGUUAUUAUACCGUCAAGGACUGUG
```

Argument values can be read from the console, also known as *standard input* in Unix and Linux. Command line arguments are stored in a special array named @ARGV (argument values) and, thus, `$ARGV[0]` contains the first argument, `$ARGV[1]` contains the second argument, and `$ARGV[$#ARGV]` contains the last argument. The sample script would be as follows when the string argument is read from the command line:

```
my $rna = $ARGV[0];
my $protein = rna_to_protein($rna);
print "$rna translates to $protein\n";
```

The string argument can also be read from the standard input in a separate line, right after the command line, as follows.

```
my $rna = <STDIN>;
my $protein = rna_to_protein($rna);
print "$rna translates to $protein\n";
```

Here, STDIN (standard input) is an example of a *file handle*, and <STDIN> can be replaced by <> for brevity. File handles will be further discussed in the third script. Reading the argument from the standard input in a separate line has the advantage that the Perl script can then be invoked upon as many input strings as desired, with each input string in a separate line, by just reading the input string inside a loop.

After reading and translating to protein all the input strings, the end of the input is signaled by just pressing Control-D (uppercase or lowercase) on Unix or Linux.

```
while (my $rna = <STDIN>) {
  my $protein = rna_to_protein($rna);
  print "$rna translates to $protein\n";
}
```

Besides, using a long conditional is not the only solution to the messenger RNA to protein translation problem in the codon_to_amino_acid subroutine. In fact, the circular genetic code table can also be encoded as a list of codons and corresponding amino acids, using a hash variable:

```
my %codon2aa = (
  "UUU" => "F", # Phe
  "UUC" => "F", # Phe
  "UUA" => "L", # Leu
  "UUG" => "L", # Leu
  "UCU" => "S", # Ser
  "UCC" => "S", # Ser
  "UCA" => "S", # Ser
  "UCG" => "S", # Ser
  "UAU" => "Y", # Tyr
  "UAC" => "Y", # Tyr
  "UAA" => "-", # stop
  "UAG" => "-", # stop
  "UGU" => "C", # Cys
  "UGC" => "C", # Cys
  "UGA" => "-", # stop
  "UGG" => "W", # Trp
  "CUU" => "L", # Leu
  "CUC" => "L", # Leu
  "CUA" => "L", # Leu
  "CUG" => "L", # Leu
  "CCU" => "P", # Pro
  "CCC" => "P", # Pro
```

```
"CCA" => "P",  # Pro
"CCG" => "P",  # Pro
"CAU" => "H",  # His
"CAC" => "H",  # His
"CAA" => "Q",  # Gln
"CAG" => "Q",  # Gln
"CGU" => "R",  # Arg
"CGC" => "R",  # Arg
"CGA" => "R",  # Arg
"CGG" => "R",  # Arg
"AUU" => "I",  # Ile
"AUC" => "I",  # Ile
"AUA" => "I",  # Ile
"AUG" => "M",  # Met
"ACU" => "T",  # Thr
"ACC" => "T",  # Thr
"ACA" => "T",  # Thr
"ACG" => "T",  # Thr
"AAU" => "N",  # Asn
"AAC" => "N",  # Asn
"AAA" => "K",  # Lys
"AAG" => "K",  # Lys
"AGU" => "S",  # Ser
"AGC" => "S",  # Ser
"AGA" => "R",  # Arg
"AGG" => "R",  # Arg
"GUU" => "V",  # Val
"GUC" => "V",  # Val
"GUA" => "V",  # Val
"GUG" => "V",  # Val
"GCU" => "A",  # Ala
"GCC" => "A",  # Ala
"GCA" => "A",  # Ala
"GCG" => "A",  # Ala
"GAU" => "D",  # Asp
"GAC" => "D",  # Asp
"GAA" => "E",  # Glu
"GAG" => "E",  # Glu
"GGU" => "G",  # Gly
"GGC" => "G",  # Gly
"GGA" => "G",  # Gly
"GGG" => "G"   # Gly
);
```

Hash variables hold lists of values. A hash is just an advanced form of an array, in which the values are accessed by a scalar *key* instead of an array position. For instance, a hash variable can be declared and have the start codon and the three stop codons assigned as a list of keys and values,

```perl
my %codon = (
  "AUG" => "start",
  "UAA" => "stop",
  "UAG" => "stop",
  "UGA" => "stop"
);
```

or it can be declared first and then have each of the four codons stored in the hash,

```perl
my %codon;
$codon{"AUG"} = "start";
$codon{"UAA"} = "stop";
$codon{"UAG"} = "stop";
$codon{"UGA"} = "stop";
```

While the value of the hash is the whole list of keys and values, individual elements can be accessed by a key, as in the previous assignment instructions, or by means of a for each loop,

```perl
foreach my $key (keys %codon) {
  print "$key $codon{$key}\n";
}
```

where the values appear in no particular order, or with the values sorted by the hash key,

```perl
foreach my $key (sort keys %codon) {
  print "$key $codon{$key}\n";
}
```

Within the **rna_to_protein** subroutine, the encoding of the genetic code table is then replaced by the hash of codons and corresponding amino acids,

```perl
    if (defined $codon2aa{$codon}) {
      $aa = $codon2aa{$codon};
    } else {
      $aa = "*";
    }
    $protein .= $aa;
```

where the **defined** function returns true if there is a value for the $codon key in the %codon2aa hash and false otherwise, when either there is no value for the key in the hash or the value is **undef**, a special scalar that denotes undefined values in Perl.

The second Perl script, also stored in the `sample.pl` file, will then look as follows.

```perl
#!/usr/bin/perl -w
use strict;

while (my $rna = <STDIN>) {
  my $protein = rna_to_protein($rna);
  print "$rna translates to $protein\n";
}

my %codon2aa = (
  ...
)

sub rna_to_protein {
  my $rna = shift;
  my $protein;
  my $aa;
  my $i = 0;
  while ($i < length($rna) - 2 &&
      substr($rna,$i,3) ne "AUG") { # start codon
    $i++;
  }
  $i += 3; # skip the start codon
  while ($i < length($rna) - 2 &&
      substr($rna,$i,3) ne "UAA" &&
      substr($rna,$i,3) ne "UAG" &&
      substr($rna,$i,3) ne "UGA") {
    my $codon = substr($rna,$i,3);
    if (defined $codon2aa{$codon}) {
      $aa = $codon2aa{$codon};
    } else {
      $aa = "*";
    }
    $protein .= $aa;
    $i += 3;
  }
  return $protein;
}
```

Third Script

The messenger RNA sequences to be translated to protein could be stored in a text file and then input to the Perl script straight from the file, instead of being input one after the other from the console. This behavior can be achieved with the previous Perl script, by using the input redirection operator in the Unix or Linux command line to have the standard input read from a file:

```
$ perl sample.pl < sample.dna
```

The Perl script can also be modified to read the input strings from a text file, with the file name passed as an argument:

```
my $file = $ARGV[0];
open IN, "<$file" or die "Cannot open file: $!";
while (my $rna = <IN>) {
  my $protein = rna_to_protein($rna);
  print "$rna translates to $protein\n";
}
close IN;
```

The following command would then invoke the Perl interpreter on the sample script with the name of the text file containing all the input strings as the first argument:

```
$ perl sample.pl sample.dna
```

All input and output take place in Perl using a file handle, which is just an unquoted string that has been associated with a particular file by means of an `open` instruction and remains valid until finished by a `close` instruction. The default file handles, which need not be opened or closed, are STDIN (standard input) for input from the console, STDOUT (standard output) for output to the console window, and STDERR (standard error) for output of error and warning messages, also to the console window.

A file can be opened for either reading, writing, or appending data. In the `open` instruction, the file name is preceded by the intuitive symbols < for reading, > for writing, and >> for appending. The symbol for reading can be omitted, since files are open for reading by default. Opening for writing an already existing file results in the loss of any previous contents of the file, while opening a file for appending results in the addition of new contents to the end of the file. Failure to open a file results in the `open` instruction returning false right after setting the special variable $! to reflect the system error. In this case, the `die` instruction will output an error message to the standard error and exit immediately.

Once a file has been opened for reading and associated with the IN file handle, for instance, the <IN> input operator is readily available, and it behaves much like <STDIN>. On the other hand, once a file has been opened for writing or appending, the `print` instruction can be used to write to the file by adding the file handle before the values to be written:

```
open OUT, ">$file" or die "Cannot open file: $!";
```

```
foreach my $key (sort keys %codon2aa) {
  print OUT "$key␣$codon2aa{$key}\n";
}
close OUT;
```

The third Perl script, stored again in the `sample.pl` file, will then look as follows.

```
#!/usr/bin/perl -w
use strict;

my $file = $ARGV[0];
open IN, "<$file" or die "Cannot␣open␣file:␣$!";
while (my $rna = <IN>) {
  my $protein = rna_to_protein($rna);
  print "$rna␣translates␣to␣$protein\n";
}
close IN;

my %codon2aa = (
  ...
)

sub rna_to_protein {
  ...
}
```

A.2   Overview of Perl

After the brief introduction by way of sample scripts in the previous section, let us focus now on a few more advanced aspects of the language which were not covered in these sample scripts.

Passing References as Arguments

Scalar values can be passed to a subroutine as arguments, by listing them right after the subroutine name. Array and hash variables, however, hold lists of values, and passing them to a subroutine as arguments is problematic for several reasons. First of all, the lists of values get messed up when passing two or more array or hash variables to a subroutine, because all the arguments are passed as a single list of values and stored in the special temporary array named @_ and then there is no way to tell them apart within the subroutine.

Second, if some or all of the values are modified within the subroutine, the modified values cannot be passed back. Third, and most important, passing a long list of values can be slow, as they are copied one by one to the @_ array.

These problems are solved by passing to the subroutine a reference to the memory location referenced by the array or hash variable, instead of passing the actual list of values.

A reference to a variable is created by just adding the backslash symbol in front of the variable name, and the reference to the variable can be passed to a subroutine as an argument.

```
my @start = ("AUG");
my @stop = ("UAA", "UAG", "UGA");
my $protein = rna_to_protein($rna,\@start,\@stop);
```

The reference to the variable is a scalar value, which can be *dereferenced* within the subroutine to get back the list of values stored in the memory location referenced by the variable. A reference to an array or hash variable is dereferenced by enclosing the reference inside @{} and %{}, respectively.

```
sub rna_to_protein {
    my ($rna,$start,$stop) = @_;
    my @start = @{$start};
    my @stop = @{$stop};
    ...
}
```

Perl also allows for *anonymous* references to lists of values, which avoid the need for storing first the list of values in an array or hash variable and still can be passed to subroutines as arguments. An anonymous reference to an array variable is created by enclosing the list of values inside square brackets, instead of parentheses.

```
my $start = ["AUG"];
my $stop = ["UAA", "UAG", "UGA"];
my $protein = rna_to_protein($rna,$start,$stop);
```

An anonymous reference to an hash variable, on the other hand, is created by enclosing the list of keys and values inside braces, instead of parentheses.

```
my $codon = {
    "AUG" => "start",
    "UAA" => "stop",
    "UAG" => "stop",
    "UGA" => "stop"
};
my $protein = rna_to_protein($rna,$codon);
```

Anonymous references are already references and, thus, there is no need for adding any backslash symbol when passing them to a subroutine.

Using Modules and Packages

Subroutines written for one script can be made available to other scripts by placing them in a *module*, a text file with the extension pm (perl module) containing Perl code. For instance, any subroutines written in a text file named sample.pm are readily available in any script *using* this module, that is, in any script containing the following header line:

```
use sample;
```

This is all right when the module and the script reside in the same directory. Otherwise, the Perl interpreter will not be able to find the sample.pm file, and the path to this file has to be given in the script header. For instance,

```
use lib "stuff/moo";
use sample;
```

The module file itself has to end with a true value, which indicates the successful loading of the module in the script for further use. The usual convention is to put the following last line in the module:

```
1;
```

Now, when using several modules in a Perl script, the variable and subroutine names declared in one module might clash with the names declared in another module. Even in the same module, there might be two different reverse_complement subroutines, one for DNA and the other one for RNA nucleotides.

Name clashes can be avoided by keeping the *name space* of one subroutine apart from the name space of the other subroutine, putting them in separate *packages* within the module. For instance, declaring packages named DNA and RNA for the two versions of the reverse_complement subroutine, the sample.pm file would look as follows.

```
package DNA;

sub reverse_complement {
  my $seq = shift;
  $seq = reverse $seq;
  $seq =~ tr/ACGT/TGCA/;
  return $seq;
}

package RNA;

sub reverse_complement {
  my $seq = shift;
  $seq = reverse $seq;
  $seq =~ tr/ACGU/UGCA/;
```

```
  return $seq;
}


1;
```

The two versions of the `reverse_complement` subroutine are readily available in any script using this module, although the subroutine name now has to be prefixed by the package name:

```
use sample;

my $dna = "TTGATTACCTTATTTGATCATTACACATTGTACGCTTGTG";
my $rc_dna = DNA::reverse_complement($dna);

my $rna = "GGGUGCUCAGUACGAGAGGAACCGCACCC";
my $rc_rna = RNA::reverse_complement($rna);
```

A.3 Perl Quick Reference Card

Most of the basic functions available in Perl, including all the Perl built-in functions used in this book, are summarized next, in a kind of inset quick reference card. They are grouped into subroutines, packages, and modules; arithmetic functions; conversion functions; string functions; list functions; array functions; hash functions; search and replace functions; and input output functions.

More detailed information on any of these built-in functions can be obtained with the Unix or Linux command `perldoc -f` followed by the function name.

Subroutines, Packages, and Modules

```
bless REF, CLASSNAME
```

Turns the object referenced by REF into an object in the CLASSNAME package.

```
package NAMESPACE
```

Declares the remainder of the current block as being in NAMESPACE.

```
ref EXPR
```

Returns a string valued `"SCALAR"`, `"ARRAY"`, `"HASH"` (among other values) if EXPR is a reference, and the empty string otherwise. If the referenced object has been *blessed* into a package, then that package name is returned instead.

`sub NAME BLOCK`

Defines NAME as a subroutine with code BLOCK.

`use MODULE VERSION LIST`

Imports the LIST of subroutines and variables into the current package from MODULE using Perl version VERSION or later.

Arithmetic Functions

`abs EXPR`

Returns the absolute value of EXPR.

`atan2 Y, X`

Returns the arc tangent of Y/X in the range $-\pi$ to π.

`cos EXPR`

Returns the cosine of EXPR (expressed in radians).

`exp EXPR`

Returns e (the natural logarithm base) to the power of EXPR.

`int EXPR`

Returns the integer portion of EXPR.

`log EXPR`

Returns the natural logarithm (base e) of EXPR.

`rand EXPR`

Returns a random fractional number greater than or equal to 0 and less than the value of EXPR. If EXPR is omitted, returns a value between 0 and 1.

`sin EXPR`

Returns the sine of EXPR (expressed in radians).

`sqrt EXPR`

Returns the square root of EXPR.

`srand EXPR`

Sets the random number seed for the **rand** function.

`time`

Returns the number of non-leap seconds since January 1, 1970. Suitable for feeding to `gmtime` and `localtime`.

Conversion Functions

`chr EXPR`

Returns the character represented by EXPR in the character set.

`gmtime EXPR`

Converts EXPR as returned by the time function to an eight-element list (seconds, minutes, hours, day of the month, month, number of years since 1900, day of the week, day of the year) with the time localized for the standard Greenwich time zone.

`hex EXPR`

Interprets EXPR as a hexadecimal string and returns the corresponding value.

`localtime EXPR`

Converts a time as returned by the `time` function to a nine-element list with the time analyzed for the local time zone. See `gmtime` for the first eight elements. The last element is true if the specified time occurs during daylight savings time and false otherwise.

`oct EXPR`

Interprets EXPR as an octal string and returns the corresponding value.

`ord EXPR`

Returns the numeric value of the first character of EXPR.

`vec EXPR, OFFSET, BITS`

Treats the string in EXPR as a bit vector made up of elements of width BITS, and returns the value of the element specified by OFFSET as an unsigned integer.

String Functions

`chomp LIST`

Removes any line endings from each of the elements of LIST. Returns the total number of characters removed.

`chop LIST`

Chops off the last character from each of the elements of LIST. Returns the last character chopped.

`index STR, SUBSTR, POSITION`

Returns the position of the first occurrence of SUBSTR in STR at or after POSITION.

`lc EXPR`

Returns a lowercased version of EXPR.

`lcfirst EXPR`

Returns the value of EXPR with the first character lowercased.

`length EXPR`

Returns the length in characters of the value of EXPR.

`q/STRING/`

Returns STRING as a single-quoted, literal string.

`qq/STRING/`

Returns STRING as a double-quoted, interpolated string.

`rindex STR , SUBSTR , POSITION`

Returns the position of the last occurrence of SUBSTR in STR at or before POSITION.

`sprintf FORMAT , LIST`

Returns the elements of LIST as a string formatted according to FORMAT.

`substr EXPR , OFFSET , LENGTH , REPLACEMENT`

Extracts a substring of LENGTH characters starting at position OFFSET out of EXPR and returns it. Removes from EXPR the characters designated by OFFSET and LENGTH, and replaces them with REPLACEMENT, if the latter is supplied.

`uc EXPR`

Returns an uppercased version of EXPR.

`ucfirst EXPR`

Returns the value of EXPR with the first character uppercased.

List Functions

`grep EXPR , LIST`

Evaluates EXPR for each element of LIST and returns the list value consisting of those elements for which EXPR evaluated to true. In scalar context, returns the number of times EXPR was true.

`join EXPR , LIST`

Joins the separate strings of LIST into a single string with fields separated by the value of EXPR, and returns the string.

`map EXPR , LIST`

Evaluates EXPR for each element of LIST and returns the list value composed of the results of each such evaluation. In scalar context, returns the total number of elements generated.

`qw/STRING/`

Splits STRING into a list of strings using embedded white space as delimiter, and returns that list.

`reverse LIST`

Returns a list value consisting of the elements of LIST in the opposite order. In scalar context, concatenates the elements of LIST and returns a string value with all characters in the opposite order.

`sort LIST`

Sorts LIST and returns the sorted list value.

Array Functions

`delete $ARRAY[EXPR]`

Deletes the elements specified by EXPR from ARRAY, and returns a list with the deleted elements.

`pop ARRAY`

Pops the last value of ARRAY and returns it, shortening the array by one element.

`push ARRAY, LIST`

Pushes the values of LIST onto the end of ARRAY, and returns the new number of elements in the array.

`shift ARRAY`

Shifts off the first value of ARRAY and returns it, shortening the array by one element.

`splice ARRAY, OFFSET, LENGTH, LIST`

Removes the elements designated by OFFSET and LENGTH from ARRAY, and replaces them with the elements of LIST, if any.

`unshift ARRAY, LIST`

Inserts the elements of LIST to the front of ARRAY, and returns the new number of elements in the array.

Hash Functions

`defined $HASH{EXPR}`

Returns true if the element with key EXPR in HASH has a value other than undef, and false otherwise.

`delete $HASH{EXPR}`

Deletes the elements with key specified by EXPR from HASH, and returns a list with the deleted elements.

`each HASH`

Returns a two-element list consisting of the key and value for the next element of HASH. After all values of the hash have been returned, an empty list is returned.

`exists $HASH{EXPR}`

Returns true if the element with key EXPR in HASH has ever been initialized, even if the corresponding value is undef, and returns false otherwise.

`keys HASH`

Returns a list consisting of all the keys of HASH. In scalar context, returns the number of keys.

`values HASH`

Returns a list consisting of all the values of HASH. In scalar context, returns the number of values.

Search and Replace Functions

`m/PATTERN/`

Searches a string for a match to PATTERN, and in scalar context returns true if it succeeds and false otherwise.

`pos SCALAR`

Returns the offset of where the last m//g search left off for the variable in question.

`s/PATTERN/REPLACEMENT/`

Searches a string for a match to PATTERN and if found, replaces it with REPLACEMENT and returns the number of substitutions made. Otherwise, it returns false.

`split /PATTERN/, EXPR, LIMIT`

Splits the string EXPR into a list of strings and returns that list. Anything matching PATTERN is taken to be a delimiter separating the fields. A positive LIMIT gives the maximum number of fields the EXPR will be split into.

`study SCALAR`

Takes extra time to study SCALAR in anticipation of doing many pattern matches on the string before it is next modified.

`tr/LIST/REPLACEMENT/`

Transliterates all occurrences of the characters found in LIST with the corresponding character in REPLACEMENT, and returns the number of characters replaced or deleted.

Input Output Functions

`close FILEHANDLE`

Closes the file or pipe associated with the file handle.

`die LIST`

Prints the value of LIST to STDERR and exits with the current value of $! (error number).

`eof FILEHANDLE`

Returns true if the next read on FILEHANDLE will return end of file, or if FILEHANDLE is not open.

`getc FILEHANDLE`

Returns the next character from the input file attached to FILEHANDLE, or **undef** at end of file.

`print FILEHANDLE LIST`

Prints LIST to the output channel attached to FILEHANDLE, and returns true if successful.

`printf FILEHANDLE FORMAT, LIST`

Equivalent to **print** FILEHANDLE **sprintf** FORMAT, LIST.

`read FILEHANDLE, SCALAR, LENGTH, OFFSET`

Reads LENGTH characters of data into variable SCALAR at OFFSET from the specified FILEHANDLE, and returns the number of characters actually read.

`select FILEHANDLE`

Returns the currently selected file handle. Sets the current default file handle for output if FILEHANDLE is supplied.

`warn LIST`

Prints the value of LIST to STDERR just like `die` but does not exit.

Bibliographic Notes

Perl was developed by Larry Wall (`http://www.wall.org/~larry/`) around 1987 as a general-purpose Unix scripting language and has since been quite influential in computational biology (Stein 1996).

General books on Perl include (Christiansen and Torkington 2003; Cozens 2005; Foy 2007; Schwartz et al. 2006; 2008; Wall et al. 2000), among many others. More specialized books on Perl for computational biology include (Birney et al. 2009; Dwyer 2003; Jamison 2003; LeBlanc and Dyer 2007; Tisdall 2001; 2003). See also (Tregar 2002).

Beside these sources, there is a comprehensive Perl manual available with any Perl distribution, including detailed descriptions and examples of almost every aspect of the Perl language. The manual is split into as many as 135 Unix or Linux man pages, which can be read with `man perl` and `perldoc perl` on the Unix or Linux command line.

Alternative Perl implementations for some of the algorithms presented in this book can be found as part of the BioPerl project (Stajich et al. 2002; Birney et al. 2009).

Appendix B

Elements of R

A brief introduction to R is given in this appendix by way of sample scripts that solve a simple computational biology problem using different methods. These scripts are then dissected in order to explain basic aspects of the R language, followed by an overview of more advanced aspects of the language, not covered in the sample scripts. This is all summarized for convenience in an R quick reference card.

B.1 R Scripts

R is an interpreted scripting language. An R program is a *script* containing a series of instructions, which are interpreted when the program is run instead of being compiled first into machine instructions and then assembled or linked into an executable program, thus avoiding the need for separate compilation and linking. The R language is actually integrated in a software environment for statistical computing and graphics.

There are R distributions available for almost every computing platform, and free distributions can be downloaded from `http://www.r-project.org/`. The actual mechanism of running a script using an R interpreter depends on the particular operating system, the common denominator being the Unix or Linux command line. Assuming one of the R scripts shown further below was already written using a text editor and stored in a file named `sample.R` (where `R` is the standard file extension for R scripts), the following command invokes the R interpreter on the sample script:

```
$ R --vanilla < sample.R
```

The following command, on the other hand, invokes the R interpreter on the sample script from within the R software environment:

```
> source("sample.R")
```

When launching the R software environment, a new window with the name "R Console" will open and a welcome message like the following one will show in the console window. When invoking the R interpreter from a console window, the same welcome message will show:

```
R version 2.7.1 (2008-06-23)
Copyright (C) 2008 The R Foundation for Statistical Computing
ISBN 3-900051-07-0

R is free software and comes with ABSOLUTELY NO WARRANTY.
You are welcome to redistribute it under certain conditions.
Type 'license()' or 'licence()' for distribution details.

R is a collaborative project with many contributors.
Type 'contributors()' for more information and
'citation()' on how to cite R or R packages in publications.

Type 'demo()' for some demos, 'help()' for on-line help, or
'help.start()' for an HTML browser interface to help.
Type 'q()' to quit R.

>
```

The simple computational biology problem at hand consists of translating to protein a messenger RNA sequence stored (in 5′ to 3′ direction) in a text file. See Appendix A for a detailed discussion of this problem.

Let us assume first the input messenger RNA sequence is already stored in a variable rna and, thus, readily available for translation to protein. The translation method will consist of skipping any nucleotides before the first start codon and then translating the rest of the sequence to protein using the standard genetic code, until the first stop codon. As an example, translating sequence GUCGCCAUGAUGGUGGUUAUUAUACCGUCAAGGACUGUGUGACUA to protein involves skipping GUCGCC and then translating AUGAUGGUGGUUAUUAUACCGUCAAGGACUGUGUGA to MVVIIPSRTV.

First Script

In an R script, the messenger RNA sequence stored in the variable rna can be translated to a protein sequence, stored in a variable protein, and output to the console window, as shown in the following script.

```
rna <- "GCCAAUGACUAAGGCCUAAAGA"
protein <- rna.to.protein(rna)
cat(rna,"translates␣to",protein)
```

Well, this is not quite the situation. The R interpreter cannot run this short script because the rna.to.protein function has not been defined yet. Nevertheless, let us dissect this script in order to discuss some basic elements of R programming.

The R script contains instructions to translate the rna sequence and output the resulting protein sequence. The first line makes the character string

"GCCAAUGACUAAGGCCUAAAGA" the value stored in the memory location referenced by the variable rna. In other words, it assigns the character string to the variable, and a variable can be assigned different values at the same or different places in an R script. Notice that variables need not be declared beforehand in R. Unlike variables, the character string is a constant whose value cannot change unless the R script is changed.

Variables are thus named references to memory locations, and there are scalar, vector, matrix, array, data frame, and list variables in R. Scalar variables hold scalar values such as integer, real, and complex numbers, or logical (Boolean) values. Vector, matrix, array, data frame, and list variables, on the other hand, hold lists of values. Vector variables hold one-dimensional vectors, matrix variables hold two-dimensional matrices, and array variables hold multi-dimensional matrices. Data frame variables hold matrices in which the columns may contain values of different classes, and list variables hold vectors of values not necessarily belonging to the same class. Character strings are not scalar values in R, and they are represented as character vectors.

Almost any name can be used as a variable name, as long as it consists of uppercase or lowercase letters, digits, or dots and begins with a letter, with a few exceptions. Some variable names have a special meaning, such as TRUE, FALSE, NULL, Inf, NaN, and NA, which cannot be used to declare new variables. Beside these conditions, variable names are case sensitive and thus rna, Rna, and RNA all reference different memory locations when used in the same R script.

A value is assigned to a variable in an assignment instruction or statement that resembles an equation, with the variable name to the left and the value to the right of the left arrow sign.

In the second line of the R script, the rna.to.protein function is invoked upon the value of the variable rna, and the resulting value is assigned to the variable protein. Before discussing functions, let us see how the protein sequence stored in the variable protein finds its way to the console window.

A cat statement outputs the concatenated values of one or more constants or variables to the console window, also known as *standard output* in Unix and Linux. The value returned by a function call can also be output in this way. In a cat statement, the values to be output are separated by commas, as in

```
cat(rna,"translates␣to",protein)
```

or

```
cat(rna,"translates␣to",rna.to.protein(rna))
```

which both output the value of the variable rna, followed by the string constant " translates to ", followed by either the value of the variable protein or the result of the function call rna.to.protein(rna) on the console window.

A `print` statement, on the other hand, outputs the value of a single constant or variable, which does not need to be a scalar, to the console window.

String constants must be enclosed in either double or single quotes in R, but unlike Perl, a string enclosed in double quotes is not subject to variable interpolation.

Invoking the R interpreter on the sample script will result in the following output,

```
GUCGCCAUGAUGGUGGUUAUUAUACCGUCAAGGACUGUGUGACUA trans
lates to MVVIIPSRTV
```

where this long output line might be broken at a different position depending on the actual console window size.

Let us come back to discussing functions now. As in any programming language, there are various functions readily available in R, which take one or more arguments as input and give a single result as output. Examples include numeric functions such as `abs` (absolute value), `exp` (raise to a power), `trunc` (integer part of a real number), `log` (natural logarithm), and `sqrt` (square root), and string functions such as `tolower` (convert to lowercase), `toupper` (uppercase), and `nchar` (number of characters), among many others.

Functions are named pieces of code that perform a specific task and are evaluated when called upon particular argument values. Functions need not return a value, although they often do, and the `rna.to.protein` function will take a string (a messenger RNA sequence) as input and return a string (a protein sequence) as output.

A function begins with the function name followed by a left arrow, `function`, any arguments enclosed in parentheses, and the function code enclosed in braces, such as in

```
rna.to.protein <- function (rna) {
  ...
}
```

In this function, the input messenger RNA sequence, stored in a string variable `rna`, will be translated to protein and stored in a string variable `protein` by first skipping any nucleotides before the first start codon and then translating the rest of the sequence to protein using the standard genetic code, until the first stop codon. The translation of each codon to amino acid will be done in turn in a `codon.to.amino.acid` function, to be defined later. The code of the function at hand is as follows.

```
rna.to.protein <- function (rna) {
  protein <- ""
  i <- 1
  while (i < nchar(rna)-1 &&
      substr(rna,i,i+2) != "AUG") { # start codon
    i <- i+1
```

```
}
i <- i+3 # skip the start codon
while (i < nchar(rna)-1 &&
     substr(rna,i,i+2) != "UAA" &&
     substr(rna,i,i+2) != "UAG" &&
     substr(rna,i,i+2) != "UGA") {
  codon <- substr(rna,i,i+2)
  protein <- paste(protein,
    codon.to.amino.acid(codon),sep="")
  i <- i+3
}
return(protein)
}
```

Values such as the actual sequence stored in the string variable **rna** can be passed to a function as arguments, by listing them right after the function name.

Recall that vector variables hold lists of values. For instance, a vector variable named **stop** can have the three stop codons assigned as a list of values by means of the **c** (combine) function that puts the arguments in a vector,

```
stop <- c("UAA", "UAG", "UGA")
```

or it can be created first and then have each of the three stop codons stored at a different position in the vector, starting with position 1,

```
stop <- vector(mode="character",length=3)
stop[1] <- "UAA"
stop[2] <- "UAG"
stop[3] <- "UGA"
```

or have each of the three stop codons pushed into the vector,

```
stop <- vector(mode="character")
stop <- c(stop,"UAA","UAG","UGA")
```

or have each of the stop codons pushed in reverse order into the vector,

```
stop <- vector(mode="character")
stop <- c("UGA",stop)
stop <- c("UAG",stop)
stop <- c("UAA",stop)
```

This illustrates another aspect of the R software environment. The R interpreter can be invoked on a script by means of the **source** command, but commands can also be typed at the R prompt. In both cases, the result of the computation is output on the console window. The [ ] preceding each output line indicates the position in the list of the first value in the line and is useful when the output is a long list of values:

```
[1]  "AAA"  "AAC"  "AAG"  "AAT"
[5]  "ACA"  "ACC"  "ACG"  "ACT"
[9]  "AGA"  "AGC"  "AGG"  "AGT"
[13] "ATA"  "ATC"  "ATG"  "ATT"
[17] "CAA"  "CAC"  "CAG"  "CAT"
[21] "CCA"  "CCC"  "CCG"  "CCT"
[25] "CGA"  "CGC"  "CGG"  "CGT"
[29] "CTA"  "CTC"  "CTG"  "CTT"
[33] "GAA"  "GAC"  "GAG"  "GAT"
[37] "GCA"  "GCC"  "GCG"  "GCT"
[41] "GGA"  "GGC"  "GGG"  "GGT"
[45] "GTA"  "GTC"  "GTG"  "GTT"
[49] "TAA"  "TAC"  "TAG"  "TAT"
[53] "TCA"  "TCC"  "TCG"  "TCT"
[57] "TGA"  "TGC"  "TGG"  "TGT"
[61] "TTA"  "TTC"  "TTG"  "TTT"
```

While the value of the vector variable is the whole list, individual elements can be accessed by position, as in the previous assignment instructions, by set of positions, or by range of positions; for instance, to access the second and third values in the vector,

```
> stop[c(2,3)]
[1] "UAG" "UGA"
```

the second through third values in the vector,

```
> stop[2:3]
[1] "UAG" "UGA"
```

or all but the first value in the vector,

```
> stop[-1]
[1] "UAG" "UGA"
```

Now, with the string value passed as an argument and stored in the `rna` variable, the previous function translates it to protein and returns the string value of the `protein` variable. The first step consists of skipping any nucleotides before the first start codon, and the scalar variable `i` will hold the initial position in the `rna` string of the first start codon, which is position 7 after skipping six nucleotides in the following sample sequence:

```
GUCGCCAUGAUGGUGGUUAUUAUACCGUCAAGGACUGUGUGACUA
.UCGCCAUGAUGGUGGUUAUUAUACCGUCAAGGACUGUGUGACUA
..CGCCAUGAUGGUGGUUAUUAUACCGUCAAGGACUGUGUGACUA
...GCCAUGAUGGUGGUUAUUAUACCGUCAAGGACUGUGUGACUA
....CCAUGAUGGUGGUUAUUAUACCGUCAAGGACUGUGUGACUA
.....CAUGAUGGUGGUUAUUAUACCGUCAAGGACUGUGUGACUA
......AUGAUGGUGGUUAUUAUACCGUCAAGGACUGUGUGACUA
123456789...
```

The position i of the first start codon in the **rna** string can be found by starting with i <- 1 and increasing the value of i as many times as needed, until the codon substr(rna,i,i+2) in the string **rna** that begins at position i and ends at position i+2 is a start codon. This is achieved by continuing to increase i by one, by means of the assignment i <- i+1, while substr(rna,i,i+2) is not a start codon, that is, until substr(rna,i,i+2) is a start codon:

```
i <- 1
while (substr(rna,i,i+2) != "AUG") { # start codon
  i <- i+1
}
```

Comments can be placed starting with a hash symbol either at the end of a line or on one or more separate lines, to make the code easier to understand. The substring function **substr** returns a substring of a given string starting at a given position and ending at another given position, and, thus, substr(rna,i,i+2) is the substring of **rna** of length 3 starting at position i and ending at position i+2, that is, the codon starting at position i of string **rna**.

Scalar values can be compared by means of various operators, and the result is always a Boolean value. The equality test == returns true if the values are equal and false otherwise. Testing for the opposite, not equal, is done with != and returns true if the values are different and false if they are equal.

Testing if two values are greater than, or greater than or equal to, each other is done with > and >=, respectively. Similarly, testing if two values are less than, or less than or equal to, each other is done with < and <=, respectively. The comparison is done in lexicographical order when the values are not scalar.

Beware of the finite representation of real numbers and the possibility for rounding error when comparing numeric values. R provides the **all.equal** function for testing near equality, which returns true if the values are nearly equal and a report of the differences otherwise. Testing exact equality, on the other hand, can be done with the **identical** function, which returns true if the values are equal and false otherwise, where isTRUE(x) is an abbreviation of identical(TRUE,x).

```
> pi1 <- (81+(19^2)/22)^(1/4)
> pi2 <- 63*(17+15*sqrt(5))/(25*(7+15*sqrt(5)))
> options(digits=18)
> pi1
[1] 3.141592652582646
> pi2
[1] 3.141592653805688
> pi1 == pi2
[1] FALSE
```

```
> identical(TRUE,all.equal(pi1,pi2))
[1] TRUE
> all.equal(223/71,22/7)
[1] "Mean relative difference: 0.000640614990390815"
```

There are several ways of looping across a block of R code. In a *for* loop, a block of code is executed once upon each of the values in a given list, without a need for the explicit position of the values in the list. For instance, by looping over the list stored in the **stop** vector, the block of code is executed once for each of the three **codon** values:

```
> for (codon in stop) print(codon)
[1] "UAA"
[1] "UAG"
[1] "UGA"
```

The list of values in a for loop can also be a list of position numbers, which are then used to access the values in the list:

```
> for (i in 1:length(stop)) print(stop[i])
[1] "UAA"
[1] "UAG"
[1] "UGA"
```

In a *while* loop, a block of code is executed while a specific condition is met: if the expression controlling the loop evaluates to true, the block of code is executed once and then it continues to execute in a loop until the expression evaluates to false, but if the expression initially evaluates to false, the block of code is not executed at all. For instance, by setting a variable to the initial value i <- 1 before the loop and then executing the block of code as long as the test expression i <= length(stop) evaluates to true, where the variable is increased with i <- i+1 within the block of code, the block of code is also executed a number of times equal to the length of the **stop** vector:

```
i <- 1
while (i <= length(stop)) {
  print(stop[i])
  i <- i+1
}
```

Finally, in a *repeat* loop, a block of code is executed until an explicit **break** instruction is encountered within the block. For instance, by setting a variable to the initial value i <- 1 before the loop and then executing the block of code until the test expression i > length(stop) evaluates to true, in which case a **break** instruction is executed, where the variable is increased with i <- i+1 within the block of code, the block of code is also executed a number of times equal to the length of the **stop** vector:

```
i <- 1
```

```
repeat {
  print(stop[i])
  i <- i+1
  if (i > length(stop)) break
}
```

Considering again the `rna.to.protein` function, the previous while loop for skipping any nucleotides before the first start codon was correct only if the `rna` string contained at least one start codon. Otherwise, the loop would eventually fall off beyond the end of the string, and `substr(rna,i,i+2)` would then return an empty string while the loop continued forever. In order to skip nucleotides within the `rna` string only, an additional condition is needed in the expression controlling the loop: the position `i` must be less than or equal to `nchar(rna)-2`, the starting position of the last codon in the string. This condition is equivalent to `i < nchar(rna)-1`, thus leading to the following code:

```
i <- 1
while (i < nchar(rna)-1 &&
    substr(rna,i,i+2) != "AUG") { # start codon
  i <- i+1
}
```

In the expression controlling this while loop, two expressions are combined by means of the Boolean `&&` (and) binary operator, which returns true if the two expressions evaluate to true and returns false otherwise, that is, if at least one of the expressions evaluates to false. Further, the Boolean `||` (or) binary operator returns true if at least one of the expressions evaluates to true and returns false otherwise, that is, if the two expressions evaluate to false, and the Boolean `!` (not) unary operator returns true if the expression evaluates to false and returns false otherwise, that is, if the expression evaluates to true.

The order of the expressions matters here, because they are always evaluated from the left to right. Therefore, if the test for a start codon is put before the test for the end of the string,

```
i <- 1
while (substr(rna,i,i+2) != "AUG" &&
    i < nchar(rna)-1) { # start of last codon
  i <- i+1
}
```

the execution will not continue forever, but in the case the `rna` string does not contain any start codon, the loop will indeed fall off the end of the string, and `substr(rna,i,i+2)` will then return only two nucleotides, instead of a codon of three nucleotides:

```
. . . . . . . . . . . . . . . . . . . . . . . . . . . . . . . . . . . . . . . . . . . . . . . . . . UGACUA
. . . . . . . . . . . . . . . . . . . . . . . . . . . . . . . . . . . . . . . . . . . . . . . . . . .GACUA
```

```
..........................................ACUA
...........................................CUA
............................................UA
```

Once the nucleotides before the first start codon have been skipped, the start codon itself has to be skipped as well,

```
i <- i+3 # skip the start codon
```

The second step consists of translating to protein the rest of the string, until the first stop codon. In the second while loop of the function, each `codon` is translated to amino acid, the amino acid is added to the end of the `protein`, the codon is skipped, and the loop is repeated as long as there are still enough nucleotides in the string and the codon is not any of the stop codons.

```
while (i < nchar(rna)-1 &&
    substr(rna,i,i+2) != "UAA" &&
    substr(rna,i,i+2) != "UAG" &&
    substr(rna,i,i+2) != "UGA") {
  codon <- substr(rna,i,i+2)
  protein <- paste(protein,
    codon.to.amino.acid(codon),sep="")
  i <- i+3
}
```

The `paste` function returns the string concatenation of two or more character vectors, here `protein` and `codon.to.amino.acid(codon)`, separated by the empty string. During the execution of this second while loop, the `protein` string grows from an initial empty string to the 10 amino acids MVVIIPSRTV coded by the 30 nucleotides AUGGUGGUUAUUAUACCGUCAAGGACUGUG, as the 10 codons are translated to protein and added one after the other to the end of the string:

```
..AUGAUGGUGGUUAUUAUACCGUCAAGGACUGUGUGA..
.....AUGGUGGUUAUUAUACCGUCAAGGACUGUGUGA..   M
........GUGGUUAUUAUACCGUCAAGGACUGUGUGA..   MV
...........GUUAUUAUACCGUCAAGGACUGUGUGA..   MVV
..............AUUAUACCGUCAAGGACUGUGUGA..   MVVI
.................AUACCGUCAAGGACUGUGUGA..   MVVII
....................CCGUCAAGGACUGUGUGA..   MVVIIP
.......................UCAAGGACUGUGUGA..   MVVIIPS
..........................AGGACUGUGUGA..   MVVIIPSR
.............................ACUGUGUGA..   MVVIIPSRT
................................GUGUGA..   MVVIIPSRTV
```

The final value of the `protein` variable is the string (character vector) returned by the function:

```
rna.to.protein <- function (rna) {
```

```
...
    return(protein)
}
```

In the absence of an explicit **return** instruction, the function would return the value of the last expression that was evaluated, which in this case is the empty string, when `substr(rna,i,i+2)` is equal to one of the stop codons.

Now, a codon can be translated to protein by looking it up in the genetic code table, and the 64 entries of the genetic code table can be encoded in a series of 64 tests within the `codon.to.amino.acid` function. Reading the circular genetic code table in clockwise order, for instance, leads to the following function code:

```
codon.to.amino.acid <- function (codon) {
    if      (codon == "UUU") return("F") # Phe
    else if (codon == "UUC") return("F") # Phe
    else if (codon == "UUA") return("L") # Leu
    else if (codon == "UUG") return("L") # Leu
    else if (codon == "UCU") return("S") # Ser
    else if (codon == "UCC") return("S") # Ser
    else if (codon == "UCA") return("S") # Ser
    else if (codon == "UCG") return("S") # Ser
    else if (codon == "UAU") return("Y") # Tyr
    else if (codon == "UAC") return("Y") # Tyr
    else if (codon == "UAA") return("-") # stop
    else if (codon == "UAG") return("-") # stop
    else if (codon == "UGU") return("C") # Cys
    else if (codon == "UGC") return("C") # Cys
    else if (codon == "UGA") return("-") # stop
    else if (codon == "UGG") return("W") # Trp
    else if (codon == "CUU") return("L") # Leu
    else if (codon == "CUC") return("L") # Leu
    else if (codon == "CUA") return("L") # Leu
    else if (codon == "CUG") return("L") # Leu
    else if (codon == "CCU") return("P") # Pro
    else if (codon == "CCC") return("P") # Pro
    else if (codon == "CCA") return("P") # Pro
    else if (codon == "CCG") return("P") # Pro
    else if (codon == "CAU") return("H") # His
    else if (codon == "CAC") return("H") # His
    else if (codon == "CAA") return("Q") # Gln
    else if (codon == "CAG") return("Q") # Gln
    else if (codon == "CGU") return("R") # Arg
    else if (codon == "CGC") return("R") # Arg
    else if (codon == "CGA") return("R") # Arg
    else if (codon == "CGG") return("R") # Arg
```

```
  else if (codon == "AUU") return("I") # Ile
  else if (codon == "AUC") return("I") # Ile
  else if (codon == "AUA") return("I") # Ile
  else if (codon == "AUG") return("M") # Met
  else if (codon == "ACU") return("T") # Thr
  else if (codon == "ACC") return("T") # Thr
  else if (codon == "ACA") return("T") # Thr
  else if (codon == "ACG") return("T") # Thr
  else if (codon == "AAU") return("N") # Asn
  else if (codon == "AAC") return("N") # Asn
  else if (codon == "AAA") return("K") # Lys
  else if (codon == "AAG") return("K") # Lys
  else if (codon == "AGU") return("S") # Ser
  else if (codon == "AGC") return("S") # Ser
  else if (codon == "AGA") return("R") # Arg
  else if (codon == "AGG") return("R") # Arg
  else if (codon == "GUU") return("V") # Val
  else if (codon == "GUC") return("V") # Val
  else if (codon == "GUA") return("V") # Val
  else if (codon == "GUG") return("V") # Val
  else if (codon == "GCU") return("A") # Ala
  else if (codon == "GCC") return("A") # Ala
  else if (codon == "GCA") return("A") # Ala
  else if (codon == "GCG") return("A") # Ala
  else if (codon == "GAU") return("D") # Asp
  else if (codon == "GAC") return("D") # Asp
  else if (codon == "GAA") return("E") # Glu
  else if (codon == "GAG") return("E") # Glu
  else if (codon == "GGU") return("G") # Gly
  else if (codon == "GGC") return("G") # Gly
  else if (codon == "GGA") return("G") # Gly
  else if (codon == "GGG") return("G") # Gly
  else return("*")
}
```

This long instruction is called a *conditional* and allows for the conditional execution of a block of code, depending on whether or not a given expression evaluates to true. In the simplest form of the conditional, a block of code is executed if an expression evaluates to true:

```
if (codon == "AUG") {
  cat("stop␣codon")
}
```

Another common form of the conditional provides for the alternative execution of two blocks of code, depending on the outcome of the evaluation of a given expression. A block of code is executed if an expression evaluates to true. Otherwise, if the expression evaluates to false, another block of code is executed:

```
if (codon == "AUG") {
  cat("stop codon")
} else {
  i <- i + 3
}
```

The whole R script, stored in the **sample.R** file, will look as follows.

```
rna <- "GCCAAUGACUAAGGCCUAAAGA"
protein <- rna.to.protein(rna)
cat(rna,"translates to",protein)

rna.to.protein <- function (rna) {
  ...
}

codon.to.amino.acid <- function (codon) {
  ...
}
```

Second Script

The previous R script would be more flexible if the **rna** string could be input from the console, instead of being hardwired in the script, since this would allow for using exactly the same script over and over again, to translate to protein any given messenger RNA sequence.

An R script can be invoked upon particular argument values, much like a function, and these argument values can be read from the console, also known as *standard input* in Unix and Linux, by means of the **scan** function:

```
rna <- scan(file=stdin(),what="character",n=1)
protein <- rna.to.protein(rna)
cat(rna,"translates to",protein)
```

Here, **stdin** (standard input) is an example of a *file handle*, and **stdin()** can be replaced by "" for brevity. File handles will be further discussed in the third script. The string (character vector) argument is read from the standard input in a separate line, right after the command line. For instance,

```
> source("sample.R")
1: GUCGCCAUGAUGGUGGUUAUUAUACCGUCAAGGACUGUGUGACUA
Read 1 item
GUCGCCAUGAUGGUGGUUAUUAUACCGUCAAGGACUGUGUGACUA trans
lates to MVVIIPSRTV
```

Reading the argument from the standard input in a separate line has the advantage that the R script can then be invoked upon as many input strings as desired, with each input string in a separate line, by just reading the input string inside a loop.

After reading and translating to protein all the input strings, the end of the input is signaled by just entering an empty line.

```
repeat {
  rna <- scan(file="",what="character",n=1)
  if (length(rna) == 0) break
  protein <- rna.to.protein(rna)
  cat(rna,"translates␣to",protein)
}
```

Beside all this, using a long conditional is not the only solution to the messenger RNA to protein translation problem in the `codon.to.amino.acid` function. In fact, the circular genetic code table can also be encoded as a list of codons and corresponding amino acids, using a vector variable:

```
codon2aa <- c(
  "UUU" = "F",  # Phe
  "UUC" = "F",  # Phe
  "UUA" = "L",  # Leu
  "UUG" = "L",  # Leu
  "UCU" = "S",  # Ser
  "UCC" = "S",  # Ser
  "UCA" = "S",  # Ser
  "UCG" = "S",  # Ser
  "UAU" = "Y",  # Tyr
  "UAC" = "Y",  # Tyr
  "UAA" = "-",  # stop
  "UAG" = "-",  # stop
  "UGU" = "C",  # Cys
  "UGC" = "C",  # Cys
  "UGA" = "-",  # stop
  "UGG" = "W",  # Trp
  "CUU" = "L",  # Leu
  "CUC" = "L",  # Leu
  "CUA" = "L",  # Leu
  "CUG" = "L",  # Leu
  "CCU" = "P",  # Pro
  "CCC" = "P",  # Pro
  "CCA" = "P",  # Pro
  "CCG" = "P",  # Pro
  "CAU" = "H",  # His
  "CAC" = "H",  # His
```

```
"CAA" = "Q", # Gln
"CAG" = "Q", # Gln
"CGU" = "R", # Arg
"CGC" = "R", # Arg
"CGA" = "R", # Arg
"CGG" = "R", # Arg
"AUU" = "I", # Ile
"AUC" = "I", # Ile
"AUA" = "I", # Ile
"AUG" = "M", # Met
"ACU" = "T", # Thr
"ACC" = "T", # Thr
"ACA" = "T", # Thr
"ACG" = "T", # Thr
"AAU" = "N", # Asn
"AAC" = "N", # Asn
"AAA" = "K", # Lys
"AAG" = "K", # Lys
"AGU" = "S", # Ser
"AGC" = "S", # Ser
"AGA" = "R", # Arg
"AGG" = "R", # Arg
"GUU" = "V", # Val
"GUC" = "V", # Val
"GUA" = "V", # Val
"GUG" = "V", # Val
"GCU" = "A", # Ala
"GCC" = "A", # Ala
"GCA" = "A", # Ala
"GCG" = "A", # Ala
"GAU" = "D", # Asp
"GAC" = "D", # Asp
"GAA" = "E", # Glu
"GAG" = "E", # Glu
"GGU" = "G", # Gly
"GGC" = "G", # Gly
"GGA" = "G", # Gly
"GGG" = "G" # Gly
)
```

Vector variables hold lists of values, which are accessed by position. However, the values in a vector can be accessed by a scalar *key* instead of a position in the vector. For instance, a vector variable can be created and have the start codon and the three stop codons assigned as a list of keys and values,

```
codon = c(
```

```
  "AUG" = "start",
  "UAA" = "stop",
  "UAG" = "stop",
  "UGA" = "stop"
)
```

or it can be created first and then have each of the four codons stored in the vector,

```
codon <- vector(mode="character")
codon["AUG"] <- "start"
codon["UAA"] <- "stop"
codon["UAG"] <- "stop"
codon["UGA"] <- "stop"
```

While the value of the vector variable is the whole list of keys and values, individual elements can be accessed by key, as in the previous assignment instructions, or by means of a for loop,

```
for (key in names(codon)) {
  cat(c(key,codon[key],"\n"))
}
```

where the values appear in no particular order, or with the values sorted by the vector key,

```
for (key in sort(names(codon))) {
  cat(c(key,codon[key],"\n"))
}
```

Within the `rna.to.protein` function, the encoding of the genetic code table is then replaced by the vector of codons and corresponding amino acids,

```
    if (is.na(codon2aa[codon])) {
        aa <- "*"
    } else {
        aa <- codon2aa[codon]
    }
```

where the `is.na` function returns false if there is a value for the `codon` key in the `codon2aa` vector and true otherwise, when either there is no value for the key in the vector or the value is `NA` (not available), a special scalar constant that denotes missing values in R.

The second R script, also stored in the `sample.R` file, will then look as follows.

```
repeat {
  rna <- scan(file="",what="character",n=1)
  if (length(rna) == 0) break
  protein <- rna.to.protein(rna)
```

```
    cat(rna,"translates␣to",protein)
}

codon2aa <- c(
    ...
)

rna.to.protein <- function (rna) {
    protein <- ""
    i <- 1
    while (i < nchar(rna)-1 &&
        substr(rna,i,i+2) != "AUG") { # start codon
        i <- i+1
    }
    i <- i+3 # skip the start codon
    while (i < nchar(rna)-1 &&
        substr(rna,i,i+2) != "UAA" &&
        substr(rna,i,i+2) != "UAG" &&
        substr(rna,i,i+2) != "UGA") {
        codon <- substr(rna,i,i+2)
        if (is.na(codon2aa[codon])) {
            aa <- "*"
        } else {
            aa <- codon2aa[codon]
        }
        protein <- paste(protein,aa,sep="")
        i <- i+3
    }
    return(protein)
}
```

Third Script

The messenger RNA sequences to be translated to protein could be stored in a text file and then input to the R script straight from the file, instead of being input one after the other from the console. In fact, the previous R script can be easily modified to read the input strings from a text file, with the file name passed as an argument:

```
rna.file <- commandArgs(trailingOnly=TRUE)[1]

RNA <- scan(file=rna.file,what="character")
for (rna in RNA) {
    protein <- rna.to.protein(rna)
    cat(rna,"translates␣to",protein,"\n")
```

```
}
```

The following command would then invoke the R interpreter on the sample script with the name of the text file containing all the input strings as the first argument:

```
$ R --slave --vanilla --args "sample.rna" < sample.R
```

The `scan` instruction is used to read data into a vector or list from the console or from a file, while `read.table` can be used to read data matrices.

All input and output takes place in R using a *connection*, which is just a variable that has been associated with a particular file by means of a `file` instruction, becomes valid when initiated by an `open` instruction, and remains valid until finished by a `close` instruction. The default connections, which need not be opened or closed, are `stdin` (standard input) for input from the console, `stdout` (standard output) for output to the console window, and `stderr` (standard error) for output of error and warning messages, also to the console window.

A file can be opened for either reading, writing, or appending data. In the `open` and `file` instructions, the file name is followed by `"r"` for reading, `"w"` for writing, and `"a"` for appending. Opening for writing an already existing file results in the loss of any previous contents of the file, while opening a file for appending results in the addition of new contents to the end of the file.

Once a file has been associated with a connection and opened for reading, the `scan` instruction can be used to read data from the file by adding the connection before the values to be read. On the other hand, once a file has been opened for writing or appending, the `cat` instruction can be used to write to the file by adding the connection after the values to be written:

```
OUT <- file("sample.out")
open(OUT,open="w")
for (key in sort(names(codon2aa))) {
  cat(c(key,codon2aa[key],"\n"),file=OUT)
}
close(OUT)
```

The connection can also be opened by the `file` instruction:

```
OUT <- file("sample.out",open="w")
```

The third Perl script, stored again in the `sample.R` file, will then look as follows.

```
rna.to.protein <- function (rna) {
  ...
}

codon2aa <- c(
  ...
```

```
)

rna.file <- commandArgs(trailingOnly=TRUE)[1]

RNA <- scan(file=rna.file,what="character")
for (rna in RNA) {
  protein <- rna.to.protein(rna)
  cat(rna,"translates to",protein,"\n")
}
```

B.2   Overview of R

After the brief introduction by way of sample scripts in the previous section, let us focus now on a few more advanced aspects of the language which were not covered in these sample scripts.

Conditional Selection

Vector variables hold lists of values, and these values can be accessed by position. For instance, codon AGC is stored in position 46 of the codon2aa vector, and it can be extracted using the square bracket operator,

```
> codon2aa[46]
AGC
"S"
```

Vector values can also be accessed by a scalar key, using again the square bracket operator. A list of one or more keys and values can be extracted using single square brackets,

```
> codon2aa["AGC"]
AGC
"S"
```

and a single value can be extracted using double square brackets,

```
> codon2aa[["AGC"]]
[1] "S"
```

A list of keys and values can also be extracted by position, enclosing a list of positions inside square brackets. For instance, UCU, UCC, UCA, UCG, AGU, AGC all code for Ser (S), and these codons are stored in positions 5, 6, 7, 8, 45, 46 of the codon2aa vector,

```
> codon2aa[c(5:8,45,46)]
UCU UCC UCA UCG AGU AGC
"S" "S" "S" "S" "S" "S"
```

The square bracket operator also allows for the conditional selection of values in a vector, matrix, array, or list of values, by enclosing a Boolean expression inside single square brackets; for instance,

```
> codon2aa[codon2aa == "S"]
UCU UCC UCA UCG AGU AGC
"S" "S" "S" "S" "S" "S"
```

These six values are the only ones throughout the codon2aa vector for which the codon2aa == "S" expression evaluates to true:

```
> codon2aa == "S"
   UUU   UUC   UUA   UUG   UCU   UCC   UCA   UCG
 FALSE FALSE FALSE FALSE  TRUE  TRUE  TRUE  TRUE
   UAU   UAC   UAA   UAG   UGU   UGC   UGA   UGG
 FALSE FALSE FALSE FALSE FALSE FALSE FALSE FALSE
   CUU   CUC   CUA   CUG   CCU   CCC   CCA   CCG
 FALSE FALSE FALSE FALSE FALSE FALSE FALSE FALSE
   CAU   CAC   CAA   CAG   CGU   CGC   CGA   CGG
 FALSE FALSE FALSE FALSE FALSE FALSE FALSE FALSE
   AUU   AUC   AUA   AUG   ACU   ACC   ACA   ACG
 FALSE FALSE FALSE FALSE FALSE FALSE FALSE FALSE
   AAU   AAC   AAA   AAG   AGU   AGC   AGA   AGG
 FALSE FALSE FALSE FALSE  TRUE  TRUE FALSE FALSE
   GUU   GUC   GUA   GUG   GCU   GCC   GCA   GCG
 FALSE FALSE FALSE FALSE FALSE FALSE FALSE FALSE
   GAU   GAC   GAA   GAG   GGU   GGC   GGA   GGG
 FALSE FALSE FALSE FALSE FALSE FALSE FALSE FALSE
```

Computing with Vectors, Matrices, and Arrays

Most of the operations on scalar values can also be applied to vectors, matrices, and arrays, as long as they are of the same length. This is a powerful feature of R indeed, which may avoid the need for looping through a list of values. For instance, the mean of a list of values stored in a vector, $\bar{x} = \sum x_i/n$, can be computed as their sum divided by the number of values, using basic arithmetical operations on the whole vector:

```
> x <- c(0:100)
> sum(x)
[1] 5050
> length(x)
[1] 101
```

```
> sum(x)/length(x)
[1] 50
```

The standard deviation of a list of values, $\sigma = \sqrt{\sum(x_i - \bar{x})^2/(n-1)}$, can also be computed using basic arithmetical operations on the whole vector:

```
> x.mean <- sum(x)/length(x)
> sqrt(sum((x-x.mean)^2)/(length(x)-1))
[1] 29.30017
```

Being that R is a software environment for statistical computing, these basic statistical functions are already available as **mean** and **sd** (standard deviation):

```
> mean(x)
[1] 50
> sd(x)
[1] 29.30017
```

The genetic code table can also be represented as a three-dimensional array, with one dimension for each of the three nucleotides in a codon. The amino acids are listed with the 64 codons sorted in reverse order, from the third nucleotide back to the first nucleotide: AAA, CAA, GAA, UAA, ACA, CCA, GCA, UCA, ..., AUU, CUU, GUU, UUU.

```
> t <- c("K","Q","E","-","T","P","A","S","R","R","G",
   "-","I","L","V","L","N","H","D","Y","T","P","A","S",
   "S","R","G","C","I","L","V","F","K","Q","E","-","T",
   "P","A","S","R","R","G","W","M","L","V","L","N","H",
   "D","Y","T","P","A","S","S","R","G","C","I","L","V",
   "F")
> acgu <- c("A","C","G","U")
> acgu <- list(acgu,acgu,acgu)
> c2aa <- array(t,dim=c(4,4,4),dimnames=acgu)
```

Array values, like vector and matrix values, can be accessed by position and also by a scalar key, using the square bracket notation:

```
> c2aa[1,3,2]
[1] "S"
> c2aa["A","G","C"]
[1] "S"
```

An empty position indicates the selection of all entries in the corresponding dimension; for instance, the amino acids coded by the four codons that start with the two nucleotides AG,

```
> c2aa["A","G",]
  A   C   G   U
"R" "S" "R" "S"
```

or the 16 codons that start with the nucleotide A,

```
> c2aa["A",,]
    A   C   G   U
A  "K" "N" "K" "N"
C  "T" "T" "T" "T"
G  "R" "S" "R" "S"
U  "I" "I" "M" "I"
```

A matrix is just a two-dimensional array, and the bracket notation can still be used for selecting individual values or whole rows or columns from a matrix. For instance, the 16 codons that start with the nucleotide A in the c2aa array can also be seen as a matrix with four rows and four columns, one for each nucleotide.

```
> mat <- c2aa["A",,]
> mat["G","C"]
[1] "S"
> mat["G",]
    A   C   G   U
   "R" "S" "R" "S"
```

In the previous matrix, AGU and AGC both code for Ser (S). Their positions in the matrix can be revealed by means of the which function, which returns the array indices for which a given expression evaluates to true:

```
> which(mat == "S",arr.ind=TRUE)
  row col
G   3   2
G   3   4
> mat[3,2]
[1] "S"
> mat[3,4]
[1] "S"
```

Most of the operations on scalar values can also be applied to matrices, as long as they are of the same length, with one important exception to the latter rule. Comparison operators can be applied to two matrices, but also to a scalar value and a matrix, and the result is, in both cases, a matrix.

```
> mat == "S"
      A     C     G     U
A FALSE FALSE FALSE FALSE
C FALSE FALSE FALSE FALSE
G FALSE  TRUE FALSE  TRUE
U FALSE FALSE FALSE FALSE
```

Missing Values

Vector, matrix, array, data frame, and list variables hold lists of values, including the special scalar constant NA (not available) that denotes missing values in R. This special value is often carried through, because most operations on NA yield also NA as a result.

```
> m <- matrix(c(1,2,NA),nrow=3,ncol=4,byrow=TRUE)
> 3*m
      [,1] [,2] [,3] [,4]
[1,]    3    6   NA    3
[2,]    6   NA    3    6
[3,]   NA    3    6   NA
```

Notice that the three values 1, 2, NA are recycled by the matrix function in this example in order to fill in the 12 entries of the matrix.

Default Argument Values

Default values for arguments to a function can be included in the function definition by just giving the default value after each argument, separated by an equal sign; for instance,

```
sort <- function (x, decreasing = FALSE) { ... }
```

Using Packages

Functions written for one script can be made available to other scripts by placing them in a *package*, a text file with the usual extension containing R code. For instance, any functions written in a text file named sample.R are readily available in any script *using* this package, that is, in any script containing the following header line:

```
library(sample)
```

A package in R is not only a text file containing R code, but a whole directory containing code and documentation files, including at least a DESCRIPTION file and a directory with at least one R code file; for instance,

```
sample/DESCRIPTION
sample/R/sample.R
```

The DESCRIPTION file contains basic information about the package, including at least the Package, Version, Title, Author, Maintainer, Description, and License information; for instance,

```
Package: sample
Version: 1.0
Title: Sample package
```

```
Author: Wilma Flintstone <wilma@hanna-barbera.com>
Maintainer: Betty Rubble <betty@hanna-barbera.com>
Description: Sample R package.
License: GPL version 3 or newer
```

The package can then be installed by the R CMD INSTALL command:

```
R CMD INSTALL sample
```

Now, when using several modules in an R script, the variable and function names declared in one module might clash with the names declared in another module. Even in the same module, there might be two different reverse_complement functions, one for DNA and the other one for RNA nucleotides.

Such a name clash can be avoided by defining a *generic* function for reverse complementing a sequence, together with a reverse.complement.dna function for DNA and a reverse.complement.rna function for RNA. The sample.R file in the sample directory would then look as follows.

```
reverse.complement <- function (seq)
  UseMethod("reverse.complement")

reverse.complement.dna <- function (seq) {
  rev <- paste(rev(unlist(strsplit(seq,split=""))),
    sep="",collapse="")
  chartr("ACGT","TGCA",rev)
}

reverse.complement.rna <- function (seq) {
  rev <- paste(rev(unlist(strsplit(seq,split=""))),
    sep="",collapse="")
  chartr("ACGU","UGCA",rev)
}
```

The two versions of the reverse.complement function are readily available in any script using this module, and it suffices to make the sequence of class DNA or RNA,

```
> dna <- "TTGATTACCTTATTTGATCATTACACATTGTACGCTTGTG"
> class(dna) <- "dna"
> rna <- "GGGUGCUCAGUACGAGAGGAACCGCACCC"
> class(rna) <- "rna"
```

before calling the general reverse complement function,

```
> reverse.complement(dna)
[1] "CACAAGCGTACAATGTGTAATGATCAAATAAGGTAATCAA"
> reverse.complement(rna)
[1] "GGGUGCGGUUCCUCUCGUACUGAGCACCC"
```

B.3   R Quick Reference Card

Most of the basic functions available in R, including all the R built-in functions used in this book, are summarized next, in a kind of inset quick reference card. They are grouped into arithmetic functions; conversion functions; string functions; list functions; matrix functions; selection functions; search and replace functions; and input output functions.

More detailed information on any of these built-in functions can be obtained with the function name as an argument of the `help` function in R. Beside this, a list of the available R functions for statistical computing can be obtained with the command `library(help="stats")`, and a list of the functions for graphics in R can be obtained with the command `library(help="graphics")`.

Arithmetic Functions

`abs(X)`

Returns the absolute value of X.

`acos(X)`

Returns the arc cosine of X.

`acosh(X)`

Returns the hyperbolic cosine of X.

`approx(X,Y)`

Returns a list of points that linearly interpolate the values in X and Y.

`asin(X)`

Returns the arc sine of X.

`asinh(X)`

Returns the hyperbolic sine of X.

`atan(X)`

Returns the arc tangent of X.

`atanh(X)`

Returns the hyperbolic tangent of X.

`atan2(Y,X)`

Returns the arc tangent of Y/X.

`ceiling(X)`

Returns the smallest integer not less than X.

`choose(X,Y)`

Returns the binomial coefficient of X and Y.

`cor(X,Y)`

Returns the correlation of the values in X and Y.

`cos(X)`

Returns the cosine of X radians.

`cov(X,Y)`

Returns the covariance of the values in X and Y.

`cumprod(X)`

Returns the cumulative products of the values in X.

`cumsum(X)`

Returns the cumulative sums of the values in X.

`exp(X)`

Returns e (the natural logarithm base) to the power of X.

`factorial(X)`

Returns the factorial of X.

`floor(X)`

Returns the largest integer not greater than X.

`log(X,Y)`

Returns the base Y logarithm of X. The default value for Y is e (the natural logarithm base).

`mean(X)`

Returns the arithmetic mean of the values in X.

`median(X)`

Returns the median of the values in X.

`prod(X)`

Returns the product of the values in X.

`quantile(X)`

Returns sample quantiles corresponding to the values in X.

`range(X)`

Returns the minimum and the maximum of the values in X.

`rank(X)`

Returns the rank of the values in X, that is, their positions in X sorted.

`round(X,Y)`

Returns X rounded to Y decimal places. The default value for Y is 0.

`sd(X)`

Returns the standard deviation of the values in X.

`signif(X,Y)`

Returns X rounded to Y significant digits. The default value for Y is 6.

`sin(X)`

Returns the sine of X radians.

`sqrt(X)`

Returns the square root of X.

`sum(X)`

Returns the sum of the values in X.

`tan(X)`

Returns the tangent of X radians.

`trunc(X)`

Returns the integer portion of X.

`var(X)`

Returns the variance of the values in X.

Conversion Functions

`array(X,Y)`

Returns an array with the values in X, where Y gives the largest index in each dimension.

`as.array(X)`

Interprets the values in X as numbers and returns an array with the corresponding numeric values.

`as.character(X)`

Interprets the values in X as characters and returns the corresponding character values.

`as.list(X)`

Returns a list of the values in X.

`as.logical(X)`

Interprets the values in X as logical and returns the corresponding Boolean values.

`as.matrix(X)`

Interprets the values in X as numbers and returns a matrix with the corresponding numeric values.

`as.numeric(X)`

Interprets the values in X as numbers and returns the corresponding numeric values.

`as.vector(X)`

Interprets the values in X as numbers and returns a vector with the corresponding numeric values.

`list(X)`

Returns a list of the values of X, any of which could also be a list.

`matrix(X,Y,Z)`

Returns a matrix of Y rows and Z columns with the values in X.

`vector(X,Y)`

Returns a vector of Y elements of type X initialized to the default value of X.

String Functions

`nchar`

Returns the number of characters of (the representation of) the values in X.

`paste(X,sep=Y)`

Interprets the values in X as character vectors and returns the corresponding values, concatenated and separated by Y.

`substr(X,Y,Z)`

Extracts a substring from position Y to position Z out of X and returns it.

`strsplit(X,Y)`

Splits X into substrings according to the presence of Y.

`tolower(X)`

Returns a lowercased version of the values in X.

`toupper(X)`

Returns an uppercased version of the values in X.

List Functions

`X:Y`

Generates a consecutive sequence from X to Y and returns the list value. Same as `seq(X,Y)`.

`c(X)`

Combines the values in X and returns the list value.

`length(X)`

Returns the number of values in X.

`max(X)`

Returns the maximum of the values in X.

`min(X)`

Returns the minimum of the values in X.

`order(X)`

Returns a permutation that rearranges X in sorted order.

`rep(X,Y)`

Replicates Y times the values in X and returns the resulting list value.

`rev(X)`

Returns a list value consisting of the elements of X in the opposite order.

`sapply(X,Y)`

Evaluates Y for each value in X and returns the list value composed of the results of each such evaluation.

`seq(X,Y)`

Generates a consecutive sequence from X to Y and returns the list value. Same as `X:Y`.

`sort(X)`

Sorts X and returns the sorted list value.

`unlist(X)`

Simplifies the list structure of the values in X and returns the resulting list value.

`unique(X)`

Removes duplicate elements from X and returns the list value.

`which.max(X)`

Returns the first position of the maximum of the values in X.

`which.min(X)`

Returns the first position of the minimum of the values in X.

Matrix Functions

`apply(X,Y,Z)`

Returns a vector, array, or list of values obtained by applying function Z to the rows (if Y=1), the columns (if Y=2), or both the rows and the columns (if Y=c(1,2)) of array X.

`cbind(X)`

Combines the vectors, matrices, or data frames in X by columns.

`chol(X)`

Returns the Choleski factorization of matrix X.

`crossprod(X,Y)`

Returns the cross-product of matrices X and Y.

`dim(X)`

Returns the dimension of array, data frame, or matrix X.

`dimnames(X)`

Returns the dimension names of array, data frame, or matrix X.

`eigen(X)`

Returns the eigenvalues and eigenvectors of matrix X.

`ncol(X)`

Returns the number of columns of matrix X.

`nrow(X)`

Returns the number of rows of matrix X.

`rbind(X)`

Combines the vectors, matrices, or data frames in X by rows.

`solve(A,B)`

Solves the system of equations $AX = B$.

`svd(X)`

Returns the singular value decomposition of matrix X.

`t(X)`

Returns the transpose of matrix X.

Selection Functions

`X[Y]`

Extracts from X the elements indexed by Y.

`X[[Y]]`

Extracts from X a single element indexed by Y.

`attr(X,Y)`

Returns the attribute Y of X.

`X$Y`

Returns the first element of X named Y.

`subset(X,Y)`

Returns the elements of X that meet Boolean condition Y. Same as `X[Y]`.

`which(X)`

Returns a vector of the indices for which the Boolean condition X is true.

Search and Replace Functions

`grep(X,Y)`

Searches for matches to X within Y.

`gsub(X,Y,Z)`

Searches for all occurrences of X within Z, and replaces them by Y.

`match(X,Y)`

Returns a vector of the positions of the first matches of the values in X as a prefix of the values of Y.

`pmatch(X,Y)`

Returns a vector of the positions of the unique partial matches of the values in X as a prefix of the values of Y.

`sub(X,Y,Z)`

Searches for the first occurrence of X within Z, and replaces it by Y.

Input Output Functions

`cat(X,file=Y,sep=Z)`

Prints the values in X to the connection or file Y, concatenated and separated by Z.

`format(X)`

Returns the values in X formatted for pretty printing.

`print(X)`

Prints the values in X to the standard output.

`scan(file=X,what=Y,sep=Z)`

Reads data of type Y separated by Z from the connection or file X.

`sink(file=X)`

Sends any further output to the connection or file X.

`source(X)`

Evaluates input from file X.

`unlink(X)`

Deletes the directories and files in X.

`write(X,file=Y)`

Writes the values in X to the connection or file Y.

Bibliographic Notes

The S language (Becker et al. 1988; Chambers 1998; Venables and Ripley 2000) was originally developed by John M. Chambers at Bell Labs, and he was honored the 1998 ACM Software System Award for "the S system, which has forever altered how people analyze, visualize, and manipulate data." There is more information about S available at http://cm.bell-labs.com/stat/S/. R is an open source implementation of S, the rationale for their names being perhaps that R is one step ahead of S.

General books on R include (Chambers 2008; Maindonald and Braun 2008; Murrell 2005; Spector 2008). More specialized books on R for computational biology include (Deonier et al. 2005; Gentleman et al. 2005; Gentleman 2008; Hahne et al. 2008; Paradis 2006). See also (Siegmund and Yakir 2007).

Beside these sources, a comprehensive R manual is available with any R distribution, including detailed descriptions and examples of almost every aspect of the R language. The documentation can be browsed in HTML format by invoking the command help.start() from within the R software environment.

Alternative R implementations for some of the algorithms presented in this book can be found in the seqinr (Sequences in R) package (Charif and Lobry 2007), in the APE (Analysis of Phylogenetics and Evolution) package (Paradis 2006), and in the Biostrings package of the Bioconductor project (Gentleman et al. 2005; Hahne et al. 2008). The latter includes the DNASuffixArray and LongestCommonPrefix functions as part of an efficient implementation of suffix arrays for DNA sequences.

References

M. I. Abouelhoda, S. Kurtz, and E. Ohlebusch. The enhanced suffix array and its applications to genome analysis. In *Proc. 2nd International Workshop on Algorithms in Bioinformatics*, volume 2452 of *Lecture Notes in Bioinformatics*, pages 449–463. Springer, 2002.

M. I. Abouelhoda, S. Kurtz, and E. Ohlebusch. Replacing suffix trees with enhanced suffix arrays. *Journal of Discrete Algorithms*, 2(1):53–86, 2004.

R. Ahlswede, B. Balkenhol, C. Deppe, and M. Fröhlich. A fast suffix-sorting algorithm. In *General Theory of Information Transfer and Combinatorics*, volume 4123 of *Lecture Notes in Computer Science*, pages 719–734. Springer, 2006.

B. L. Allen and M. A. Steel. Subtree transfer operations and their induced metrics on evolutionary trees. *Annals of Combinatorics*, 5(1):1–13, 2001.

U. Alon. *An Introduction to Systems Biology: Design Principles of Biological Circuits*. Chapman & Hall/CRC, 2006.

S. F. Altschul. Amino acid substitution matrices from an information theoretic perspective. *Journal of Molecular Biology*, 219(3):555–565, 1991.

S. F. Altschul, W. Gish, W. Miller, E. W. Myers, and D. J. Lipman. Basic local alignment search tool. *Journal of Molecular Biology*, 215(3):403–410, 1990.

A. Amir and D. Keselman. Maximum agreement subtree in a set of evolutionary trees: Metrics and efficient algorithms. *SIAM Journal on Computing*, 26(6):1656–1669, 1997.

A. Amir and G. M. Landau, editors. *Proc. 12th Annual Symp. Combinatorial Pattern Matching*, volume 2089 of *Lecture Notes in Computer Science*. Springer, 2001.

A. Apostolico and Z. Galil, editors. *Combinatorial Algorithms on Words*, volume 12 of *NATO Advanced Science Institutes Series F, Computer and Systems Sciences*. Springer, 1985.

A. Apostolico and R. Giancarlo. Sequence alignment in molecular biology. *Journal of Computational Biology*, 5(2):173–196, 1998.

A. Apostolico and C. Guerra. The longest common subsequence problem revisited. *Algorithmica*, 2(1–4):316–336, 1987.

A. Apostolico and J. Hein, editors. *Proc. 8th Annual Symp. Combinatorial Pattern Matching*, volume 1264 of *Lecture Notes in Computer Science*. Springer, 1997.

A. Apostolico and M. Takeda, editors. *Proc. 13th Annual Symp. Combinatorial Pattern Matching*, volume 2373 of *Lecture Notes in Computer Science*. Springer, 2002.

A. Apostolico, M. Crochemore, Z. Galil, and U. Manber, editors. *Proc. 3rd Annual Symp. Combinatorial Pattern Matching*, volume 644 of *Lecture Notes in Computer Science*. Springer, 1992.

A. Apostolico, M. Crochemore, Z. Galil, and U. Manber, editors. *Proc. 4th Annual Symp. Combinatorial Pattern Matching*, volume 684 of *Lecture Notes in Computer Science*. Springer, 1993.

A. Apostolico, M. Crochemore, and K. Park, editors. *Proc. 16th Annual Symp. Combinatorial Pattern Matching*, volume 3537 of *Lecture Notes in Computer Science*. Springer, 2005.

M. Arenas, G. Valiente, and D. Posada. Characterization of phylogenetic reticulate networks based on the coalescent with recombination. *Molecular Biology and Evolution*, 25(12):2517–2520, 2008.

R. A. Baeza-Yates, E. Chávez, and M. Crochemore, editors. *Proc. 14th Annual Symp. Combinatorial Pattern Matching*, volume 2676 of *Lecture Notes in Computer Science*. Springer, 2003.

H.-J. Bandelt and A. Dress. Reconstructing the shape of a tree from observed dissimilarity data. *Advances in Applied Mathematics*, 7(3):309–343, 1986.

M. Baroni, C. Semple, and M. Steel. Hybrids in real time. *Systematic Biology*, 55(1):46–56, 2006.

R. A. Becker, J. M. Chambers, and A. R. Wilks. *The New S Language: A Programming Environment for Data Analysis and Graphics*. Wadsworth & Brooks/Cole, 1988.

M. A. Bender and M. Farach-Colton. The LCA problem revisited. In *Proc. 4th Latin American Symp. Theoretical Informatics*, volume 1776 of *Lecture Notes in Computer Science*, pages 88–94. Springer, 2000.

M. A. Bender, M. Farach-Colton, G. Pemmasani, S. Skiena, and P. Sumazin. Lowest common ancestors in trees and directed acyclic graphs. *Journal of Algorithms*, 57(2):75–94, 2005.

D. A. Benson, I. Karsch-Mizrachi, D. J. Lipman, J. Ostell, and D. L. Wheeler. GenBank. *Nucleic Acids Research*, 36(D):25–30, 2008.

E. Birney, S. Markel, and J. E. Stajich. *Using BioPerl*. Cambridge University Press, 2009. In press.

F. R. Blattner, G. Plunkett, C. A. Bloch, N. T. Perna, V. Burland, M. Riley, J. Collado-Vides, J. D. Glasner, C. K. Rode, G. F. Mayhew, J. Gregor, N. W. Davis, H. A. Kirkpatrick, M. A. Goeden, D. J. Rose, B. Mau, and Y. Shao. The complete genome sequence of *Escherichia coli* K-12. *Science*, 277(5331):1453–1462, 1997.

J. Bluis and D.-G. Shin. Nodal distance algorithm: Calculating a phylogenetic tree comparison metric. In *Proc. 3rd IEEE Symp. BioInformatics and BioEngineering*, pages 87–94. IEEE Press, 2003.

G. S. Brodal, R. Fagerberg, and C. N. S. Pedersen. Computing the quartet distance between evolutionary trees in time $O(n \log n)$. *Algorithmica*, 38:377–395, 2003.

D. Bryant, J. Tsang, P. Kearney, and M. Li. Computing the quartet distance between evolutionary trees. In *Proc. 11th Annual ACM-SIAM Symposium on Discrete Algorithms*, pages 285–286. ACM/SIAM, 2000.

H. Bunke and J. Csirik. Parametric string edit distance and its application to pattern recognition. *IEEE Transactions on Systems, Man and Cybernetics*, 25(1):202–206, 1995.

F. Burkhardt and S. Smith, editors. *The Correspondence of Charles Darwin*, volume 2, 1837–1843 of *The Correspondence of Charles Darwin*. Cambridge University Press, 1987.

G. Cardona, M. Llabrés, F. Rosselló, and G. Valiente. A distance metric for a class of tree-sibling phylogenetic networks. *Bioinformatics*, 24(13):1481–1488, 2008a.

G. Cardona, F. Rosselló, and G. Valiente. A Perl package and an alignment tool for phylogenetic networks. *BMC Bioinformatics*, 9:175, 2008b.

G. Cardona, F. Rosselló, and G. Valiente. Extended Newick: It is time for a standard representation of phylogenetic networks. *BMC Bioinformatics*, 9:532, 2008c.

G. Cardona, F. Rosselló, and G. Valiente. Tripartitions do not always discriminate phylogenetic networks. *Mathematical Biosciences*, 211(2):356–370, 2008d.

G. Cardona, M. Llabrés, F. Rosselló, and G. Valiente. Metrics for phylogenetic networks I:

Generalizations of the Robinson-Foulds metric. *IEEE/ACM Transactions on Computational Biology and Bioinformatics*, 6(1):46–61, 2009a.

G. Cardona, M. Llabrés, F. Rosselló, and G. Valiente. Metrics for phylogenetic networks II: Nodal and triplets metrics. *IEEE/ACM Transactions on Computational Biology and Bioinformatics*, 2009b. In press.

G. Cardona, F. Rosselló, and G. Valiente. Comparison of tree-child phylogenetic networks. *IEEE/ACM Transactions on Computational Biology and Bioinformatics*, 2009c. In press.

L. L. Cavalli-Sforza and A. W. F. Edwards. Phylogenetic analysis: Models and estimation procedures. *American Journal of Human Genetics*, 19(3):233–257, 1967.

A. Cayley. On the theory of the analytical forms called trees. *Philosophical Magazine*, 13:19–30, 1857.

A. Cayley. On the analytical forms called trees. *American Journal of Mathematics*, 4(1):266–268, 1881.

J. M. Chambers. *Programming with Data: A Guide to the S Language*. Springer, 1998.

J. M. Chambers. *Software for Data Analysis: Programming with R*. Springer, 2008.

D. Charif and J. R. Lobry. SeqinR 1.0-2: A contributed package to the R project for statistical computing devoted to biological sequences retrieval and analysis. In U. Bastolla, M. Porto, H. E. Roman, and M. Vendruscolo, editors, *Structural Approaches to Sequence Evolution: Molecules, Networks, Populations*, Biological and Medical Physics, Biomedical Engineering, pages 207–232. Springer, 2007.

C. Choy, J. Jansson, K. Sadakane, and W.-K. Sung. Computing the maximum agreement of phylogenetic networks. *Theoretical Computer Science*, 335(1):93–107, 2005.

C. Christiansen, T. Mailund, C. N. S. Pedersen, M. Randers, and M. S. Stissing. Fast calculation of the quartet distance between trees of arbitrary degrees. *Algorithms for Molecular Biology*, 1:16, 2006.

T. Christiansen and N. Torkington. *Perl Cookbook*. O'Reilly, 2nd edition, 2003.

J. Cohen. Bioinformatics: An introduction for computer scientists. *ACM Computing Surveys*, 36(2):122–158, 2004.

R. Cole, M. Farach-Colton, R. Hariharan, T. M. Przytycka, and M. Thorup. An $O(n \log n)$ algorithm for the maximum agreement subtree problem for binary trees. *SIAM Journal on Computing*, 30(5):1385–1404, 2000.

S. Cozens. *Advanced Perl Programming*. O'Reilly, 2nd edition, 2005.

D. E. Critchlow, D. K. Pearl, and C. Qian. The triples distance for rooted bifurcating phylogenetic trees. *Systematic Biology*, 45(3):323–334, 1996.

M. Crochemore and D. Gusfield, editors. *Proc. 5th Annual Symp. Combinatorial Pattern Matching*, volume 807 of *Lecture Notes in Computer Science*. Springer, 1994.

M. Crochemore and M. Paterson, editors. *Proc. 10th Annual Symp. Combinatorial Pattern Matching*, volume 1645 of *Lecture Notes in Computer Science*. Springer, 1999.

M. Crochemore and W. Rytter. *Text Algorithms*. Oxford University Press, 1994.

M. Crochemore and W. Rytter. *Jewels of Stringology*. World Scientific, 2003.

M. Crochemore, C. Hancart, and T. Lecroq. *Algorithms on Strings*. Cambridge University Press, 2007.

G. Csárdi and T. Nepusz. The igraph software package for complex network research. *InterJournal Complex Systems*, 1695, 2006.

B. DasGupta, X. He, T. Jiang, M. Li, J. Tromp, and L. Zhang. On distances between phylogenetic trees. In *Proc. 8th Annual ACM-SIAM Symposium on Discrete Algorithms*, pages 427–436. ACM/SIAM, 1997.

W. H. E. Day. Optimal algorithms for comparing trees with labeled leaves. *J. Classification*, 2 (1):7–28, 1985.

M. O. Dayhoff, R. Schwartz, and B. C. Orcutt. A model of evolutionary change in proteins. In M. O. Dayhoff, editor, *Atlas of Protein Sequence and Structure*, volume 5, supplement 3, pages 345–358. National Biomedical Research Foundation, Washington, DC, 1978.

R. C. Deonier, S. Tavaré, and M. S. Waterman. *Computational Genome Analysis: An Introduction*. Springer, 2005.

Y. Deville, D. Gilbert, J. van Helden, and S. J. Wodak. An overview of data models for the analysis of biochemical pathways. *Briefings in Bioinformatics*, 4(3):246–259, 2003.

M.-M. Deza and E. Deza. *Dictionary of Distances*. Elsevier Science, 2006.

R. Durbin, S. Eddy, A. Krogh, and G. Mitchison. *Biological Sequence Analysis: Probabilistic Models of Proteins and Nucleic Acids*. Cambridge University Press, 1998.

R. A. Dwyer. *Genomic Perl: From Bioinformatics Basics to Working Code*. Cambridge University Press, 2003.

S. R. Eddy. Where did the BLOSUM62 alignment score matrix come from? *Nature Biotechnology*, 22(8):1035–1036, 2004a.

S. R. Eddy. What is dynamic programming? *Nature Biotechnology*, 22(7):909–910, 2004b.

G. Estabrook, F. R. McMorris, and C. Meacham. Comparison of undirected phylogenetic trees based on subtrees of four evolutionary units. *Systematic Zoology*, 34(2):193–200, 1985.

P. A. Evans. Finding common subsequences with arcs and pseudoknots. In *Proc. 10th Annual Symp. Combinatorial Pattern Matching*, volume 1645 of *Lecture Notes in Computer Science*, pages 270–280. Springer, 1999.

M. Farach. Optimal suffix tree construction with large alphabets. In *Proc. 38th Annual Symposium on Foundations of Computer Science*, pages 137–143. IEEE Computer Science Press, 1997.

M. Farach-Colton, editor. *Proc. 9th Annual Symp. Combinatorial Pattern Matching*, volume 1448 of *Lecture Notes in Computer Science*. Springer, 1998.

J. Felsenstein. Evolutionary trees from DNA sequences: A maximum likelihood approach. *Journal of Molecular Evolution*, 17(6):368–376, 1981.

J. Felsenstein. *Inferring Phylogenies*. Sinauer Associates, 2004.

P. Ferragina and G. M. Landau, editors. *Proc. 19th Annual Symp. Combinatorial Pattern Matching*, volume 5029 of *Lecture Notes in Computer Science*. Springer, 2008.

P. Ferragina, R. Giancarlo, V. Greco, G. Manzini, and G. Valiente. Compression-based classification of biological sequences and structures via the Universal Similarity Metric: Experimental assessment. *BMC Bioinformatics*, 8:252, 2007.

A. Firth, T. Bell, A. Mukherjee, and D. Adjeroh. A comparison of BWT approaches to string pattern matching. *Software Practice and Experience*, 35(13):1217–1258, 2005.

B. D. Foy. *Mastering Perl*. O'Reilly, 2007.

Z. Galil and E. Ukkonen, editors. *Proc. 6th Annual Symp. Combinatorial Pattern Matching*, volume 937 of *Lecture Notes in Computer Science*. Springer, 1995.

G. Gati. Further annotated bibliography on the isomorphism disease. *Journal of Graph Theory*, 3(2):95–109, 1979.

R. Gentleman. *R Programming for Bioinformatics*, volume 12 of *Chapman & Hall/CRC Computer Science & Data Analysis*. Chapman & Hall/CRC, 2008.

R. Gentleman, V. Carey, W. Huber, R. Irizarry, and S. Dudoit, editors. *Bioinformatics and Computational Biology Solutions Using R and Bioconductor*. Springer, 2005.

R. Giancarlo and D. Sankoff, editors. *Proc. 11th Annual Symp. Combinatorial Pattern Matching*, volume 1848 of *Lecture Notes in Computer Science*. Springer, 2000.

R. Giegerich. A systematic approach to dynamic programming in bioinformatics. *Bioinformatics*, 16(8):665–677, 2000.

R. Giegerich and S. Kurtz. From Ukkonen to McCreight and Weiner: A unifying view of linear-time suffix tree construction. *Algorithmica*, 19(3):331–353, 1997.

W. Goddard, E. Kubicka, G. Kubicki, and F. R. McMorris. The agreement metric for labeled binary trees. *Mathematical Biosciences*, 123(2):215–226, 1994.

G. H. Gonnet, R. A. Baeza-Yates, and T. Snider. New indices for text: PAT trees and PAT arrays. In W. B. Frakes and R. A. Baeza-Yates, editors, *Information Retrieval: Data Structures and Algorithms*, pages 66–82. Prentice-Hall, 1992.

O. Gotoh. An improved algorithm for matching biological sequences. *Journal of Molecular Biology*, 162(3):705–708, 1982.

O. Gotoh. Multiple sequence alignment: Algorithms and applications. *Advances in Biophysics*, 36(1):159–206, 1999.

R. E. Green, A.-S. Malaspinas, J. Krause, A. W. Briggs, P. L. F. Johnson, C. Uhler, M. Meyer, J. M. Good, T. Maricic, U. Stenzel, K. Prüfer, M. Siebauer, H. A. Burbano, M. Ronan, J. M. Rothberg, M. Egholm, P. Rudan, D. Brajković, Željko Kućan, I. Gušić, M. Wikström, L. Laakkonen, J. Kelso, M. Slatkin, and S. Pääbo. A complete Neandertal mitochondrial genome sequence determined by high-throughput sequencing. *Cell*, 134(3):416–426, 2008.

R. Grossi and J. S. Vitter. Compressed suffix arrays and suffix trees with applications to text indexing and string matching. *SIAM Journal on Computing*, 35(2):378–407, 2005.

D. Gusfield. *Algorithms on Strings, Trees, and Sequences: Computer Science and Computational Biology*. Cambridge University Press, 1997.

D. Gusfield, K. Balasubramanian, and D. Naor. Parametric optimization of sequence alignment. *Algorithmica*, 12(4/5):312–326, 1994.

D. Gusfield, S. Eddhu, and C. Langley. The fine structure of galls in phylogenetic networks. *INFORMS Journal on Computing*, 16(4):459–469, 2004a.

D. Gusfield, S. Eddhu, and C. Langley. Optimal, efficient reconstruction of phylogenetic networks with constrained recombination. *Journal of Bioinformatics and Computational Biology*, 2 (1):173–213, 2004b.

O. Haddrath and A. J. Baker. Complete mitochondrial DNA genome sequences of extinct birds: Ratite phylogenetics and the vicariance biogeography hypothesis. *Proc. Royal Society B: Biological Sciences*, 268(1470):939–945, 2001.

F. Hahne, W. Huber, R. Gentleman, and S. Falcon. *Bioconductor Case Studies*. Springer, 2008.

R. W. Hamming. Error detecting and error correcting codes. *The Bell System Technical Journal*, 26(2):147–160, 1950.

Y.-J. He, T. N. D. Huynh, J. Jansson, and W.-K. Sung. Inferring phylogenetic relationships avoiding forbidden rooted triplets. *Journal of Bioinformatics and Computational Biology*, 4 (1):59–74, 2006.

J. Hein. Reconstructing the history of sequences subject to gene conversion and recombination. *Mathematical Biosciences*, 98(2):185–200, 1990.

J. Hein. A heuristic method to reconstruct the history of sequences subject to recombination. *Journal of Molecular Evolution*, 36(4):396–405, 1993.

J. Hein, T. Jiang, L. Wang, and K. Zhang. On the complexity of comparing evolutionary trees. *Discrete Applied Mathematics*, 71(1–3):153–169, 1996.

S. Henikoff and J. G. Henikoff. Amino acid substitution matrices from protein blocks. *Proc. National Academy of Sciences USA*, 89(22):10915–10919, 1992.

D. S. Hirschberg. A linear space algorithm for computing maximal common subsequences. *Communications of the ACM*, 18(6):341–343, 1975.

D. S. Hirschberg. Algorithms for the longest common subsequence problem. *Journal of the ACM*, 24(4):664–675, 1977.

D. S. Hirschberg and E. W. Myers, editors. *Proc. 7th Annual Symp. Combinatorial Pattern Matching*, volume 1075 of *Lecture Notes in Computer Science*. Springer, 1996.

W.-K. Hon, T. W. Lam, K. Sadakane, W.-K. Sung, and S.-M. Yiu. A space and time efficient algorithm for constructing compressed suffix arrays. *Algorithmica*, 48(1):23–36, 2007.

D. C. Jamison. *Perl Programming for Biologists*. John Wiley & Sons, 2003.

J. Jansson and W.-K. Sung. Inferring a level-1 phylogenetic network from a dense set of rooted triplets. *Theoretical Computer Science*, 363(1):60–68, 2006.

J. Jansson and W.-K. Sung. The maximum agreement of two nested phylogenetic networks. In O. N. Terikhovsky and W. N. Burton, editors, *New Topics in Theoretical Computer Science*, chapter 4, pages 119–141. Nova Science Publishers, Hauppauge, New York, 2008.

J. Jansson, N. B. Nguyen, and W.-K. Sung. Algorithms for combining rooted triplets into a galled phylogenetic network. *SIAM Journal on Computing*, 35(5):1098–1121, 2006.

N. C. Jones and P. A. Pevzner. *An Introduction to Bioinformatics Algorithms*. MIT Press, 2004.

T. H. Jukes and C. R. Cantor. Evolution of protein molecules. In H. N. Munro, editor, *Mammalian Protein Metabolism*, volume 3, pages 21–123. Academic Press, New York, 1969.

J. Kärkkäinen, P. Sanders, and S. Burkhardt. Linear work suffix array construction. *Journal of the AÇM*, 53(6):918–936, 2006.

M. Kasahara and S. Morishita. *Large-Scale Genome Sequence Processing*. World Scientific, 2006.

T. Kasai, G. Lee, H. Arimura, S. Arikawa, and K. Park. Linear-time longest-common-prefix computation in suffix arrays and its applications. In *Proc. 12th Annual Symp. Combinatorial Pattern Matching*, volume 2089 of *Lecture Notes in Computer Science*, pages 181–192. Springer, 2001.

D. K. Kim, J. S. Sim, H. Park, and K. Park. Constructing suffix arrays in linear time. *Journal of Discrete Algorithms*, 3(2–4):126–142, 2005.

M. Kimura. A simple method for estimating evolutionary rates of base substitutions through comparative studies of nucleotide sequences. *Journal of Molecular Evolution*, 16(2):111–120, 1980.

H. Kitano. Computational systems biology. *Nature*, 420(6912):206–210, 2002a.

H. Kitano. Systems biology: A brief overview. *Science*, 295(5560):1662–1664, 2002b.

P. Ko and S. Aluru. Space efficient linear time construction of suffix arrays. *Journal of Discrete Algorithms*, 3(2–4):143–156, 2005.

J. Koolman and K.-H. Roehm. *Color Atlas of Biochemistry*. Georg Thieme Verlag, 2nd edition, 2005.

M. Kreitman. Nucleotide polymorphism at the alcohol dehydrogenase locus of *Drosophila melanogaster*. *Nature*, 304(5925):412–417, 1983.

M. Křivánek. Computing the nearest neighbor interchange metric for unlabeled binary trees is NP-complete. *Journal of Classification*, 3(1):55–60, 1986.

S.-Y. Le, R. Nussinov, and J. V. Maizel. Tree graphs of RNA secondary structures and their comparisons. *Computers and Biomedical Research*, 22(5):461–473, 1989.

M. D. LeBlanc and B. D. Dyer. *Perl for Exploring DNA*. Oxford University Press, 2007.

C.-M. Lee, L.-J. Hung, M.-S. Chang, C.-B. Shen, and C.-Y. Tang. An improved algorithm for the maximum agreement subtree problem. *Information Processing Letters*, 94(5):211–216, 2005.

C. Y. Lee. An algorithm for path connection and its applications. *IRE Trans. Electronic Computers*, 10(3):346–365, 1961.

V. I. Levenshtein. Binary codes capable of correcting deletions, insertions, and reversals. *Doklady Physics*, 10(8):707–710, 1966.

M. Levitt and S. Lifson. Refinement of protein conformations using a macromolecular energy minimization procedure. *Journal of Molecular Biology*, 46(2):269–279, 1969.

M. Lewenstein and G. Valiente, editors. *Proc. 17th Annual Symp. Combinatorial Pattern Matching*, volume 4009 of *Lecture Notes in Computer Science*. Springer, 2006.

R. B. Lyngsø, Y. S. Song, and J. Hein. Minimum recombination histories by branch and bound. In *Proc. 5th International Workshop on Algorithms in Bioinformatics*, volume 3692 of *Lecture Notes in Bioinformatics*, pages 239–250. Springer, 2005.

B. Ma and K. Zhang, editors. *Proc. 18th Annual Symp. Combinatorial Pattern Matching*, volume 4580 of *Lecture Notes in Computer Science*. Springer, 2007.

W. P. Maddison. Gene trees in species trees. *Systematic Biology*, 46(3):523–536, 1997.

T. Mailund, G. S. Brodal, R. Fagerberg, C. N. S. Pedersen, and D. Phillips. Recrafting the neighbor-joining method. *BMC Bioinformatics*, 7:29, 2006.

J. Maindonald and J. Braun. *Data Analysis and Graphics Using R: An Example-based Approach*. Cambridge University Press, 2nd edition, 2008.

V. Mäkinen. Compact suffix array: A space-efficient full-text index. *Fundamenta Infomaticae*, 56(1–2):191–210, 2003.

U. Manber and G. Myers. Suffix arrays: A new method for on-line string searches. *SIAM Journal on Computing*, 22(5):935–948, 1993.

M. A. Maniscalco and S. J. Puglisi. An efficient, versatile approach to suffix sorting. *Journal of Experimental Algorithmics*, 12(1):1.2, 2008.

G. Manzini. Two space saving tricks for linear time LCP array computation. In *Proc. 9th Scandinavian Workshop on Algorithm Theory*, volume 3111 of *Lecture Notes in Computer Science*, pages 372–383. Springer, 2004.

E. M. McCreight. A space-economical suffix tree construction algorithm. *Journal of the ACM*, 23(2):262–272, 1976.

E. F. Moore. The shortest path through a maze. In *Proc. Int. Symp. Theory of Switching*, pages 285–292. Harvard University Press, 1959.

G. W. Moore, M. Goodman, and J. Barnabas. An iterative approach from the standpoint of the additive hypothesis to the dendogram problem posed by molecular data sets. *Journal of Theoretical Biology*, 38(3):423–457, 1973.

B. M. E. Moret, L. Nakhleh, T. Warnow, C. R. Linder, A. Tholse, A. Padolina, J. Sun, and R. Timme. Phylogenetic networks: Modeling, reconstructibility, and accuracy. *IEEE/ACM*

Transactions on Computational Biology and Bioinformatics, 1(1):13–23, 2004.

M. M. Morin and B. M. E. Moret. NETGEN: Generating phylogenetic networks with diploid hybrids. *Bioinformatics*, 22(15):1921–1923, 2006.

P. Murrell. *R Graphics*, volume 6 of *Computer Science & Data Analysis*. Chapman & Hall/CRC, 2005.

E. W. Myers and W. Miller. Optimal alignments in linear space. *Computer Applications in the Biosciences*, 4(1):11–17, 1988.

G. Navarro. A guided tour to approximate string matching. *ACM Computing Surveys*, 33(1): 31–88, 2001.

G. Navarro and M. Raffinot. *Flexible Pattern Matching in Strings*. Cambridge University Press, 2002.

S. B. Needleman and C. D. Wunsch. A general method applicable to the search for similarities in the amino acid sequence of two proteins. *Journal of Molecular Biology*, 48(3):443–453, 1970.

D. L. Nelson and M. M. Cox. *Lehninger Principles of Biochemistry*. W. H. Freeman, 5th edition, 2008.

C. Notredame. Recent evolutions of multiple sequence alignment algorithms. *PLOS Computational Biology*, 3(8):1405–1408, 2007.

B. Ø. Palsson. *Systems Biology: Properties of Reconstructed Networks*. Cambridge University Press, 2006.

G. Pandey, V. Kumar, and M. Steinbach. *Computational Approaches for Protein Function Prediction*. Wiley Series on Bioinformatics: Computational Techniques and Engineering. Wiley, 2008.

E. Paradis. *Analysis of Phylogenetics and Evolution with R*. Springer, 2006.

J. R. Parrish, J. Yu, G. Liu, J. A. Hines, J. E. Chan, B. A. Mangiola, H. Zhang, S. Pacifico, F. Fotouhi, V. J. DiRita, T. Ideker, P. Andrews, and R. L. Finley, Jr. A proteome-wide protein interaction map for *Campylobacter jejuni*. *Genome Biology*, 8(7):R130.1–R130.19, 2007.

N. D. Pattengale, E. J. Gottlieb, and B. M. Moret. Efficiently computing the Robinson-Foulds metric. *Journal of Computational Biology*, 14(6):724–735, 2007.

W. R. Pearson and D. J. Lipman. Improved tools for biological sequence comparison. *Proc. National Academy of Sciences USA*, 85(8):2444–2448, 1988.

D. Penny and M. D. Hendy. The use of tree comparison metrics. *Systematic Zoology*, 34(1): 75–82, 1985.

P. A. Pevzner. *Computational Molecular Biology: An Algorithmic Approach*. MIT Press, 2000.

S. S. Rao. Time-space trade-offs for compressed suffix arrays. *Information Processing Letters*, 82(6):307–311, 2002.

R. C. Read and D. G. Corneil. The graph isomorphism disease. *Journal of Graph Theory*, 1(4): 339–363, 1977.

F. Restle. A metric and an ordering on sets. *Psychometrika*, 24(3):207–220, 1959.

D. F. Robinson and L. R. Foulds. Comparison of phylogenetic trees. *Mathematical Biosciences*, 53(1/2):131–147, 1981.

D. F. Robinson and L. R. Foulds. Comparison of labeled trees with valency three. *Journal of Combinatorial Theory*, 11:105–119, 1971.

A. Rzhetsky and M. Nei. A simple method for estimating and testing minimum-evolution trees. *Molecular Biology and Evolution*, 9(5):945–967, 1992.

S. C. Sahinalp, S. Muthukrishnan, and U. Dogrusüz, editors. *Proc. 15th Annual Symp. Combinatorial Pattern Matching*, volume 3109 of *Lecture Notes in Computer Science*. Springer, 2004.

N. Saitou and M. Nei. The neighbor-joining method: A new method for reconstructing phylogenetic trees. *Molecular Biology and Evolution*, 4(4):406–425, 1987.

F. Sanger and M. Dowding, editors. *Selected Papers of Frederick Sanger*. World Scientific, 1996.

F. Sanger, S. Nicklen, and A. R. Coulson. DNA sequencing with chain-terminating inhibitors. *Proc. National Academy of Sciences USA*, 74(12):5463–5467, 1977.

W. R. Schmitt and M. S. Waterman. Linear trees and RNA secondary structure. *Discrete Applied Mathematics*, 51(3):317–323, 1994.

K.-B. Schürmann and J. Stoye. An incomplex algorithm for fast suffix array construction. *Software Practice and Experience*, 37(3):309–329, 2007.

R. L. Schwartz, B. D. Foy, and T. Phoenix. *Intermediate Perl*. O'Reilly, 2006.

R. L. Schwartz, T. Phoenix, and B. D. Foy. *Learning Perl*. O'Reilly, 5th edition, 2008.

C. Semple and M. Steel. Unicyclic networks: Compatibility and enumeration. *IEEE/ACM Transactions on Computational Biology and Bioinformatics*, 2(1):84–91, 2006.

B. A. Shapiro. An algorithm for comparing multiple RNA secondary structures. *Computer Applications in the Biosciences*, 4(3):387–393, 1988.

B. A. Shapiro and K. Zhang. Comparing multiple RNA secondary structures using tree comparisons. *Computer Applications in the Biosciences*, 6(4):309–318, 1990.

F. Shi. Suffix arrays for multiple strings: A method for on-line multiple string searches. In *Proc. 2nd Asian Computing Science Conf. Concurrency and Parallelism, Programming, Networking, and Security*, volume 1179 of *Lecture Notes in Computer Science*, pages 11–22. Springer, 1996.

P. W. Shor. A new proof of Cayley's formula for counting labeled trees. *Journal of Combinatorial Theory Series A*, 71(1):154–158, 1995.

D. Siegmund and B. Yakir. *The Statistics of Gene Mapping*. Springer, 2007.

N. J. A. Sloane and S. Plouffe. *The Encyclopedia of Integer Sequences*. Academic Press, 1995.

T. F. Smith and M. S. Waterman. How alike are two trees? *American Mathematical Monthly*, 87(7):552–553, 1980.

T. F. Smith and M. S. Waterman. Identification of common molecular subsequences. *Journal of Molecular Biology*, 147(1):195–197, 1981.

B. Smyth. *Computing Patterns in Strings*. Addison-Wesley, 2003.

P. H. A. Sneath and R. R. Sokal, editors. *Numerical Taxonomy: The Principles and Practice of Numerical Classification*. W. H. Freeman, 1973.

P. Spector. *Data Manipulation with R*. Springer, 2008.

J. E. Stajich, D. Block, K. Boulez, S. E. Brenner, S. A. Chervitz, C. Dagdigian, G. Fuellen, J. G. Gilbert, I. Korf, H. Lapp, H. Lehvaslaiho, C. Matsalla, C. J. Mungall, B. I. Osborne, M. R. Pocock, P. Schattner, M. Senger, L. D. Stein, E. Stupka, M. D. Wilkinson, and E. Birney. The BioPerl toolkit: Perl modules for the life sciences. *Genome Research*, 12(10):1611–1618, 2002. URL http://www.bioperl.org.

M. A. Steel and D. Penny. Distributions of tree comparison metrics: Some new results. *Systematic Biology*, 42(2):126–141, 1993.

M. A. Steel and T. Warnow. Kaikoura tree theorems: Computing the maximum agreement subtree. *Information Processing Letters*, 48(2):77–82, 1993.

L. D. Stein. How Perl saved the Human Genome Project. *The Perl Journal*, 1(2), 1996.

G. A. Stephen. *String Searching Algorithms*. World Scientific, 1998.

K. Strimmer and V. Moulton. Likelihood analysis of phylogenetic networks using directed graphical models. *Molecular Biology and Evolution*, 17(6):875–881, 2000.

K. Strimmer, C. Wiuf, and V. Moulton. Recombination analysis using directed graphical models. *Molecular Biology and Evolution*, 18(1):97–99, 2001.

J. A. Studier and K. J. Keppler. A note on the neighbor-joining algorithm of Saitou and Nei. *Molecular Biology and Evolution*, 5(6):729–731, 1988.

R. E. Tarjan. Depth-first search and linear graph algorithms. *SIAM Journal on Computing*, 1(2):146–160, 1972.

J. Tisdall. *Beginning Perl for Bioinformatics*. O'Reilly, 2001.

J. Tisdall. *Mastering Perl for Bioinformatics*. O'Reilly, 2003.

S. Tregar. *Writing Perl Modules for CPAN*. Apress, 2002.

E. Ukkonen. Constructing suffix trees on-line in linear time. In J. van Leeuwen, editor, *Proc. IFIP 12th World Computer Congress*, pages 484–492. Elsevier, 1992.

G. Valiente. *Algorithms on Trees and Graphs*. Springer, 2002.

G. Valiente. A fast algorithmic technique for comparing large phylogenetic trees. In *Proc. 12th Int. Symp. String Processing and Information Retrieval*, volume 3772 of *Lecture Notes in Computer Science*, pages 371–376. Springer, 2005.

W. N. Venables and B. D. Ripley. *S Programming*. Springer, 2000.

S. Vinga and J. Almeida. Alignment-free sequence comparison: A review. *Bioinformatics*, 19(4):513–523, 2003.

M. Vingron and M. S. Waterman. Sequence alignment and penalty choice: Review of concepts, case studies and implications. *Journal of Molecular Biology*, 235(1):1–12, 1994.

R. A. Wagner and M. J. Fischer. The string-to-string correction problem. *Journal of the ACM*, 21(1):168–173, 1974.

L. Wall, T. Christiansen, and J. Orwant. *Programming Perl*. O'Reilly, 3rd edition, 2000.

L. Wang, K. Zhang, and L. Zhang. Perfect phylogenetic networks with recombination. *Journal of Computational Biology*, 8(1):69–78, 2001.

M. S. Waterman. Secondary structure of single-stranded nucleic acids. In G.-C. Rota, editor, *Studies in Foundations and Combinatorics*, volume 1 of *Advances in Mathematics: Supplementary Studies*, pages 167–212. Academic Press, 1978.

M. S. Waterman. *Introduction to Computational Biology: Maps, Sequences and Genomes*. Chapman & Hall/CRC, 1995.

P. Weiner. Linear pattern matching algorithms. In *Proc. 14th Annual IEEE Symposium on Switching and Automata Theory*, pages 1–11. IEEE Computer Science Press, 1973.

D. Weininger. SMILES, a chemical language and information system. 1. Introduction to methodology and encoding rules. *Journal of Chemical Information and Computer Sciences*, 28(1):31–36, 1988.

W. T. Williams and H. T. Clifford. On the comparison of two classifications of the same set of elements. *Taxon*, 20(4):519–522, 1971.

S. M. Woolley, D. Posada, and K. A. Crandall. A comparison of phylogenetic network methods using computer simulation. *Plos ONE*, 4(3):e1913, 2008.

K. A. Zaretskii. Constructing a tree on the basis of a set of distances between the hanging vertices. *Uspekhi Matematicheskikh Nauk*, 20(6):90–92, 1965.

Index

alignment, 95
alternative splicing, 5
amino acid, 8

backtracking, 262

cladogram, 122
classification tree, 118
cluster map, 262, 266
codon, 7, 8
combinatorial pattern matching, 3
common ancestor, 122, 138, 148, 150, 215
common prefix, 63, 80
common semi strict ancestor, 215, 235, 249, 254, 269
common subsequence, 67, 69, 77, 80, 84
common subtree, 161
common suffix, 72
computational biology, 4
counting
    graphs, 182, 195
    secondary structures, 33
    sequences, 22, 33
    trees, 115, 125

dendrogram, 122
distance
    in a graph, 214
    in a tree, 138, 144, 150
distance between graphs
    error rate, 243
    nodal distance, 234
    path multiplicity distance, 220
    tripartition distance, 228
    triplets distance, 269
distance between sequences

alignment free distance, 49
    correlation of k-mer frequencies, 50
    edit distance, 86, 96
    Hamming distance, 86
    Levenshtein distance, 88, 96
distance between trees
    nodal distance, 144, 151
    partition distance, 140, 151
    quartets distance, 175
    triplets distance, 172
dynamic programming, 71, 89, 92, 96, 98, 103, 109, 162, 167

exon, 5
extended Newick format, 193, 249

gap, 95, 109
gene, 5, 8
gene transfer, 194
generalized suffix array, 74, 77, 80, 84
generating
    graphs, 198
    sequences, 35
    trees, 126
global alignment, 98
graph
    directed, 183
    directed acyclic, 184, 188, 264
    isomorphism, 262, 267
    traversal
        bottom-up, 212
        breadth-first, 183
        depth-first, 183
    undirected, 181

hybridization, 194

hybridization cycle, 191

intron, 5

k-mer, 43
k-mer frequency, 49

local alignment, 103

maximum agreement subgraph, 259, 262
maximum agreement subtree, 161, 172

Newick format, 123, 240
node
    height, 212
    hybrid, 188
    strict descendant, 214, 228
    tree, 188
nucleotide, 4, 7, 8

open reading frame, 5, 8
outgroup, 122

path
    elementary, 155, 156, 159, 238, 239, 241, 247, 249, 260
    in a graph, 211, 214, 220, 234, 247, 249
    in a tree, 137, 148
pattern, 3
pattern matching, 3
phylogenetic network, 184, 188
    fully resolved, 188
    galled-tree, 191, 195
    time-consistent, 191, 198
    tree-child, 191, 198
    tree-sibling, 191
    unicyclic, 195
phylogenetic tree, 118, 122, 238
    fully resolved, 122
prefix, 53, 102
primary structure, 29, 31, 119
protein, 5, 8
protein interaction network, 187

pseudo-knot, 119

quartet, 159, 175

reading frame, 7, 8
recombination, 194
recombination cycle, 191
regular expression, 3, 55
reticulation, 188

secondary structure, 3, 30, 33, 118, 119, 121
sequence, 4, 29
    arc annotated, 119
    reverse complement, 31
    simultaneous traversal, 225, 231
    traversal, 26
    word composition, 43
subgraph
    bottom-up, 247, 260, 263
    induced, 247, 248, 259
    top-down, 247
subsequence, 53, 54, 60
subtree, 155
    bottom-up, 155
    induced, 155, 156, 159, 253, 254, 269
    top-down, 155
suffix, 53, 102
suffix array, 56, 60
    binary search, 60, 63

transcription, 7
translation, 7
tree, 115
    ordered, 121
    rooted, 117
    traversal
        postorder, 118, 124, 240
        preorder, 118
    unrooted, 115
triplet, 156, 172, 253, 254, 269

Printed and bound by CPI Group (UK) Ltd, Croydon, CR0 4YY

25/10/2024

01779208-0001